Extraction '90

Institution of Chemical Engineers,
Rugby, UK.

Hemisphere Publishing Corporation
A Member of the Taylor & Francis Group
New York Washington Philadelphia London

Extraction '90

Members of the Institution of Chemical Engineers should order as follows:

Worldwide	Institution of Chemical Engineers, Davis Building, 165–171 Railway Terrace, RUGBY, Warwickshire CV21 3HQ, U.K.
Australia only	R. M. Wood, School of Chemical Engineering and Industrial Chemistry, University of New South Wales, PO Box 1, Kensington, NSW, Australia 2033.

Non members' orders should be directed as follows:

UK, Eire and Australia	Institution of Chemical Engineers, Davis Building, 165–171 Railway Terrace, RUGBY, Warwickshire CV21 3HQ, U.K.
or	Taylor & Francis Ltd., Rankine Road, BASINGSTOKE, Hampshire RG24 0PR, U.K.
Rest of the World	Taylor & Francis Inc., 1900 Frost Road, Suite 101, Bristol, PA 19007, U.S.A.

Extraction '90

A five-day symposium organised by the Dounreay Nuclear Power Development Establishment and the Institution of Chemical Engineers (Scottish Branch) and held at Dounreay Nuclear Power Establishment, Caithness, 10–14 September 1990.

Organising Committee

W. Batey (Chairman)	UKAEA
G. Davies	UMIST
R. Edge	University of Strathclyde
S. Ortiz	Imperial College
J. D. Thornton	Consultant
B. Waldie	Heriot-Watt University
L. Weatherley	Heriot-Watt University

INSTITUTION OF CHEMICAL ENGINEERS

SYMPOSIUM SERIES No. 119
EFCE Event No. 421
EFCE Publication No. 82
ISBN 0 85295 254 6

Preface

Over the past few years there has been a change in liquid-liquid solvent extraction from a process used mainly for metal recovery to more diverse areas such as pharmaceutical drug production.

This Conference, organised by the Scottish Branch of the IChemE in association with AEA Technology Dounreay, will address the recent advances in solvent extraction, both industrial and academic, carried out in both the nuclear and non-nuclear fields

W. Batey
(Chairman)

Contents

Extraction Fundamentals

Extraction from Dilute Solutions

Sorption

Extraction of Biological Products

Processes and Equipment

THE BREAKUP OF A JET OF CHLOROBENZENE IN WATER

Elhag, E.A.* and Edge, R.M.+

The breakup of a jet of chlorobenzene in an aqueous continuous phase is described in detail. The effect on the breakup of the oscillation of the jet is discussed and it is explained how the oscillations lead to a dispersion containing a range of drop sizes. The use of existing theoretical equations for predicting the range of drop sizes is discussed in the light of the above findings.

INTRODUCTION

When drops are formed at very low flowrates from a nozzle into a continuous liquid phase a uniform size of drop is produced. Under these conditions the drops are formed at the nozzle and the mechanism of formation is well understood. When the flowrate is increased a critical flowrate is reached at which a jet of liquid is formed from the nozzle. At flowrates above this critical flowrate the drops are formed from the end of the jet and a uniform drop size is rarely produced. At most flowrates a wide range of drop sizes is found and no satisfactory explanation of this has been given. Also, all of the theoretical models which have been used to describe drop formation from a jet of liquid predict that a uniform size of drop will be produced for a given system at a given flowrate.

It is useful at this point to briefly consider the above theoretical models. These are based on a model which was discussed by Rayleigh in a paper published in 1879(1). In this model it is assumed that the formation of drops from the end of a jet of liquid is similar to the formation of drops from an infinite cylinder of liquid which is breaking [Figure 1]. A mathematical model for the latter process is obtained by assuming that the primary disturbance, which leads to the breakup, can be represented by a series of wave-like disturbances and that the fastest growing disturbance will break the cylinder into drops. This model predicts that the cylinder will divide into segments which are of uniform size. Subsequent workers have refined this model and extended its use to a wider range of systems.

+ Chemical Engineering Division, University of Strathclyde, Glasgow, G1 1XJ.

* current address: Industrial Research & Consulting Centre, PO Box 268, Khartoum, Sudan.

However, they also predict that single segments will detach to form drops with a uniform size. This is contrary not only to the findings concerning drop size distribution in the resulting dispersion but is also in disagreement with the observations of previous workers that two or more segments can detach to form drops which have large volumes[2,3].

The aim of the present paper is to describe the drop-formation process and to explain how this process produces a range of drop sizes; also to discuss how the theoretical models proposed by previous authors should be used to predict the drop sizes found in the dispersion produced from a nozzle.

APPARATUS

The apparatus is shown in Figure 2. The column consisted of a 15 cm x 10 cm nominal-bore glass cross-piece with the 10 cm arms vertical and the 15 cm arms horizontal. The 15 cm arms were blanked off by schlieren quality glass windows through which the jet was photographed. The bottom of the column was reduced in size and contained a PTFE plug tap to allow drainage of the dispersed phase which collected at the bottom of the column. The jets of liquid were formed from the end of stainless steel tubes. The tubes had a length of 2 m and the end of each tube was machined so that it was sharp and circular. Three sizes of stainless steel tube were used giving nozzle inside diameters of 0.104 cm, 0.213 cm and 0.292 cm. The dispersed phase was fed to the stainless steel tubes from a constant head tank and the flowrate was controlled by a fine control valve and was measured with a rotameter which had been calibrated. Connecting tubing was made of PTFE. All parts of the apparatus which came into contact with the dispersed and continuous liquid phases were made from glass, PTFE or stainless steel. Prior to use these were cleaned by filling with chromic acid or immersion in chromic acid for 24 hours. This was followed by repeated washing with distilled water. The apparatus was housed in a constant temperature air bath which was controlled at 25°C.

The jets of liquid were photographed using a focussed-shadow optical system and a Fastax high-speed camera operating at between 500 and 1000 frames s^{-1}. Details of the optical arrangement were given an earlier paper by Edge and Grant[4].

MATERIALS

Double distilled water was used for the continuous phase and chlorobenzene was used for the dispersed phase. The chlorobenzene was of Analar quality and was distilled immediately prior to use in a 30 plate glass Oldershaw column.

PROCEDURE

Jets were formed from each of the three nozzles over a range of flowrates. The jets were photographed in focussed shadow using a high-speed cine camera and 16 mm Tri X cine film. The films were analysed frame by frame.

RESULTS AND DISCUSSION

The terminology used when discussing the results is shown in Figure 1a. Before discussing the mechanism for jet breakup it is useful to group the results into four groups. These groups are based on the regions shown when either the mean length of the jet immediately prior to breakup or immediately after breakup is plotted against the mean velocity of the jet liquid through the nozzle.

Mean Length of the Jet at Breakup

The length of the jet was measured both immediately prior to the breakaway of a drop (L) and immediately after the breakaway (R). At most of the flowrates studied, between 10 and 20 breakaways were examined and $\bar{L}$ and $\bar{R}$, the mean values of L and R, were determined.

Figure 3a shows the results obtained for $\bar{L}$ at various flowrates for each of the nozzles studied. Results for $\bar{R}$ are given for one nozzle [Figure 3b]; the results for the other two nozzles show similar trends and are given by Elhag[5]. Figure 3 can be divided into four regions and these will be referred to as the low, moderate, intermediate and high nozzle-velocity regions.

In the low nozzle-velocity region the average jet length immediately prior to breakup increases only slightly with increase in nozzle-velocity up to the critical velocity at which there is a sudden increase in jet length. This marks the boundary between the low and moderate nozzle-velocity regions. For a range of nozzle velocities above this critical nozzle-velocity, the mean jet length at breakup has an approximately linear relationship with nozzle-velocity. This linear relationship is in agreement with the findings of previous workers who used other liquids[6-9]. This is referred to as the moderate nozzle-velocity region. Immediately above this region is a region in which the mean jet length does not show a clear dependence on nozzle-velocity and this is referred to as the intermediate nozzle-velocity region. This extends up to the nozzle-velocity at which the maximum jet length occurs. Above this region is the high nozzle-velocity region in which the jet length decreases with increase in nozzle-velocity.

The results for jet length immediately after the breakaway of a drop show similar trends except that in the low nozzle-velocity region the average jet length increases with increase in nozzle-velocity and this increase is approximately linear. This linear increase is in agreement with the results of Meister and Sheele[9] for other systems.

Description of the Breakup Process

1) Low Nozzle-Velocity Region

At most flowrates in the low nozzle-velocity region an alternating pattern of breakup is observed and two sizes of drop are produced. The mechanism by which the two sizes of drop are produced is shown in Figure 4. This shows sketches taken from the projected cine film. Figure 4(a) shows the jet at a time when two well developed segments are present. At this stage the jet is almost free from oscillation and the first segment detaches by a force-balance type mechanism which is similar to that observed at very low nozzle-velocities. This produces the larger size of drop. During this detachment the second segment of the drop is drawn out towards the first segment so that immediately after the break has occurred the second segment has a distorted shape with a forward protrusion. This protrusion is drawn back into the jet causing it to oscillate. Also, a disturbance runs over the surface of the jet [Figure 4(b)] and when it reaches the region immediately in front of the original second neck a new neck is observed to form in this region [Figure 4(c)]. Due to its oscillation the current first segment becomes more oblate and its rear is drawn in towards the segment, thereby enhancing the necking process. The jet breaks at neck B to produce the smaller size of drop. Following this break the jet oscillates but at this stage there is no neck which is sufficiently well formed [Figure 4(d)] and the oscillations die out without a break occurring. The jet then grows until its shape is as shown in Figure 4(a) and the process repeats itself.

Therefore, at a given flowrate in the low nozzle-velocity region, two sizes of drop are produced alternately. The larger size of drop is produced by a force-balance type of mechanism which is similar to that observed when a drop detaches from a nozzle at very low nozzle-velocities except that the drop now detaches from the end of the jet instead of from the nozzle. The smaller size of drop is produced by an oscillation-enhanced mechanism. This oscillation-enhanced mechanism will be seen to be the predominant mechanism at higher nozzle velocities.

Moderate Nozzle Velocity Region

At flowrates immediately above the critical flowrate for sudden jet lengthening the breakup mechanism has many similarities to the breakup mechanism in the low nozzle-velocity region. The main difference is that immediately prior to the detachment of a segment by the force-balance type mechanism the jet has three well developed segments instead of the two observed in the low nozzle-velocity region [Figure 5(a)]. This explains the sudden increase in L and R at the critical nozzle velocity [Figure 3]. Following detachment of the first segment by a force-balance type of mechanism the next segment tries to detach by an oscillation-enhanced mechanism which is similar to that found in the low nozzle-velocity region [Figure 5(b)]. However, it fails to detach and the disturbance is seen to run over the surface of the next segment and this segment starts to oscillate. The jet breaks at what is now its second neck [Figure 5(c)] by an oscillation-enhanced mechanism and a double drop detaches. Following this detachment there are no well developed necks on the surface of the jet and the oscillations of the jet do not lead to a further breakup of the jet. The oscillations die out and the jet grows in length and segmentalises so that subsequently it contains three well developed segments. The process then repeats itself. In this way the drops which are formed alternate between large single segment drops which are formed by a force-balance type mechanism, and double segment drops, which are formed by an oscillation-enhanced mechanism.

With the 0.104 cm diameter nozzle at one flowrate, which was above but close to the critical flowrate, an unusual mechanism was observed. All of the resulting drops contained two segments and the breakup mechanism was of the force-balance type [Figure 6]. Figure 6(a) shows the jet immediately after a double segment drop has detached. At this stage it contains two well developed segments. The first segment tries to detach by an oscillation-enhanced mechanism but fails to do so. The disturbance is seen to run over the surface of the second segment which also starts to oscillate. When the disturbance is part way along the surface of the segment a new neck is formed in the region where the disturbance has reached and this grows to produce a well formed neck which divides the original second segment into two parts, thereby producing a jet with three well developed segments [Figure 6(b)]. The length of the jet increases and the segments grow in size. During this period the oscillations die out. Subsequently, the region between the third neck and the nozzle becomes extended and a new neck forms in this region so that the jet contains four segments. The next breakup occurs at the second neck by a force-balance type of mechanism which is little affected by jet oscillations which are now small. The process is then repeated. This mechanism was only observed with this nozzle and at one flowrate. However, it is probable that a similar breakup mechanism, giving only double segment drops, will be observed with the other nozzles provided the necessary flowrates are achieved. These should be close to the critical flowrate for sudden jet lengthening.

At higher nozzle-velocities in this region drops are formed by an increased number of mechanisms. One of the most important of these is an oscillation-enhanced mechanism which produces a single segment drop. This mechanism is similar to that shown in Figure 4 (b and c) for the low nozzle-velocity region except that following the breakaway of a single segment drop the jet contains several well formed segments and because of this the next breakaway of a drop may also be by an oscillation-enhanced mechanism. In the low nozzle-velocity region an oscillation-enhanced breakaway is always followed by a breakaway of the force-balance type [Figure 4(a)]. The oscillation-enhanced type of breakaway can occur either at the original first neck or at the new neck formed in front of the original neck. This makes some difference to the size of drop produced, particularly when the new neck is formed some way in front of the original first neck. In some instances breaks occurs at both necks and a small satellite drop is produced in addition to the single-segment drop.

An important additional mechanism is observed in this region. This is when the jet fails to break by an oscillation-enhanced mechanism at the first neck[Figure 7(a)] and the first neck moves with the disturbance over the surface of the second segment [Figure 7(b)] so that effectively liquid passes from the second segment into the first. When this neck has moved part way towards the second neck it grows [Figure 7(c)] and the first segment, which now contains a large fraction of the liquid which was previously in the second segment, detaches to form a drop which is intermediate in volume between a single segment drop formed by an oscillation enhanced mechanism and a double segment drop formed by an oscillation enhanced mechanism [Figure 7(d)]. After the drop has broken away the remaining liquid of the second segment enters the original third segment to produce an enlarged new first segment. However, this new first segment does not remain enlarged and by the time the next break occurs it is approximately normal size.

When the jet fails to break by one of the above oscillation-enhanced mechanisms, other mechanisms are observed. Occasionally triple segment drops are detached by an oscillation-enhanced mechanism. After detachment these sometimes divide into two drops by breaking at one of the necks, thereby producing a double segment drop and a single segment drop. Sometimes the oscillations of the jet following a breakup lead to an oscillation-enhanced breakup during the second oscillation of the jet segments. Although the second oscillation is of lower amplitude than the first the necks are more developed and this can lead to a breakup at either the first or second neck. In this case the segments are attached to the jet for a longer period than when the break occurs during the first oscillation. This allows additional liquid to enter the segments and the resulting drops are larger than those produced at the first oscillation.

As the nozzle-velocity is increased the average length of the jet before breakup increases [Figure 3] and the number of well defined segments on the jet increases. At the higher nozzle-velocities in this region there are always present segments which are sufficiently well developed that they could be detached by an oscillation-enhanced mechanism. However, periodically a break fails to occur by an oscillation-enhanced mechanism and a long growth period ensues followed by a force-balance type breakup. This is then followed by a number of oscillation-enhanced breakups of the types discussed above.

Intermediate Nozzle-Velocity Region

In this region there is no longer a linear relationship between $\overline{L}$ and nozzle-velocity, and as the nozzle- velocity is increased the jet length varies in an irregular fashion [Figure 3]. Some jets contain a large number of well defined segments whereas others contain few segments and have a long undisturbed section. Jets having a long, undisturbed section are generally longer than those containing many segments. In all cases the jet has an undisturbed section near to the nozzle and necks are first observed some way along the jet. The necks travel towards the end of the jet.

Drops are formed by a range of mechanisms which are similar to those described above for the moderate nozzle-velocity region. However, as the flowrate is increased the proportion of single segment drops formed by an oscillation-enhanced mechanism increased until, for each nozzle at a particular nozzle-velocity, only single segment drops are formed. This is the much discussed and well studied[3,10] nozzle-velocity which produces the maximum interfacial area per unit volume of dispersion. In this case, prior to each breakup the first neck and the new neck formed at the position of the disturbance are close to each other. The half-period of segment oscillation, the period between the arrival of successive necks at the end of the jet and the time for the disturbance to travel over the surface of the first segment, are well matched so that breaks occur in a regular manner to give only single segment drops by an oscillation-enhanced mechanism. This produces a dispersion which contains small drops of a uniform size.

As the flowrate is increased above the value for maximum interfacial area per unit volume of dispersion, drops are once again formed by the other mechanisms and the proportion of single segment drops formed by the oscillation-enhanced mechanism decreases. The segments at the end of the jet become distorted and the major axis of these segments is no longer along the axis of the jet. Due to this distortion it becomes increasingly difficult to categorise the breakup as the flowrate is increased. However, most drops are formed by oscillation-enhanced mechanisms.

High Nozzle-Velocity Region

In this region a jet contains few segments and has a long section, stretching from the nozzle, which is free from disturbance. The segments at the end of the jet are distorted and it is difficult to categorise the breakup. However, most drops are formed by oscillation-enhanced mechanisms.

Summarising, it can be seen that drops are formed from a jet of liquid by a number of mechanisms. At nozzle-velocities below the critical value for sudden jet lengthening drops are formed alternately by two mechanisms. The first is a force-balance type of mechanism and the second is an oscillation-enhanced mechanism. At nozzle velocities above the critical value drops are mainly formed by oscillation-enhanced mechanisms. In this case the basic mechanism is for the jet oscillations to promote the separation of single segments of the jet. At the nozzle velocity which gives the maximum interfacial area per unit volume of dispersion all of the drops are formed by this basic mechanism. At other flowrates the period between arrival of the necks at the end of the jet and the half-period of segment oscillation are not sufficiently well matched for the jet to break consistently at the first neck. Drops are then formed by some or all of the following oscillation-enhanced mechanisms

1) Oscillation-enhanced break-away of two segments and occasionally three segments.

2) Oscillation-enhanced division of the second segment to produce a drop which is intermediate in size between a single segment drop and a double segment drop.

3) Oscillation-enhanced breakup at the second oscillation of the segments leading to a large single segment or a large double segment drop.

In addition, when the oscillations of the jet do not induce a break and the oscillations are damped out, the next breakup occurs by a force-balance type mechanism of the sort observed in the low nozzle velocity region.

If the theoretical models for jet breakup discussed by Rayleigh and subsequent workers are now examined using the above findings it will be seen that these models predict the normal volume of a segment of the jet as the segment reaches the end of the jet. This is equal to the volume of a single-segment drop produced by an oscillation-enhanced mechanism. This drop is the smallest found in the dispersion. This explains why the theoretical predictions for drop size made by Scheele and Meister[3,11] and by Bright[12] agree better with the experimental minimum drop size than with the mean drop size. It is of interest that in liquid-gas systems there is a similar finding and the theoretical predictions of Tyler[13] show better agreement with the minimum drop size.

There remains the task of obtaining a measure of the range of drop sizes in the dispersion. It will be noted that most of the drops in the dispersion have volumes intermediate between single-segment drops and double-segment drops. If it is assumed that double-segment drops have twice the volume of single segment drops then most of the drops in the dispersion will have volumes intermediate between the theoretical volume and twice that volume.

In a subsequent paper, tables will be presented giving individual drop sizes and mechanisms of formation for chlorobenzene jets breaking in a continuous aqueous phase and the experimental results for drop size will be compared with the sizes predicted theoretically.

CONCLUSIONS

It has been shown that a jet of liquid breaks into drops by a number of mechanisms. At nozzle velocities below the critical nozzle velocity for sudden jet lengthening, single segment drops are detached alternately by two mechanisms. One is of the force-balance type and the other is oscillation-enhanced. At nozzle velocities above the critical value additional oscillation-enhanced mechanisms are observed. These produce drops containing two segments, one complete segment and part of a second segment, and occasionally three segments. The volumes of these segments are increased if the break occurs at the second oscillation of the segments instead of at the first oscillation.

Visual observations of the breakup process show that the theoretical models discussed by Rayleigh and subsequent workers should be used to predict the volume of the single segment drops produced by an oscillation-enhanced mechanism. These are the smallest drops found in the dispersion. The volumes of most of the individual drops in the dispersion should lie between the theoretical volume and twice this volume.

REFERENCES

1. RAYLEIGH, Lord
   1879. Proc. Royal Soc., 29, 71.

2. CHRISTIANSEN, R.M. and HIXEN, A.N.
   1957. Ind. Eng. Chem., 49, 1017.

3. MEISTER, B.J. and SCHEELE, G.F.
   1969. AIChE J., 15(5), 700.

4. EDGE, R.M. and GRANT, C.D.
   1971. Chem. Eng. Sci., 26, 1001.

5. ELHAG, E.A.
   1986. PhD thesis, University of Strathclyde, 142-147.

6. MERRINGTON, A.C. and RICHARDSON, E.C.
   1947. Proc. Royal Phys. Soc., 59, 1.

7. MOTTEM, J.W.
   1978. AIChE Eng. Symp., Ser. 74 (181), 75.

8. STROBEL, W.J. and WHITAKER, S.
   1969. AIChE J., 15(4), 527.

9. MEISTER, B.J. and SCHEELE, G.F.
   1969. AIChE J., 15(5), 689.

10. KEITH, F.W. Jr. and HIXSON, A.N.
    1955. Ind. Eng. Chem., 47, 258.

11. SCHEELE, G.F. and MEISTER, B.F.
    1968. AIChE J., 14(1), 15.

12. BRIGHT,
    1985. Chem. Eng. Res. Des., 63, 59.

13. TYLER, E.
    1933. Phil. Mag., 16, 504.

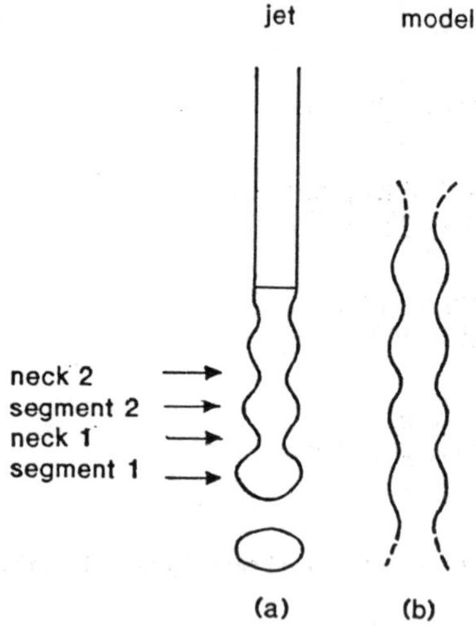

FIG 1. MODEL USED FOR BREAKUP THEORY (ALSO SHOWING TERMINOLOGY)

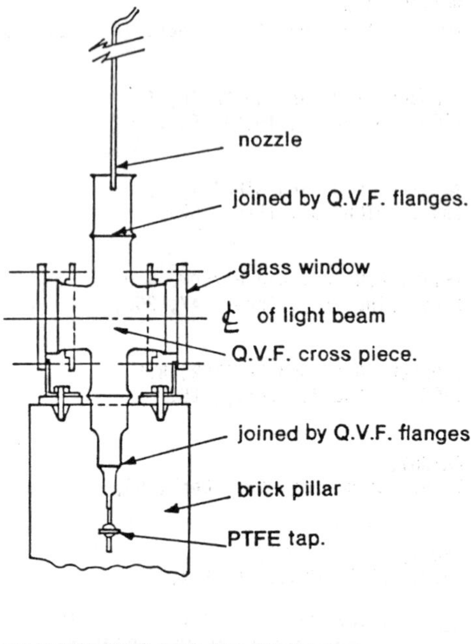

FIG. 2. THE APPARATUS AND ITS SUPPORT.

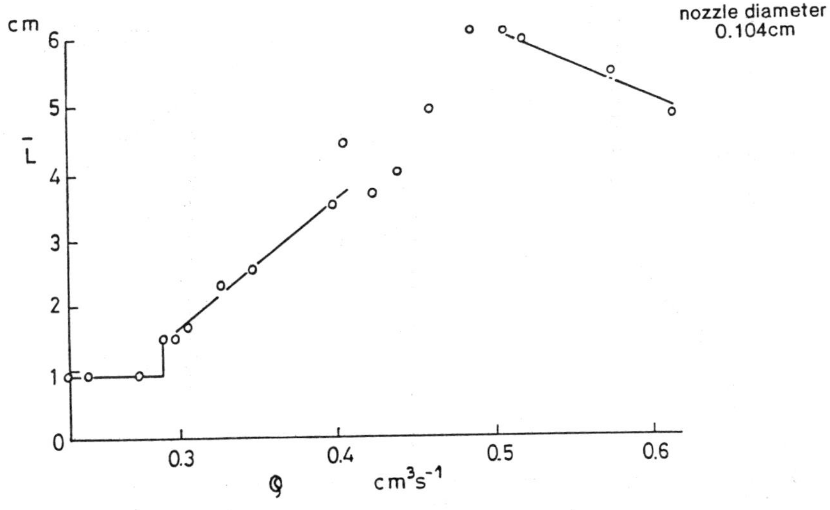

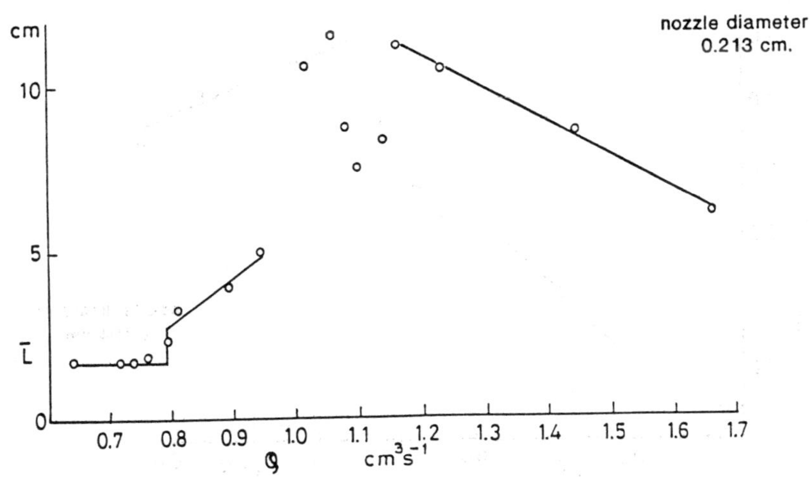

FIG.(3a) VARIATION OF $\bar{L}$, THE MEAN LENGTH OF THE JET IMMEDIATELY BEFORE BREAKUP, WITH Q, THE FLOWRATE THROUGH THE NOZZLE.

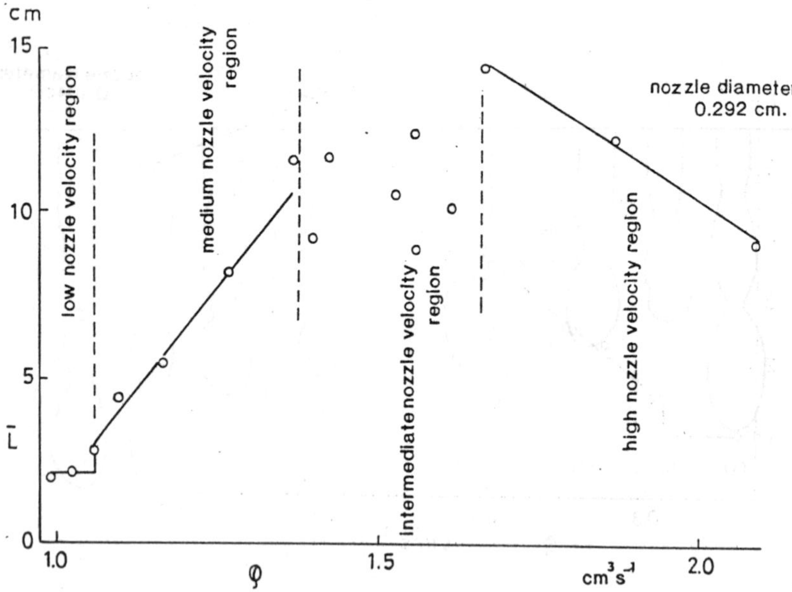

FIG. (3a) CONTINUED

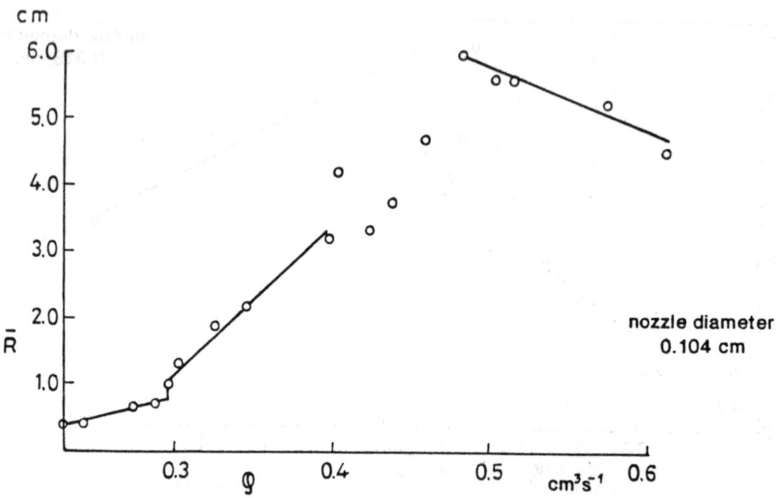

FIG (3b) VARIATION OF $\bar{R}$, THE MEAN LENGTH OF THE JET IMMEDIATELY FOLLOWING BREAKUP, WITH φ, THE FLOWRATE THROUGH THE NOZZLE.

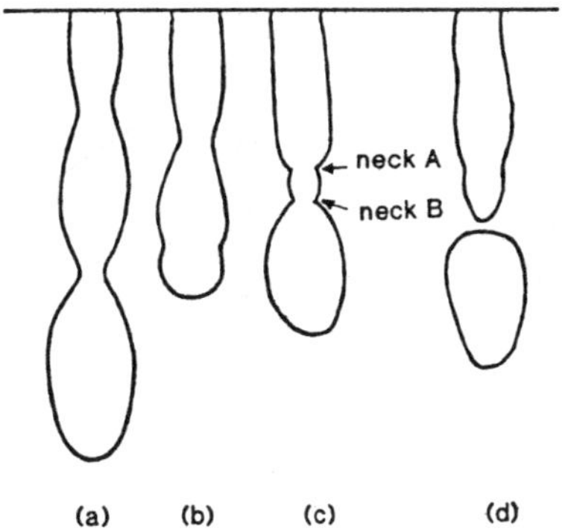

FIG. (4) JET BREAKUP IN THE LOW NOZZLE-VELOCITY REGION

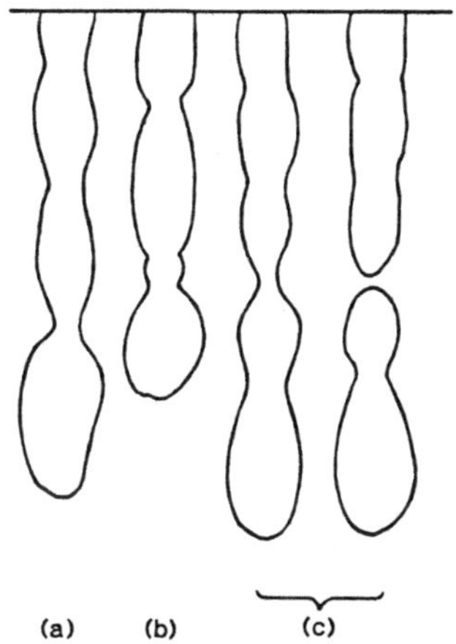

FIG. (5) JET BREAKUP AT FLOWRATES IMMEDIATELY ABOVE THAT FOR SUDDEN JET LENTHENING

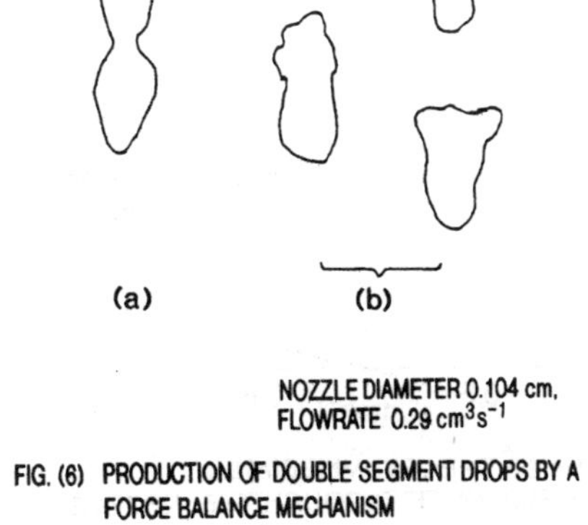

NOZZLE DIAMETER 0.104 cm,
FLOWRATE 0.29 cm^3s^{-1}

FIG. (6) PRODUCTION OF DOUBLE SEGMENT DROPS BY A FORCE BALANCE MECHANISM

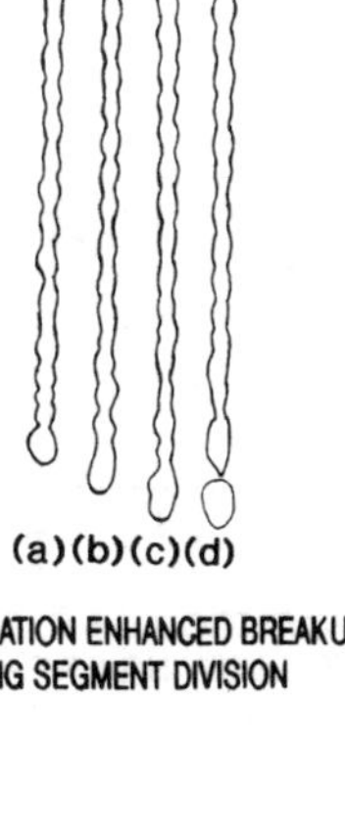

FIG. (7) OSCILLATION ENHANCED BREAKUP CAUSING SEGMENT DIVISION

TERMINAL VELOCITY RELATIONSHIPS OF TWO-PHASE DROPS

M.A.Hashim* and M.Balasubramaniam

Two-phase drops are formed when air is incorporated into dispersed phase drops. For systems in which the dispersed phase is less dense than the continuous phase, the relationship between the terminal velocity and the ratio of the air phase to the dispersed phase (β) shows maximal values. Up to these values, the drops exhibit swinging motions. These values correspond to the onset of shape oscillations and the frequency of oscillations increase with increasing β values. Correlations are proposed for the onset of shape oscillations and for the terminal velocities for both non-oscillating and oscillating swinging drops.

INTRODUCTION

Spray columns are often used in liquid-liquid extraction processes as they offer simplicity of design and low capital and operating costs. However, efforts to improve their efficiency through increasing or decreasing the size of the dispersed phase drops have met with very little success; smaller drops result in limited throughput due to their low rising velocities whereas larger drops imply a reduction in interfacial area.

Investigations by Pajak and Kaczmarski(1) and Hayakawa and Shigeta(2-4) involve the application of a new concept of two-phase drops. Two-phase drops consist of both liquid and gaseous phases and are formed when a gaseous medium is incorporated into dispersed phase drops.

In their work on air agitation in various extractors, Hashim(5) and Suppiah(6) had also observed the formation of two-phase drops and subsequently suggested the usefulness of this phenomena in increasing mass transfer rates. They had also implied that owing to the minimum energy requirements, mechanical simplicity and hence low maintenance costs, air agitated extractors could be particularly beneficial to the nuclear industry.

The two-phase drops are characterized by an accelerated rectilineal motion which is often accompanied by secondary motions such as swinging and shape oscillations which bring about an increase in mass transfer. The investigations(1-4) mentioned above were confined to limited systems and the correlations formulated

*Department of Chemical Engineering, University of Malaya, Kuala Lumpur, Malaysia

were for swinging and non-swinging drops with no distinction between oscillating and non-oscillating swinging drops. In addition, no effort has been made to correlate the maximal terminal velocity of such drops to the volumetric ratio of air to the dispersed phase (β).

In the present investigation, terminal velocities of two-phase drops are reported for systems where the dispersed phase is both denser and lighter than the continuous phase. Correlations are also developed for swinging drops characterized as oscillating and non-oscillating.

EXPERIMENTAL

Figure 1 illustrates the experimental set-up and Figure 2 provides details of the spray nozzle. The spray nozzle which consists of two concentric tubes was attached to the base of the glass column of height 1050 mm and an internal diameter 70 mm. Air was injected from the inner tube into the dispersed phase flowing in the annular space of the concentric tubes.

Glass syringes, accurate to 0.5 x 10^{-9} m^3, was used to deliver precisely a measured quantity of air, and a precision dosing pump, accurate to 0.1 x 10^{-9} m^3, was used to deliver measured amounts of the dispersed phase. The systems employed distilled water as the continuous phase and air as the gaseous phase. The solvents employed as the dispersed phase were: benzene, toluene, xylene, hexane and carbon tetrachloride and their physical properties at 25°C are shown in Table 1. The values of surface tension and interfacial tension refer to water-saturated solvents or solvent-saturated water.

A number of dispersed phase drops were pumped through the nozzle at the start of each experiment so as to saturate the continuous phase with the dispersed phase in order to minimize the effect of mass transfer. The time taken for each drop to traverse 500 mm along the column was recorded. All measurements were done in triplicates, and the average values recorded.

TABLE 1 - Physical properties of the solvents

Solvent	Density (kgm^{-3})	σ_C (x 10^3 Nm^{-1})	σ_D (x 10^3 Nm^{-1})	σ_{CD} (x 10^3 Nm^{-1})	Solvent Purity(%)
Benzene	879	62.2	28.8	35.0	99.9
Toluene	861	72.0	27.4	26.0	99.9
Xylene	861	72.8	28.4	37.8	98.9
Hexane	660	70.7	18.2	52.6	99.0
Carbon Tetrachloride	1594	69.6	26.6	44.5	99.8

RESULTS AND DISCUSSION

Relationship Between β And Terminal Velocity

The relationship between the terminal velocity of the two-phase drops and β is illustrated in Figure 3. The plots are of similar pattern, with each exhibiting a maximum and a minimum terminal velocity at particular β values. A small increase in β above the minimum results in the terminal velocity rising at a slower rate.

The above behaviour can be explained by considering the motions of the two-phase drops. For small overall drop diameter with β less than 0.3, the drops rise in a rectilineal manner. An increase in β results in increased buoyancy which should increase their terminal velocities. However, this would cause instability with respect to the shapes of the drops as was observed in the experiments where there was a periodic change of the liquid portion from prolate to oblate ellipsoids. The drops are then said to oscillate.

Oscillating drops no longer travel in a rectilineal manner but are accompanied by superimposed swinging motions about the axis normal to the direction of motion as they follow zig-zag or spiral trajectories. As energy is required for swinging and for shape changes, the amount of energy available as kinetic energy for vertical translation is reduced, resulting in a decrease in terminal velocity. Further increase in β results in the drops exhibiting more violent oscillations and other secondary motions, thus further suppressing an increase in terminal velocity. This explains the relatively constant terminal velocities at large β.

The decrease in terminal velocity immediately after the onset of shape oscillations was comparatively small for the three aromatic systems as compared with the hexane system investigated. This could be accounted for, in part, to the high interfacial tension between hexane and water. For such a system, the increase in surface energy for a given deviation from the most stable shape dictated by force balances is greater than for systems of low interfacial tension. As the driving force to eliminate shape deviations is proportional to the increase in surface free energy, oscillations tend to be more intense in the water-hexane-air system. This results in lower terminal velocity.

Figure 4 shows the relationship between the terminal velocity and β for the water-carbon tetrachloride-air system in which the density of the dispersed phase is greater than that of the continuous phase. The plot shows the expected rise in terminal velocity for β values less than 1 but it does not exhibit any turning point. The rate of increase in terminal velocity gradually decreases with increasing β values. As was observed by Pajak and Kaczmarski(1), when the dispersed phase is denser than the continuous phase, the drops begin to oscillate at β values much higher than that initiating the rise of the drops. Moreover, the oscillations tend to be less violent as gravitational force is greater in the case of carbon tetrachloride whilst buoyancy force dominates in the case of air. This results in greater stretching of the drops and helps to dampen shape changes.

Onset Of Shape Oscillations

The onset of shape oscillations at a definite β value is a characteristic of a system. Oscillations are determined to a large extent by the ability or disability of the drop to maintain its stable shape, which is in turn governed by its inertia and interfacial tension. As such, the point of transition from non-oscillating to oscillating motion (β_o) can be correlated in terms of the density of the dispersed and continuous phases, surface tension and interfacial tension. This will result in a relationship of the form

$$\beta_o = f(\rho_D/\rho_C, \sigma_C/\sigma_{CD}) = f(\gamma', \alpha) \quad (1)$$

Using the least squares method, the correlation which provides values of β_o has the form

$$\beta_o = -3.16 + 3.01\,\alpha^{\gamma'} - 4.18\,\gamma'^{1.5}\,\alpha^{\alpha-1} \quad (2)$$

with a maximum error of $\pm$ 1%.

Drag Coefficient And Internal Circulation

Figure 5 shows the relationship between drag coefficients and Reynolds numbers for two-phase drops and the line for rigid spheres is included for the purpose of comparison. It is clear that the plot for two-phase drops lies above that for rigid spheres, indicating higher drag coefficients. Deformation of drops from the spherical shape appears to cause an increase in drag coefficient. The variation of drag coefficient with Reynolds number agrees with the work of Hayakawa and Shigeta(2,3). A minimum in the plot is observed at a Reynolds number of approximately 760. Beyond this value, there is an increase in drag coefficient as shape oscillations become more intense and violent, as mentioned previously.

Figure 6 shows the plot of Reynolds numbers against Eötövs numbers. Such a plot will give an indication of the dependence of terminal velocity on viscosity ratio(7). The scatter of the points could be attributed in part to the differences in viscosity ratio, thus inferring the presence of internal circulation or to the differences in the values of β as this factor is not present in the case of single-phase drops.

As the Eötvös numbers(7) in these experiments were all smaller than 4, it can be inferred that internal circulation was absent. As wakes were repeatedly observed in the oscillating two-phase drops, this strengthens the argument that internal circulation is absent as the latter is known to delay wake formation.

Terminal Velocity Correlations

The prediction of the behaviour of two-phase drops is best facilitated by dimensional analysis.

For the case of a rigid sphere moving with a steady velocity in a gravitational field its motion is governed by an equation of the general form

$$f(g\Delta\rho, \rho_c, \mu_c, d_e, u_t, \bar{\sigma}) = 0 \qquad (3)$$

However, for a non-rigid particle subjected to various forms of secondary motions, the particle to continuous phase density ratio (γ) is a significant variable. In addition, β is another important variable as it affects the shape and terminal velocity of the particle.

If the following assumptions are made:

a. Single two-phase drop rising in a static continuous phase is free from wakes or other forms of disturbances,
b. No mass transfer between the dispersed and continuous phase,
c. Absence of wall effects,
d. Absence of internal circulation, and
e. $390<Re<967$ and $0.04<\beta<4.50$,

the motion of a two-phase drop would then be governed by a relationship of the form

$$f(\Pi_1, \Pi_2, \Pi_3, \gamma, \beta) = 0 \qquad (4)$$

This results in the dimensional groups of Reynolds number (Re), Eötvös number (Eo) and the surface tension group (S).

The equation describing the motion of a two-phase drop is thus

$$Re = k\, Eo^a\, S^b\, \beta^c\, \gamma^d \qquad (5)$$

where k, a, b, c and d are constants.

Equation (5) thus provides a method of estimating the terminal velocity of two-phase drops. The constants in Equation (5) were determined by multiple regression of the experimental data, and the results are shown below:

a. For non-oscillating swinging two-phase drops

$$Re = 1.18\, Eo^{0.25}\, S^{0.54}\, \beta^{0.02}\, \gamma^{0.26} \qquad (6)$$

b. For swinging oscillating two-phase drops

$$Re = 1.33\, Eo^{0.35}\, S^{0.55}\, \beta^{0.20}\, \gamma^{0.53} \qquad (7)$$

Figures 7 and 8 show the deviation of the experimental data from the calculated terminal velocities. The error in calculating the terminal velocity of non-oscillating two-phase drops is $\pm 10\%$ and for oscillating drops is $\pm 5\%$.

Comparing Equations (6) and (7) it is evident that there is a greater dependence on β for oscillating drops. This is probably due to the greater degree of secondary motions associated with these drops. In addition to increasing the buoyancy it also affects the effective particle interfacial tension which in turn affects the degree of oscillations and hence influences the terminal velocity of the drops.

ACKNOWLEDGEMENTS

The authors would like to express their gratitude to all persons who have contributed to this work in any way and in particular to S.S.Gan and P.S.Yap.

SYMBOLS USED

C_D = drag coefficient

d_e = equivalent diameter of particle (m)

Eo = Eötvös number

g = gravitational acceleration (ms^{-2})

Re = Reynolds number

S = surface tension group

u_t = terminal velocity (ms^{-1})

Subscripts

C = continuous phase

CD = interface between continuous and dispersed phases

D = dispersed phase

Greek letters

α = σ_C/σ_{CD}

β = volumetric ratio of air to dispersed phase

β_o = value of β at the onset of shape oscillation

γ = dispersed to continuous phase density ratio

γ' = particle to continuous phase density ratio

Δ = difference

μ = viscosity (Nsm^{-2})

ρ = density (kgm^{-3})

σ = surface tension (Nm^{-1})

$\bar{\sigma}$ = surface average interfacial tension (Nm^{-1})

Π_1, Π_2, Π_3 = dimensionless groups

REFERENCES

1. Pajak, M., and Kacmarski, K., 1984, Chem Eng. Commun. 26, 173.

2. Hayakawa, T., and Shigeta, M., 1974, J. Chem. Eng. Japan 7(2), 140.

3. Hayakawa, T., and Shigeta, M., 1973, J. Chem. Eng. Japan 6(6), 545.

4. Hayakawa, T., and Shigeta, M., 1973, J. Chem. Eng. Japan 6(3), 273.

5. Hashim, M.A., 1980, PhD Thesis, University of Birmingham, England.

6. Suppiah, S., 1980, PhD Thesis, University of Birmingham, England.

7. Clift, R., Grace, J.R., and Weber, M.E., 1978, "Bubbles, Drops and Particles", Academic Press, New York, USA.

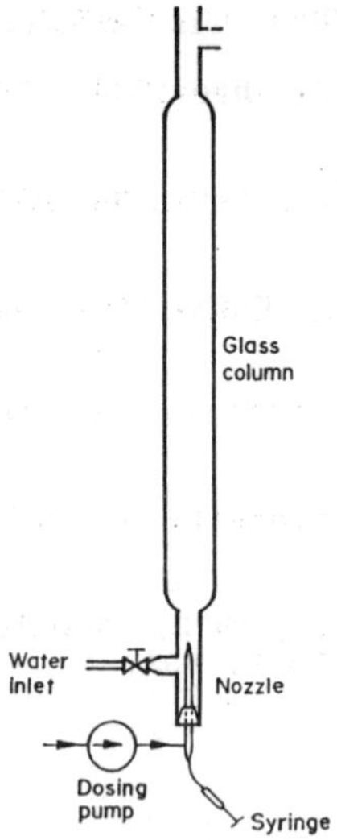

Figure 1 Schematic diagram of the experimental set-up

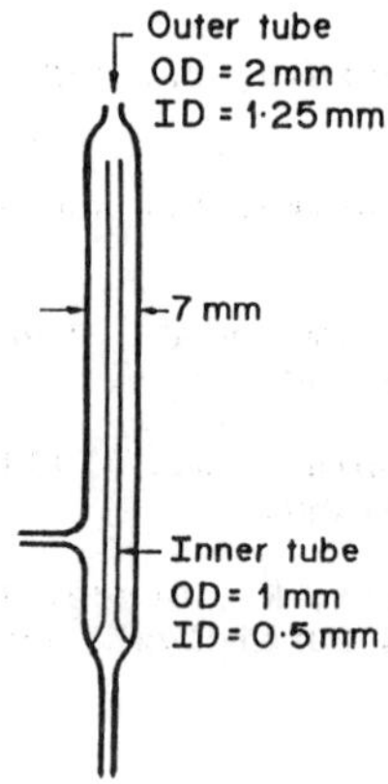

Figure 2 Spray nozzle

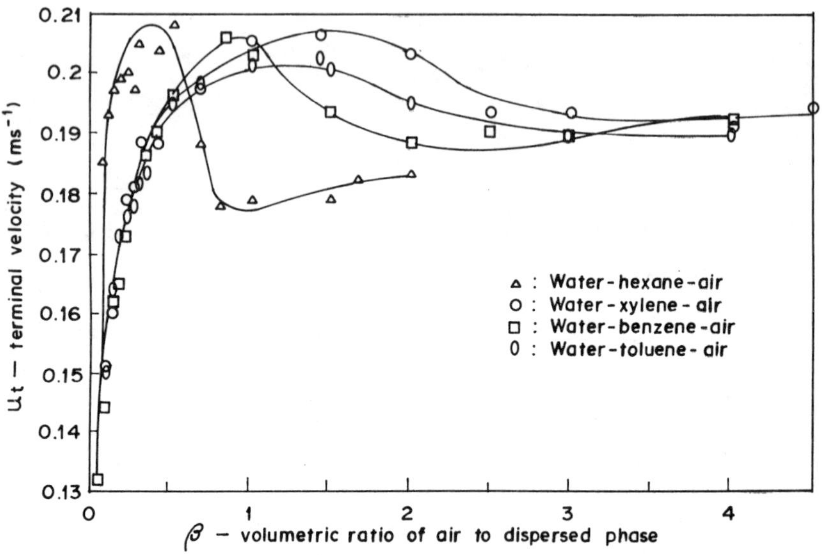

Figure 3 Relationship between terminal velocity and β

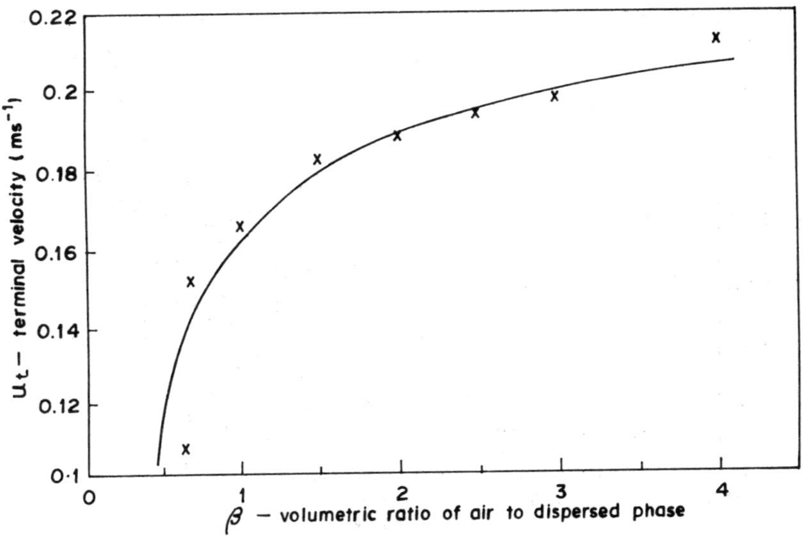

Figure 4 Relationship between terminal velocity and β for water-carbon tetrachloride-air system

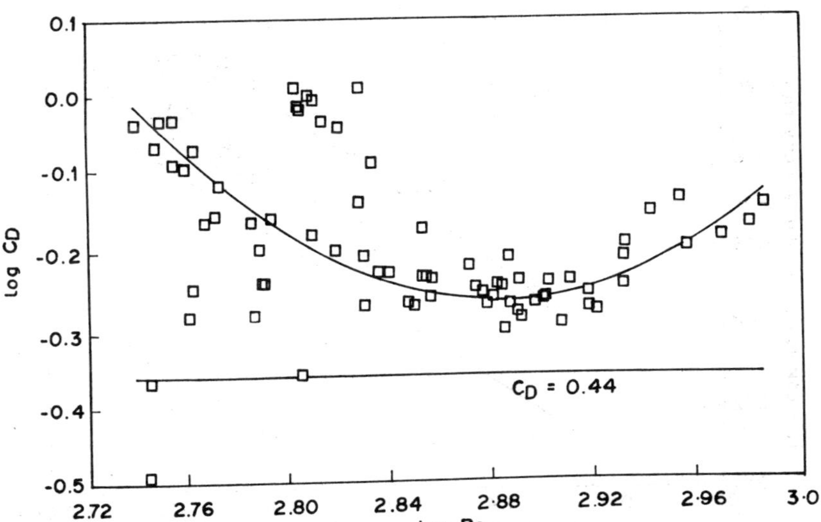

Figure 5 Dependence of drag coefficient on Reynolds number

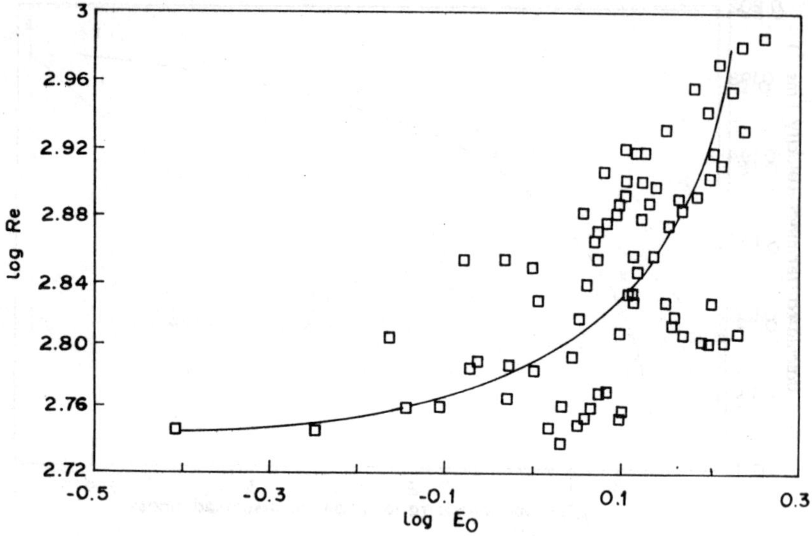

Figure 6 Effect of viscosity ratio on terminal velocity

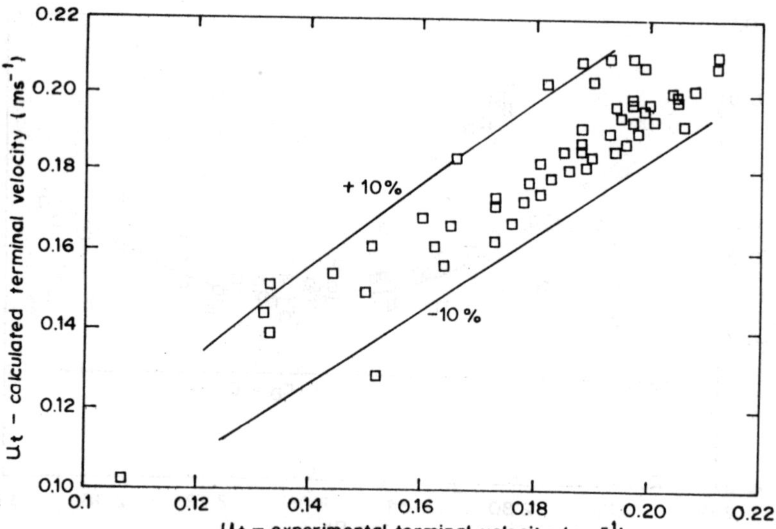

Figure 7 Comparison of calculated and experimental terminal velocity for non-oscillating two-phase drops

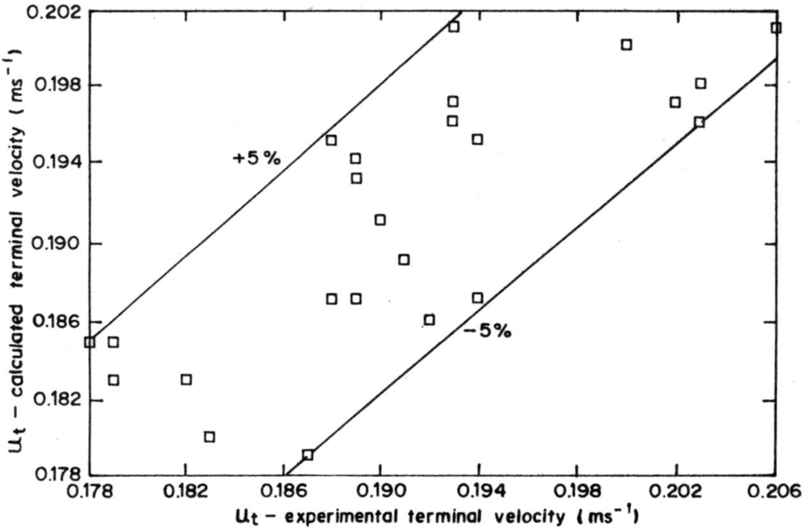

Figure 8 Comparison of calculated and experimental terminal velocity for oscillating two-phase drops

VISUALISATION OF THE MOVEMENT OF DROPS IN A PACKED VESSEL CONTAINING A LIQUID CONTINUOUS PHASE

N.M.W. AlSharbati and R.M. Edge*

A method is described for visualising the distribution of both the packing and the dispersed phase in a packed vessel. In this method the refractive indices of the three phases are matched and a fluorescent dye is added to the dispersed phase. The presence of the packing and the dispersed phase in various planes in the vessel is detected by tracking a beam of laser light in these planes and photographing the fluorescent light given off by the various phases. The choice of the liquid phases and the fluorescent dye is discussed. Details are given of the method used to free the packed vessel from air bubbles.

INTRODUCTION

Much experimental work has been carried out in the fields of liquid-liquid extraction and coalescence in packed vessels. In spite of this work, there is little information available on how the two liquid phases move in the packing and how coalescence and breakdown of the drops is occurring. Most of the available information has come from visual observations which have been made close to the wall of a vessel or close to a probe inserted into a vessel where wall effects are large.

In 1985 the Chemical Engineering Division of Strathclyde University was asked to propose methods for visualising the viscous fingering which occurs under certain circumstances when an aqueous phase is used to drive an oil phase from a bed of particles. Viscous fingering occurs in oil fields and seriously limits the amount of oil which can be recovered from the field. One of the methods which was proposed and developed (1) uses an optical technique and this technique is well suited to the study of the flow of two liquid phases through packed beds in several fields which are of interest to the chemical engineer, particularly those dealt with in this paper.

The optical technique used in the fingering work was developed from the method used by Allak and Jeffreys(2) to study the motion of drops in two-phase systems. They matched the refractive indices of the two phases and added a phototropic dye to the dispersed phase. This dye was sensitive to ultraviolet light and by illuminating regions of the system with ultraviolet light they observed the movement of drops in this region.

*Chemical Engineering Division, University of Strathclyde, Glasgow, G1 1XJ.

Matching of the refractive indices in two-phase systems is not difficult and has been used by several authors in their experimental work[3-6]. Matching in three-phase systems, where one of the phases is a solid, is more difficult. Burdett et al, who studied liquid-liquid dispersions in a packed column, achieved some success by using a packing made from silicone rubber, which has a low refractive index[7]. This made matching easier. They added the dye nigrosine to the dispersed phase and used a light absorption technique to obtain holdup measurements in their matched system.

More recently, and whilst the procedure described in the present paper was being developed, a paper was published by Chen and Wada[8] in which the authors described a method for visualising the distribution of two immiscible liquid phases in a three-dimensional bed of glass particles. Their method had many similarities to that described in the present paper except that a sheet of laser light was used to illuminate planes in the bed and no fluorescent dye was used. These authors published several photographs showing the distribution of the three phases in the bed but gave few details of the experimental methods or of the compositions of the matched phases. Their results were of preliminary work but they indicated that they intended to use the method in studies of the fingering of aqueous drives in oil-related work.

OBJECT

To develop a flow visualisation method for use in the study of the flow of a dispersed liquid phase through a packed vessel saturated with a continuous liquid phase.

APPARATUS AND PROCEDURE

Summary of the Flow Visualisation Method

A fluorescent dye is added to the dispersed phase. The refractive indices of both liquid phases are matched to that of the packing and the two phases are contacted to establish equilibrium. Light from a laser is used to illuminate a plane in the bed and the fluorescent dye in this plane is activated and emits light. Photographs of the emitted light show clearly the position of the dispersed phase in this plane. Various planes are illuminated to build up a three-dimensional picture of the dispersed phase in the bed.

Apparatus

The apparatus used in the development work is shown in Figure 1. It contained a packed column which was 46 cm long and had a 10.2 cm square cross-section. This was mounted vertically and the entry to it and the exit from it were located centrally. The column was packed with Pyrex Raschig rings which were 7 mm in diameter and 7 mm long. The liquid phases were stored in glass vessels and were fed to the column by means of peristaltic pumps which had gas filled vessels on their discharge side to smooth the flow. The apparatus was situated in a constant temperature room which was controlled at 20°C.

Selection of the Packing

In order that the visualisation method can be used, it is necessary to use a packing which is transparent to light. In the present work glass was used. Unfortunately, glass has a relatively high refractive index and this limits the range of liquids which can be used. This is particularly so if a glass with a high refractive index is chosen. Therefore, Pyrex was used in the current work. This had a refractive index of 1.471, which is low for a glass. This was annealed at 560°C to relieve stresses, thereby producing a product with a more uniform refractive index.

Selection of the Liquid Phases and the Fluorescent Dye

The main requirements for the two liquid phases are:

1) They must have the same refractive index as Pyrex glass.

2) They must be readily available, inexpensive and relatively non-toxic.

3) It is desirable that the physical properties of the liquids can be varied so that conditions met in a wide range of systems can be simulated.

A wide range of solutions was examined and the following were chosen for the development work:

Continuous Phase

The main constituent of the continuous phase was a mineral oil which was supplied by Messrs. Aldrich. This has a refractive index of 1.4778 and a viscosity of 2.01×10^{-1} kg m^{-1} s^{-1} at 20°C. This was diluted with kerosene to produce an oil with a refractive index of 1.471 at 20°C which matched the refractive index of the glass packing. This mixture had a viscosity of 4.6×10^{-2} kg m^{-1} s^{-1} and a density of 858 kg m^{-3} and contained $83.6\%_v$ mineral oil. In future work, when a viscosity other than 4.6×10^{-2} kg m^{-1} s^{-1} is required, extra components will be added to the continuous phase. Suitable components are xylene and coconut oil. By using mixtures of xylene, kerosene, mineral oil and coconut oil a range of viscosities between 1.6×10^{-3} and 1×10^{-1} kg m^{-1} s^{-1} can be obtained. Carbon tetrachloride will be added to the continuous phase when a higher density is required.

Dispersed Phase

A dispersed phase with a high refractive index is produced by adding a salt, with a high refractive index, to water. Unfortunately, many salts which have a high refractive index are insufficiently soluble to produce a solution with the required refractive index. Also, many solutions have to be rejected because of toxicity. For the present work an aqueous solution of potassium thiocyanate was chosen. Potassium thiocyanate has a relatively low toxicity, is readily available and is inexpensive. The solution contained $62.3\%_w$ of potassium thiocyanate and had a viscosity of 2.2×10^{-3} kg m^{-1} s^{-1} and a density of 1.36×10^3 kg m^{-3}. In future work, when a higher viscosity is required, sucrose will be added as a third component.

Fluorescent Dye

Fluorescein was chosen as the fluorescent dye. For this work, a laser quality dye is not required and the normal laboratory reagent can be used. This is readily available and is inexpensive. It has been used in laboratories for many years and there do not appear to be any serious health hazards associated with its use. This is not the case with many dyes currently being used in laser work. It has a good absorption for the light from an argon laser. Also, the fluorescent light from the dye and the incident light from the laser have wavelengths which are sufficiently far apart that any scattrered incident light can be filtered from the fluorescent light reaching the camera.

Thornton et al (9) have stressed that when dyes are added to liquid-liquid systems the hydrodynamics can be affected by the surface activity of the dye. In the present work, the dye was used at low concentrations. Also, the interfacial tension, as measured by the drop weight method, was the same when the dye was present as when it was absent. Although this does not prove that there was no surface activity associated with the dye it is probable that any surface activity of the dye was no greater than that associated with impurities in the presently used organic phase. The extent of this surface activity will be examined as the system is further refined.

Light Source and Optical Set-Up

It is logical to use a sheet of light produced from a laser to illuminate the plane in the bed. However, preliminary experiments indicated that to get a sheet of light with a sufficient intensity requires an expensive, water-cooled, high-powered, laser(1). Air cooled lasers with beam power of up to 200 mW are much less expensive.

For the development work, an argon ion laser with a beam power of 20 mW was used and the beam was tracked through the bed in the opposite direction to the direction of motion of the drop. This gave satisfactory photographs (Fig. 4). However, it should be noted that the instant of recording the photographic image varied throughout the bed because the laser beam took about 10 s to travel the length of the bed. In future work this time will be reduced by using a higher powered laser. The laser was set up to give blue lines only and the output at 514.5 nm was removed as this is too close to the wavelength of the light emitted by the dye. Any of the incident laser light which was scattered or reflected in the column towards the camera was removed by a filter, positioned in front of the camera lens. A sharp-cut long-pass filter giving 50% transmission at 515 nm was found to be satisfactory for this purpose.

The optical set-up is shown in Figure 1b. The beam from the laser was turned through 90° by mirror A so that it travelled parallel to the walls of the column. A second mirror, which was mounted on a P210 pneumatic rodless cylinder supplied by Messrs. Origa, tracked the beam along the length of the column so that a plane in the bed was illuminated. The light given off by the fluorescent dye in this plane was recorded on high-speed colour film using a reflex camera. The shutter of the camera was open throughout the period of tracking. By varying the position of mirror A, various planes in the bed were examined. The exposed film was push developed to achieve a speed of 3200 ASA.

PROCEDURE

It was found that when the organic continuous phase was pumped into the bed a larger number of air bubbles was trapped in the bed. These cannot be seen in normal packed beds but can easily been seen in a transparent system [Figure 2]. The presence of these bubbles is undesirable, not only because they affect the hydrodynamics of the bed but also because they scatter the laser beam as it strikes them. These bubbles cannot be removed by putting the column under partial vaccuum before charging the continuous phase. Therefore, the following procedure was developed: prior to pumping oil into the bed, the air was thoroughly purged from the packing by passing butane gas through the bed. Oil was then pumped upwards into the bed and several pore volumes were passed through it until all but a few of the bubbles had been removed and the bed was transparent. Bubbles which were not displaced by the flow slowly dissolved into the oil. The presence of the dissolved butane in the oil affected its refractive index and any oil so affected has to be displaced from the bed by further purging. The refractive index of the displaced oil was readjusted back to 1.471 and it was re-used for purging. After the bed had been saturated with oil the aqueous phase was injected downwards into the bed and passed through the bed as drops.

RESULTS AND DISCUSSION

Figures 3, 4 and 5 show photographs of the packed bed which were taken using the above technique. Figure 3 was taken immediately after the bed had been saturated with oil and the distribution of the glass packing can be clearly seen.

Figure 4 was taken whilst the aqueous phase was passing into the column. In this case the distribution of the drops of the aqueous phase in each plane of the packing can be seen in addition to the position of the glass packing.

Figure 5 was taken after the flow of the aqueous phase had ceased and after a period had elapsed so that droplets of the aqueous phase could settle from the column. It shows the droplets of the dispersed aqueous phase which are trapped in the packing.

These photographs show that the method can be used to examine both the distribution of the dispersed phase and the distribution of the packing in a packed bed. The method has the advantage that regions of the bed well away from the wall can be examined and there is no need to disturb the bed with probes or other devices.

The above method should provide much useful information concerning the flow of two phases through a packing. However, before attempting to use it, the following points should be carefully considered:

1) The method is easier to use if the drops are moving at low velocities. Low drop velocities can be achieved by increasing the viscosity of the continuous phase and by reducing the size of the packing.

2) The smaller the size of the packing the more critical is the matching of the refractive indices of the three phases and the more difficult it is to get the laser beam to penetrate deeply into the bed without adverse refraction.

REFERENCES

1) AL SHARBATI, N.M.W., ROSSELAND, E. and EDGE, R.M.
To be published.

2) ALLAK, A.M.A. and JEFFREYS,G.V.
1974. Proceedings International Solvent Extraction Conference, Lyon, 265.

3) HANDLEY, D.
1957. PhD thesis, University of Leeds.

4) HUMMEL, R.L. and POPOVICH, A.T.
1967. Chem. Engng. Sci., 22, 21.

5) DAVIES, G.A., RUSHTON, E. and THOMAS, J.A.
1973. Coalescence of droplets in dispersions.
Symp. Interfacial Phenomena, AIChE National Meeting, New Orleans.

6) DYBBS, A. and EDWARDS, R.V.
1987. Second International Conference on Laser Anemometry, Glasgow. 1.

7) BURDETT, I.D., WEBB, D.R. and DAVIES, G.A.
1981. Chem. Engng. Sci., 36, 1915.

8) CHEN, J.D. and WADA, N.
1986. SPE Formation Evaluation, 205.

9) THORNTON, J.D., ANDERSON, T.J., JAVED, K.H. and ACHWAL, S.K.
1985. AIChE J., 31, 1069.

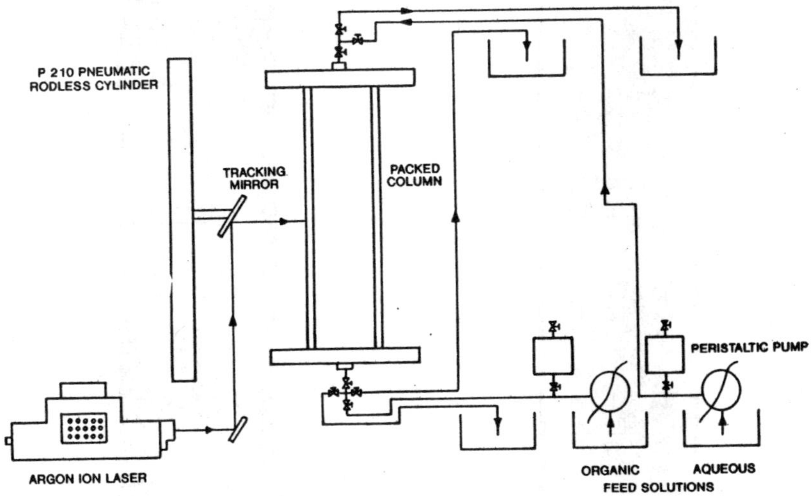

FIG. 1a APPARATUS (Front View)

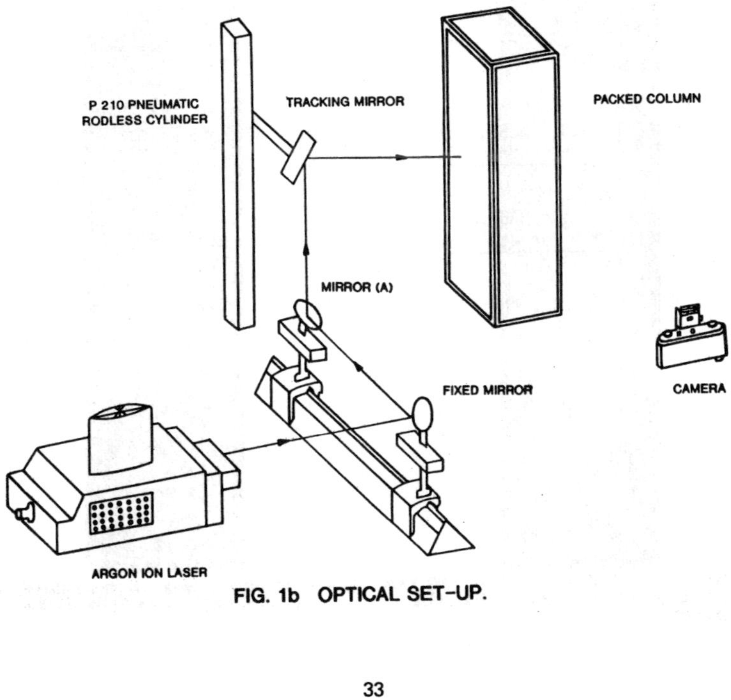

FIG. 1b OPTICAL SET-UP.

FIG. 2 PACKED BED SHOWING AIR BUBBLES TRAPPED IN THE PACKING.

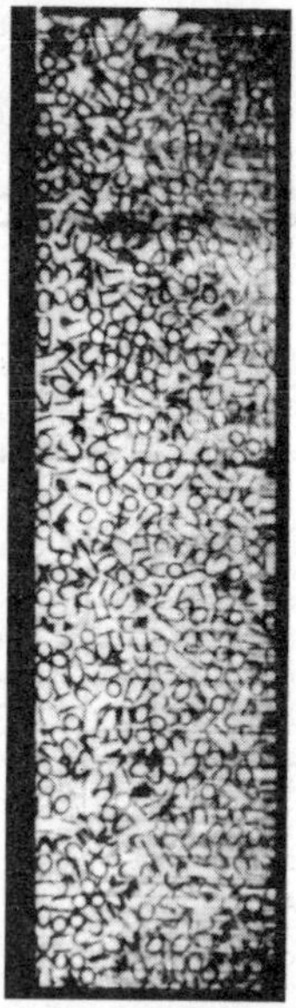

FIG. 3 PHOTOGRAPH SHOWING THE DISTRIBUTION OF THE PACKING IN A PLANE THROUGH THE COLUMN.

FIG. 4 PHOTOGRAPH SHOWING THE DROPS PASSING DOWN THE PACKED COLUMN IN A PLANE THROUGH THE COLUMN

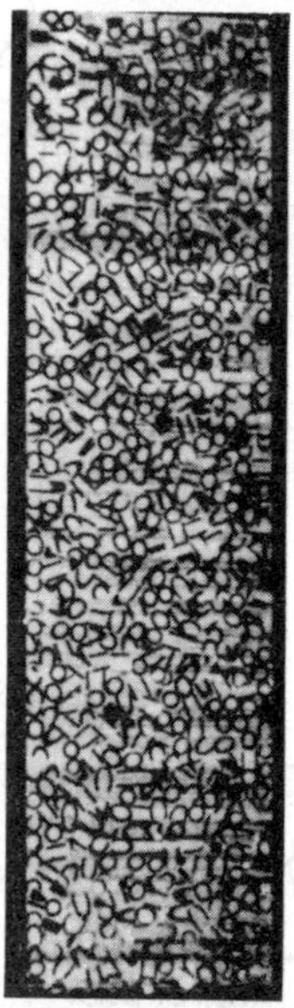

FIG. 5 PHOTOGRAPH SHOWING THE DROPS TRAPPED IN THE PACKING IN A PLANE THROUGH THE COLUMN.

PHENOL RECOVERY WITH SUPPORTED LIQUID MEMBRANES.

A.M. Urtiaga, M.I. Ortiz, A. Irabien*

Hollow fibre supported liquid membranes have been used for the recovery of phenol from aqueous solutions. Kerosene and mixtures of methyl isobutyl ketone with kerosene have been used as organic membranes. The influence of the distribution coefficient on the overall mass transfer parameter has been investigated and the boundary layer coefficient in the aqueous phase and the permeability coefficient in the membrane phase have been obtained.

INTRODUCTION

Phenol and phenolic derivatives are chemical products commonly encountered in aqueous effluents from manufacturing processes. They occur in industrial operations such as coking, oil refining, foundries and resin manufacturing.

Toxicity of phenolic effluents is well known and their treatment must be accomplished depending on the nature of the phenols, the volume and rate of discharge and the nature and volume of the receiving water. There are several methods of treating phenolic wastes, including recovery, incineration, adsorption, biological treatment and chemical oxidation. Phenols may be recovered by liquid-solvent extraction processes. This method is applicable when phenol concentration is at least 2,000 mg/l. However, high capital costs and relatively high number of operations are required, Lanoutte (1).

Liquid membranes offer a potentially attractive alternative to solvent extraction, in that they combine the processes of extraction, stripping and regeneration into a single step. In a liquid membrane the organic extractant phase is reduced to a thin liquid film separating the feed and stripping solutions. Two configurations are encountered:
a) Emulsion Liquid Membranes (ELM): The stripping and organic phases form a water in oil emulsion which is suspended in the feed solution. Several authors have studied their potential for environmental applications Salazar et al. (2) and phenol removal, Li (3), Kim et al. (4), Xijuan and Jianghong (5).

* Departamento Ingeniería Química. Facultad de Ciencias. Universidad del País Vasco, Apdo. 644, 48080 Bilbao (Spain).

b) Supported Liquid Membranes (SLM): The organic phase is adsorbed on a microporous oleophilic polymeric support separating feed and stripping solutions, Sengupta et al. (6), Urtiaga et al. (7).

Liquid membranes were first applied to gas transport (oxygen, carbon dioxide, carbon monoxide, hydrogen sulfide and others). Metal ions separation from hydrometallurgical industries has received considerable interest in order to recover the metals as well as to control pollution. Finally, biotechnical applications and recovery of other products, such as acetic acid and phenols have been reported. An extensive review including industrial applications has been published, Noble et al. (8).

Phenol is a weak acid and can diffuse through the liquid membrane in its undissociated form to react with a caustic stripping solution (NaOH) to give sodium phenolate. The acid-base reaction involved is as follows

$$C_6H_5OH + NaOH \rightleftharpoons C_6H_5ONa + H_2O$$

Sodium phenolate is a strong electrolyte non-soluble in the organic phase.

Despite its advantages, the suitability of a liquid membrane system for simultaneous extraction and stripping basicaly depends on the solute flux rate, stability and working life-time.

Mass transfer rate through the liquid membrane will be affected by factors such as solubility and diffusivity of solutes in the organic phase. Then, organic solvents with high distribution coefficients will produce high mass transfer coefficients. For removal of phenol from water, polar organic fluids are much better solvents than non-polar hydrocarbons. However, polar solvents are much more soluble in water. On the other hand, the working life-time of the liquid membrane is a function of the organic solvent stability on the polymeric support. Solubility in water may cause big losses of extractant from the pores of the support.

In this work, both factors are considered. Two extractants, kerosene -a commercialy available mixture of hydrocarbons- and methyl isobutyl ketone (MIBK) -a polar solvent- were used as liquid membranes. Due to the low distribution coefficient for phenol between water and kerosene, mixtures of both solvents were tested. We propose a first-order mass transfer model for solute transport in a recycling mode with instantaneous chemical reaction at the stripping side. The variation of the overall mass transfer coefficient with the liquid membrane composition has also been investigated. Distribution coefficients for phenol between water and mixtures of kerosene-MIBK have been measured. The water-MIBK-kerosene equilibrium has also been studied.

EXPERIMENTAL PROCEDURE

A schematic diagram of the experimental set-up and a representation of the liquid membrane structure are shown in Figure 1. The membrane module is formed by three hydrophobic microporous polypropylene hollow fibres (Accurel, ENKA, A.G.) inserted in a glass shell. The fibres were firmly epoxy-bonded to the ends of the permeation shell. The shell is a 4 mm and 25 cm long glass tube with lateral openings at both ends allowing circulation of the stripping solution from a small reservoir. Feed solution circulates through the lumen of the fibres at a constant linear velocity of 8.5 cm/s. The volumes of the feed and stripping solutions treated in each experiment were 100 cm^3 and 25 cm^3 respectively. Both inside and outside streams come from magnetically stirred tanks where they are recirculated by a peristaltic pump. The characteristics of the support material are shown in table 1.

TABLE 1- Dimensions and physical properties of the fibres.

Diameter (mm)		Pore Size (μm)	Porosity	Tortuosity
Inner	Outer			
0.6	1.0	0.4	75%	1.67

The organic solvents used as liquid membranes are kerosene, MIBK and mixtures of both solvents in proportions ranging from 100% kerosene to 100% MIBK. Impregnation was carried out by pumping the organic extractant through the fibre bores for one hour. Solvent passed rapidly through the porous walls of the fibres and was collected in the outer shell. This probably means that impregnation was completed in a few minutes. The solubility of MIBK in water made necessary to add MIBK to the feed solution in order to get equilibrium conditions with the membrane composition. The different kerosene-MIBK combinations provided a wide range of values for the distribution coefficient m_D. Starting from a phenol concentration of 5 g/l, 1 ml samples were taken at different times from the feed tank and analyzed after dilution in a PYE-UNICAM UV spectrophotometer, at λ=272 nm.

Distribution coefficients for MIBK between water and kerosene were determined by measuring equilibrium mass fraction of solute in the aqueous phase. About 20 cm^3 each of kerosene and MIBK aqueous solution were equilibrated in a 50 cm^3 erlenmeyer flask magnetically stirred for 2 hours. Both phases were allowed to settle down and separated. Analysis was performed with a Perkin-Elmer Sigma 3B Gas Cromatograph equiped with a dual-flame ionization detector and a stainless steel column (1,5 m long; 3.18 mm o.d.) packed with Porapack Q with helium as carrier gas. Four 5 μl aqueous samples were analyzed for each distribution coefficient.

The same procedure was followed to measure distribution coefficients for phenol between water and mixtures kerosene-MIBK. MIBK was added to the aqueous phenol solutions at feed initial concentration in order to get equilibrium conditions with the organic mixtures. Aqueous samples were analyzed by UV spectrophotometry.

MATHEMATICAL MODELLING

The feed solution containing the solute circulates through the lumen of the hollow fibre. The solute is permeating radially through the fibre wall by diffusion in the liquid membrane. Assuming linear concentration gradients, a steady-state macroscopic mass balance of the solute in a certain volume of the fibre yields

$$\frac{\partial C^f}{\partial t} = -u \frac{\partial C^f}{\partial z} - \frac{A}{V} K \left(C^f - C^s\right) \qquad (1)$$

C^s, the solute concentration in the outer side of the fibre, is considered to be zero due to the irreversible fast chemical reaction in the strip solution, and $\frac{\partial C^f}{\partial t} << u \frac{\partial C^f}{\partial z}$

The solute mass balance in the feed tank at a time interval dt, assuming complete mixing can be written as

$$\frac{dC}{dt} = \frac{1}{\tau} \left(C^f_{z=L} - C^f_{z=0}\right) \qquad (2)$$

Integration of this system of differential equations (1) and (2) leads to an expression of the overall mass transfer parameter as follows

$$\ln \frac{C_{T_{t=0}}}{C_T} = \frac{1}{\tau} \left(1-\exp\left(- K \frac{2\pi r L}{u\pi r^2}\right) t\right. \qquad (3)$$

In this way, experimental results can be fitted to a first order kinetic law

$$V_T \ln \frac{C_{T_{t=0}}}{C_T} = S \cdot t \qquad (4)$$

where S is a factor dependant on the geometry of the fibres, linear velocity of the fluids, feed tank volume and the mass transfer parameter of the system, K. Therefore, in case all the other parameters are held constant, the experimental overall mass transfer parameters can be calculated from the next expression

$$K = \frac{u\ r}{2\ L} \left(-\ln \left(1 - \frac{S}{Q}\right)\right) \qquad (5)$$

Typical solute concentration profile in the fibre wall when there is an irreversible fast reaction in the stripping side is shown in Figure 2. A way to define an overall mass transfer parameter K is as follows

$$R_L = K\ n(\pi d_i C^f - \pi d_o C^s) \qquad (6)$$

where R_L is the solute transport rate per unit transfer length. Analyzing transport with an overall mass transfer coefficient requires assuming that changes in concentration are limited to the small part of the feed volume near the inner fibre wall. The overall mass transfer coefficient may be related to the individual transfer coefficients through a "resistences-in-series approach" as follows:

$$R_L = k_i \; n \; \pi d_i \; (C^f - C_i^f) = \tag{7a}$$

$$= k_m \; n \; \pi d_i (C_m^f - C_m^s) = \tag{7b}$$

$$= k_s \; n \; \pi d_o \; (C_i^s - C^s) \tag{7c}$$

Noting that $C_m^f = m_D C^f$; $C_m^s = m_D C^s$ and $C^s = 0$ due to an instantaneous strip phase reaction. In the same way, the instantaneous strip reaction would eliminate the strip boundary layer resistance, so that,

$$\frac{1}{K} = \frac{1}{k_i} + \frac{di}{d_{lm}} \cdot \frac{1}{m_D . k_m} \tag{8}$$

From equation (8), the overall mass transfer coefficient K will vary with the distribution coefficient m_D of the system whenever the feed flow rate and consequently the boundary layer coefficient, are held constant.

RESULTS AND DISCUSSION

The distribution coefficient m_D is defined by

$$m_D = \frac{x_s}{x_w} \tag{9}$$

where x_s is the mass fraction of solute in the solvent and x_w is the mass fraction of solute in water.

Distribution coefficients for MIBK in the water-kerosene system were measured. Results are given in Figure 3.

Mass transfer experiments were carried out with membrane composition ranging from 100% kerosene to 100% MIBK. Figure 4 plots phenol concentration against time for the pointed out membrane compositions. It must be noticed that for a 100% MIBK membrane phenol transfer is completely stopped after 1.5 hours. Apparently MIBK has been solubilized in water causing membrane rupture. To avoid MIBK losses, this extractant has been added to the aqueous phases in order to get equilibrium conditions with the organic phase. Equilibrium data are taken from Figure 3.

Distribution coefficients for phenol in mixtures of Kerosene and MIBK are shown in Figure 5. Won and Prausnitz (9) have reported dependency of m_D on solute concentration for phenol- MIBK system. However, average distribution coefficients can be used without appreciable errors.

Hydrodynamic conditions are held constant in all the mass transfer experiments in the fibre module, allowing to study the overall mass transfer parameter dependency on the distribution coefficients m_D. Figure 7 plots $V_T \ln(C_{T_{t=0}}/C_T)$ against time for different membrane compositions. Regression lines have been drawn, observing a good fitting to the first order transport model in equation 4 . The S factors have been taken from the slopes. Equation 5 allows the calculation of the experimental values of K, which are shown in table 2.

TABLE 2 - Experimental values of the overall mass transfer coefficient for different liquid membrane compositions.

%MIBK in the aqueous phase	%MIBK in the organic phase	S factor, cm^3/h	K, cm/s	m_D
2.0	90.4	51.0	9.295×10^{-4}	37.0
1.5	61.4	48.5	8.789×10^{-4}	24.5
1.0	32.5	43.0	7.671×10^{-4}	13.0
0.75	18.0	30.0	5.219×10^{-4}	7.0
0.5	6.25	19.2	3.268×10^{-4}	1.9
0.25	2.9	13.3	2.241×10^{-4}	0.9
0.1	1.2	9.2	1.530×10^{-4}	0.5
0.0	0.0	4.9	0.810×10^{-4}	0.22

Figure 7 plots 1/K against $1/m_D$. Experimental values show a good fitting to equation 8. Values of $1/k_i$ and $1/k_m$ are obtained from the regression line. The resulting values for the boundary layer coefficient and permeability coefficient are:

$$k_i = 8.205 \times 10^{-4} \text{ cm/s}$$
$$k_m = 3.112 \times 10^{-4} \text{ cm/s}$$

The inner boundary layer coefficient is related to the hydrodynamic conditions, basically linear flow velocity and dimensions of the fibres. The permeability membrane coefficient is a property of the wall thickness, porosity and tortuosity of the fibres.

CONCLUSIONS

Phenol transfer experiments were carried out in a hollow fibre supported liquid membrane permeator. The influence of the organic phase in the flux rate was investigated through the distribution coefficient m_D. For high enough values of m_D, the maximum transport rate is accomplished when phenol transference is controlled by the boundary layer resistance in the inside of the hollow fibre, in the conditions of linear velocity tested. A good agreement was observed with the proposed first order model.

The solubility of MIBK is a factor that must be taken into account. Although the tested membrane mixtures showed a good behaviour it was necessary to use an aqueous phase containing 1.5% of MIBK to get the maximum transport rate conditions, implying a MIBK subsequent separation step for the aqueous effluent.

Acknowledgments

This work has been partially financed by the Spanish Education Ministry, under a Research Grant and EHU/UPV Project Nº 069.310.0050/88.

SYMBOLS USED

A/V	= transfer area to volume ratio in the fibre (cm^{-1})
C^f, C^s	= solute concentration in the feed and strip solutions, (g/l).
C_i^f, C_o^s	= solute concentration in the inside and outside wall of the hollow fibre, respectively (g/l).
C_m^f, C_m^s	= solute concentration in the membrane phase at the feed and strip side respectively (g/l).
C_T	= solute concentration in the feed tank (g/l).
d_i, d_o, d_{lm}	= hollow fibre inside diameter, outside diameter and the long-mean diameter (cm).
k_i	= boundary layer coefficient in the feed phase (cm/s).
k_m	= permeability coefficient of the organic membrane (cm/s).
K	= overall mass transfer parameter (cm/s).
L	= hollow fibre permeation length (cm).
m_D	= solute distribution coefficient ((g/l)/(g/l)).
n	= number of fibres in the module.
Q	= flow rate of the feed solution (cm^3/s)
R_L	= solute transfer rate per unit module length (g/cm.s).
r	= inner radius of the hollow fibre (cm).
S	= parameter defined in eqs. (4) and (5) (cm^3/s).
u	= average linear velocity in the hollow fibre (cm/s).
x_s, x_w	= mass fraction of solute in the solvent and water, respectively.
τ	= residence time in the feed tank (s).

REFERENCES

1. Lanoutte, K.H., 1977, Chem. Engng. 84, 99,

2. Salazar, E., Ortiz, M.I. and Irabien, A., 1989, "Recovery of cromium (VI) with Emulsion Liquid Membranes in mechanically stirred tanks", 7th International Conference Chemistry for the Protection of the Environment, Lublin, Poland.

3. Li, N.N., 1971, AIChEJ, 17 459.

4. Kim, K.S., Choi, S.J. and Ihm, S.K., 1983, Ind. Eng. Chem. Fundam., 22, 167.

5. Xijuan, Z. and Jianghong, L., 1987, "Industrial application of liquid membrane separation for phenolic wastewater treatment", 2nd International Conference on Separation Technology, Schloss Elman, West Germany.

6. Sengupta, A., Basu, R. and Sirkar, K.K., 1988, AICheJ 34, 1698

7. Urtiaga, A.M., Ortiz, M.I. and Irabien, A., 1989, "Phenol recovery with supported liquid membranes. Experimental study" 7th International Conference Chemistry for the Protection of the Environment, Lublin, Poland.

8. Noble, R.D., Way, J.D. and Bunge, A.L., 1988, Ion Exchange Solvent Extraction,10, 63.

9. Won, K.W. and Prausnitz, J.M., 1975, J. Chem. Thermodynamics, 7, 661.

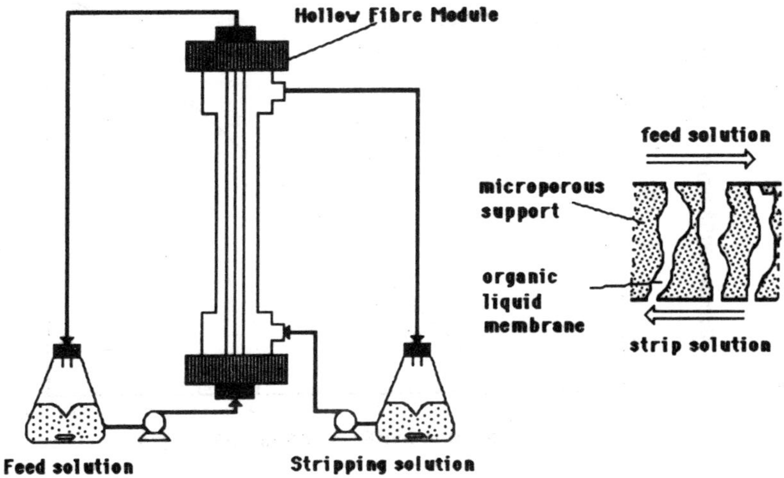

Figure 1 Schematic diagram of the experimental set-up and supported liquid membrane structure.

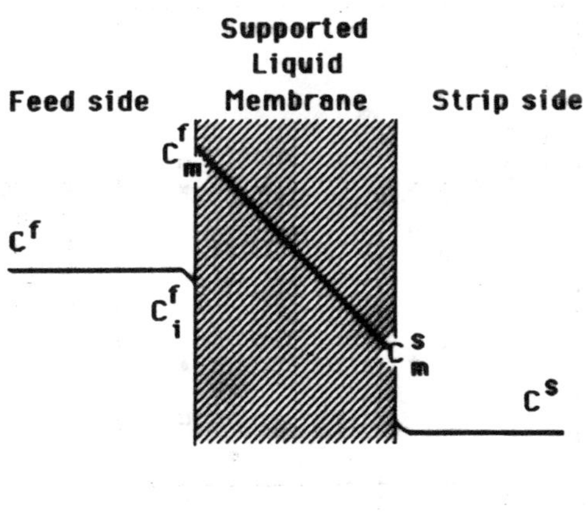

Figure 2 Typical solute concentration profile in the fibre wall

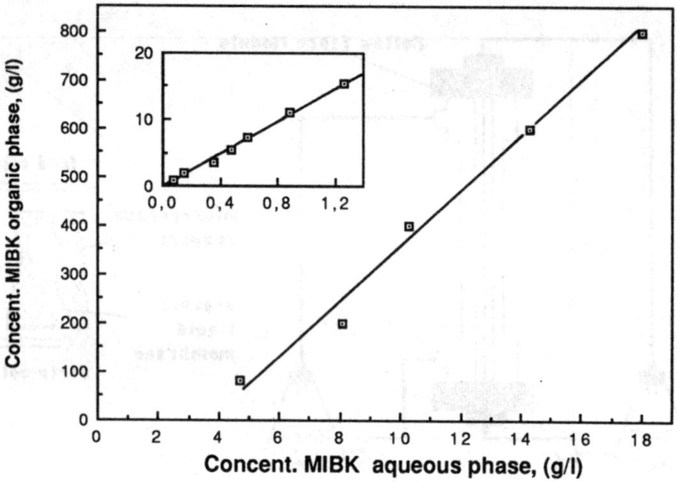

Figure 3 Equilibrium of MIBK in the water-kerosene system.

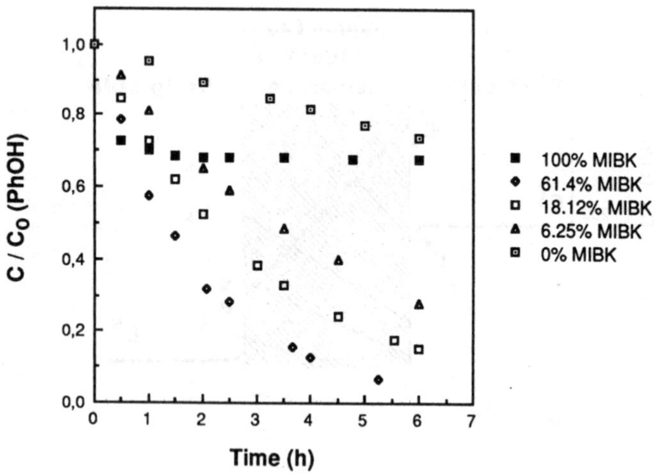

Figure 4 Phenol concentration against time for different membrane compositions.

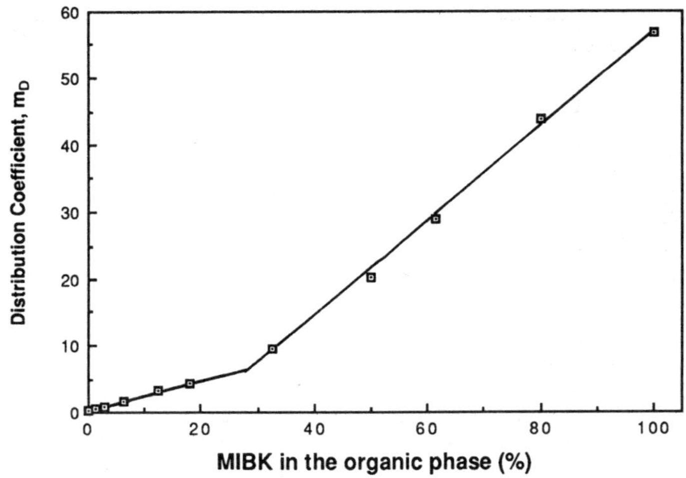

Figure 5 Distribution coefficient for phenol between water and mixtures of MIBK and kerosene.

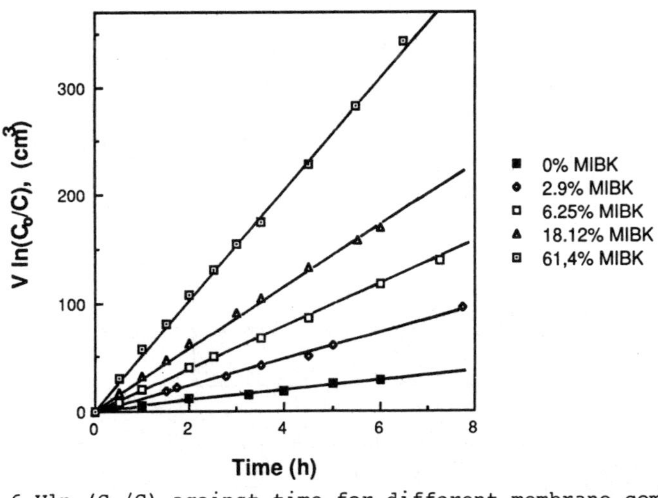

Figure 6 Vln (C_o/C) against time for different membrane compositions The S factor are the slopes of the regression lines.

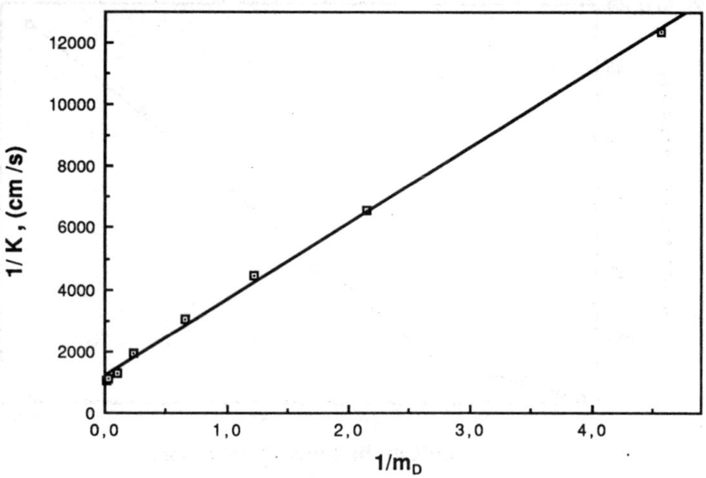

Figure 7 Inverse of the experimental overall mass transfer coefficients against the inverse of corresponding distribution coefficients.

NEW EXTRACTANTS FOR THE TREATMENT OF NUCLEAR WASTE SOLUTIONS

C. CUILLERDIER*, C. MUSIKAS*, P. HOEL*

Pentaalkyl propane diamides are studied in Fontenay-aux-Roses to extract α emitters from nitric acid solutions of nuclear wastes. Their formula is $(RR'NCO)_2$ CHR" with R and R' alkyl radicals R" alkyl or oxyalkyl one. Main features concerning the chemistry of these solvents are given. A molecule with $R=CH_3$ $R'=C_4H_9$ and $R"=C_2H_4$-O-C_6H_{13} has been selected for practical applications. The results of an experiment realised in a mixer-settler battery to treat a real solution of wastes were described.

INTRODUCTION

One of the main important direction of research concerning the nuclear fuel cycle is to decrease the cost of wastes treatment. One reason for the high cost of nuclear wastes is the existence of long lived alpha emitters such as Pu, Am, Cm, Np. Those which are stable in tetravalent state such as Pu are easy to extract with classical solvents (TBP for instance) but trivalent ones Am(III), Cm(III) need other extraction systems.

For that purpose TRUEX process has been developped in the USA, this process uses several bifunctional extractants such as N,N-diethyl dihexyl carbamoyl methylene phosphonate (DHDECMP), $((C_6H_5O)_2POCH_2CON(C_2H_5)_2)$ or octyl-phenyl N,N"-diisobutyl carbamoylmethylphosphine oxyde (CMPO) $C_6H_5(C_8H_{17})PO$ CH_2 $CON(CH_2CH(CH_3)_2)_2$ this last mixed with tributylphosphate and diluted in an aliphatic (for instance Conoco) or chlorinated diluent (for instance tetrachloroethylene) (Vandegrift (1), Leonard et al. (2)). In spite of their good efficiency towards trivalent cations extraction both in nitric or chloride medium, these extractants have many disadvantages. They are expensive and difficult to purify, they contain phosphorus and then they are not completely incinerable giving P_2O_5 or H_3PO_4 by calcination, in consequence they create new wastes. They are rather stable towards radiolysis but degradation products such as DBP, MBP can produce precipitates with Pu, Zr and must be eliminated from the degraded solvent needing then an important treatment.

*Commissariat à l'Energie Atomique BP n° 6 92265 Fontenay-aux-Roses FRANCE

For several years a new class of extractants has been studied in Fontenay-aux-Roses (Hubert (3), Musikas et al. (4)) pentaalkyl propane diamides the generic formula of which is $(RR'NCO)_2CHR''$ with R and R' alkyl R" alkyl or oxyalkyl radicals. These solvents have a chelating effect so they extract trivalent actinides and lanthanides from nitric acid solutions. As they contain no phosphorus they are completely incinerable. According to first studies, the degradation products are not noxious to extractant efficiency (Cuillerdier et al. (5)).

Many alkyl and oxyalkyl radicals can be used theoretically for R and R' CH_3 to C_8H_{17} for instance and for R" radicals from C_2H_4-O-C_6H_{13} to C_2H_4-O-C_2H_4-O-C_8H_{17} or longer are possible. For practical applications choices are made according to extractive properties and price of the molecule.

TABLE 1 - Influence of the Structure of the Solvent on D_{Am3+} in HNO_3

R" Radicals	R	R'	D_{Am3+}	Conditions
				Diluent :
H	C_6H_{13}	C_2H_5	0.11	Benzene
H	C_4H_9	C_4H_9	0.18	
H	C_3H_7	C_8H_{17}	0.45	$D_{maximum}$
H	CH_3	C_4H_9	0.55	
H	CH_3	C_8H_{17}	1.18	Solvent 0.5M
				Diluent :
C_6H_{13}	C_4H_9	CH_3	1.16	t-butyl benzene
C_2H_4-O-C_2H_5	C_4H_9	CH_3	3.28	
C_2H_4-O-C_6H_{13}	C_4H_9	CH_3	7.55	
C_2H_4-O-C_2H_4-O-C_8H_{17}	C_4H_9	CH_3	9.43	HNO_3 5M
C_2H_4-O-C_2H_4-O-$CH(C_2H_5)CH_2\ C(CH_3)_3$	C_4H_9	CH_3	10.38	Solvent 0.5M
C_2H_4-O-C_2H_4-O-C_6H_{13}	C_4H_9	CH_3	11.48	

MAIN FEATURES CONCERNING SUBSTITUTED MALONAMIDES CHEMISTRY

The most important criterion of optimization of the extractant formula is the extraction efficiency for trivalent metals. Malonamides with one CH_2 between the two C = O having a chelating effect permit extraction of M^{3+}. A proper choice of R, R' and R" gives good distribution coefficients for Am^{3+} in concentrated nitric acid solution (see TABLE 1).

The solvent used are synthesized by IRCHA (Vert le Petit (France)). The synthesis described in details in (Thiollet and Musikas (6)) has three main steps :

* preparation of the unsubstituted malonamide
* preparation of R" OH and then R" Br
* reaction between R" Br and $(R\ R'\ NCO)_2$ CH Li
* purification of the product and characterization by NMR and mass spectrometry.

For economical reasons it is interesting to choose R = CH_3 and R' = C_4H_9 because the secondary amine is a commercial product.

The nature of R" is very important because the synthesis yields and the purification problems are highly dependant from the length and the number of oxygen of R". Increasing length of R" generally increases the price of the product but enhances the extraction ability. A compromise has to be found. The better extraction is due to the decrease of the extractant basicity when increasing R" (5) then in concentrated acid medium competition between HNO_3 and M^{3+} is reduced for the bonding with the extractant (5). In addition a large alkyl or oxyalkyl radical seems to protect the molecule from hydrolysis and radiolysis (5).

Aside trivalent actinides and lanthanides (Am^{3+}, Cm^{3+}, Ln^{3+}) which are well extracted in a sufficiently concentrated nitric acid medium, tetra and hexavalent ones are extracted too (Pu(IV), Pu(VI), Np(IV), Np(VI), U(VI)) (5). Neptunium when introduced in the solution as Np(V) is completely extracted too because of oxydation, disproportionation and displacement of the equilibrium by extraction.

Stripping of trivalent actinides needs a diluted medium, Pu(IV), Pu(VI) and Np(VI) a reducing one such as 0.1M HNO_3 0.2M hydroxylamine nitrate at 45°C. Np(IV) when extracted has to be washed with oxalic acid. U(VI) is stripped with water or very diluted nitric acid.

Corrosion products, iron and zirconium are strongly extracted with actinides they can be separated with oxalic acid. Technetium with distribution coefficients near one is still an unsolved problem.

Previous studies ((5) and Cuillerdier et al. (7)) have shown that degradation of the solvent is not very important in working conditions ($t \leq 45°C$) and that extraction and back extraction of actinides are still possible with degraded solvent. No retention of plutonium or americium was observed.

BENCH SCALE EXTRACTION EXPERIMENT IN A MIXER-SETTLER BATTERY

Choice of the solvent and of the aqueous phase

The purpose of the experiment was to test the efficiency of a substituted malonamide synthesized at a pilot scale for the extraction of actinides from a real waste solution.
As said previously a compromise had to be found for the length of R" between the price of the diamide and its extractive properties. The 2 -(3-oxanonyl) NN'-dibutyl NN'-dimethyl 1,3-diamide propane, was choosen : one kilogram of that diamide was synthesized by G. Thiollet (IRCHA). This extractant was used at a 0.5M concentration diluted in tertiobutyl benzene. The choice of an aromatic diluent is due to the problem of third phase formation in contact with concentrated nitric acid solution (for the diamide described here the third phase appears roughly for 9N nitric acid when diluted in tertiobutyl benzene and only 0.9N with hydrogenated tetrapropylene).

The aqueous solution of wastes was representative of a real Purex effluent, it came from the pilot pulsed columns facility of Fontenay-aux-Roses. The composition of the concentrated effluent is given in TABLE 2. As it did not contain curium, ^{152}Eu was added for Cm^{3+} simulation. Neptunium was added too (50 mg/l of ^{237}Np and traces of ^{239}Np for analytical purposes).

TABLE 2 - Composition of the aqueous Phase used for the Mixers-Settlers Experiment

Element	Content
H^+	5 M
Pu	110 mg/l
Am	0.91 mg/l
U	5.9 g/l
Tc	134 mg/l
Fe	500 mg/l
P_{total}	270 mg/l
NO_3^-	5.5 M
α activity	50 mCi/l
$*^{239}Np$	For analysis
$*^{152}Eu$	For analysis

*Added to the solution

This experiment was the first realized with solvent prepared at a pilot scale so the qualities of the solvent had to be tested. Purity of the solvent was found 95%. The diamide was titrated by $HClO_4$ diluted in acetic acid in an acetic anhydride medium. Degradation of the solvent mainly produces carboxylic acids and secondary amines they can be titrated in non aqueous medium respectively by NaOH and by HNO_3 diluted in ethanol in an ethanol benzene mixture. The results were : amine 0.0125M ; carboxylic acid < 0.040M. However these methods are not sufficient, they cannot show the presence of tensio-active components that could create many problems during the experiment (stable emulsions, incomplete disengagements of phases ...). Settling tests have been undertaken and diamide has been compared to TBP in the same conditions. Experimental method comes from (Ginisty and Pierlas (8)). Results are given in TABLE 3 and fig (1). In acidic medium disengagement times are a little more important than for TBP. In basic medium settling is less easy especially in organic continuous phase but basic treatment of solvent, if necessary is still possible.

Extractive properties of the solvent were checked by measuring distribution coefficients of actinides (Am^{3+}) they were found equal to those obtained with the solvent prepared in a laboratory scale.

Optimization of the conditions of experiment by calculation

First of all the distribution coefficients for all the cations present in the solution have been measured. For "sensitive" element such as Am^{3+} they have been

TABLE 3 - Disengagement Time for various Conditions (Time in Second)

$C^{aq}_{H^+}$ mol.l^{-1}	"fac" % disengaged volume				"foc" % disengaged volume			
	25	50	75	100	25	50	75	100
0.1	22	34	43	78	14	23	30	51
0.1**					15	20	25	34
1	27	36	44	105	27	37	48	65
2	23	32	39	77	22	30	37	47
3	23	30	36	57	24	32	40	50
5	22	29	37	54	27	35	45	62
6	20	28	35	52	impossible to obtain			

* "fac" continuous aqueous phase * "foc" continuous organic phase
** Back-extraction medium HNO_3 0.1M - NH_2OH 0.2M at = 45°C

checked with the solvent and the real aqueous phase. All the values are given in TABLE 4 for extraction and back-extraction. Calculations values are given in TABLE 4 for extraction and back-extraction. Calculations of the theoretical concentrations profiles of Am(III), Cm(III), and HNO_3 were carried on with the program "Newsolv" (Baron (9)).

TABLE 4 - Values of the Distribution Coefficients used for the Optimization of the Flow sheet

$C^{aq}_{HNO_3}$ mol.l^{-1}	0.5	0.5*	0.6*	0.7*	1	2	3	4	5	6
D_{HNO3}	0.014	0.014	0.020	0.02	0.046	0.053	0.098	0.0115	0.123	0.132
D_U	5.53	3.09	4.29	6.32	7.38	37.19	67.9	69.9	72.0	75.0
D_{Pu}	7.0	0.009	0.015	0.036	11.54	29.2	47.1	50.0	54.3	94.0
D_{Am}	0.003	0.005	0.005	0.011	0.018	0.202	0.847	2.20	3.99	5.77
D_{Eu}	0.002	0.0005	0.0022	0.0062	0.018	0.105	0.580	1.30	2.33	3.73
D_{NpVI}	3.81				5.20	7.91	15.11	46.35	81.6	131
D_{Np}*	0.02	0.14	0.151	0.173	0.029	0.40	3.1	7.7	81.6	131
D_{Fe}	0.0016	0.0016	0.002	0.002	0.002	0.03	0.7	7.0	48	312

* Back-extraction 45°C ; HNO_3 0.1M - NH_2OH 0.2M

The main results are given in TABLE 5. The concentration of nitric acid in the organic phase is an important parameter because when the solvent is overloaded with acid, back-extraction becomes more difficult. These calculations were made for a twelve stages battery, ten stages for extraction and two stages for scrub.

TABLE 5 - Results of the Optimization Calculations

a) Scheme without Scrub 12 Stages Extraction (O/A = 1)

C_{HNO_3} feed $mol.l^{-1}$	Log DF_{Am}	Log $DF_{Eu\ or\ Cm}$	$C^{org}_{HNO_3}$ $mol.l^{-1}$
4	4.27	1.95	0.46
5	7.20	4.56	0.61
6	9.09	6.86	0.79

b) Scheme with Scrub 10 Stages Extraction, (O/A = 1) 2 Stages Scrub

$C^{aq}_{HNO_3}$ $mol.l^{-1}$ extraction	Scrub conditions O/A = 5 $C^{aq}_{HNO_3}$ $mol.l^{-1}$	O/A = 10 $C^{aq}_{HNO_3}$ $mol.l^{-1}$	$C^{org}_{HNO_3}$ $mol.l^{-1}$ (loaded solvent)	log DF_{Am}	log DF_{Eu}
4	1		0.08	2.2	0.65
	2		0.2	2.55	0.95
		1	0.21	3.12	1.34
		2	0.29	3.16	1.38
5	1		0.12	4.67	2.48
	2		0.25	4.87	2.72
		1	0.29	5.50	3.33
		2	0.37*	5.54*	3.37*
6	1		0.16	6.34	4.36
	2		0.29	6.46	4.52
		1	0.37	7.1	5.22
		2	0.46	7.12	5.25

* choosen conditions

A compromise was found between efficiency of extraction of Am^{3+} and decrease of HNO_3 concentration in the organic phase and schemes with acid scrub were preferred - (2N acicity for the scrub was taken to prevent accumulation problems). Nominal conditions were choosen as : feed acidity 5N HNO_3 ; scrub acidity 2N with O/A = 10.
The theoretical decontamination factors permit to otbain a waste having an α activity less than 50 nCi/g. The global scheme of the experiment is given in fig (2). The main purpose was to extract alpha emitters from the waste solutions ; iron was not supposed to be separated. At this point of development of the process the technetium problem is not solved, its repartition was just followed. A neptunium (IV) stripping with oxalic acid in a separating funnel was planned because, as the solution contained iron in presence of a reducing medium, Fe(II) could reduce Np as Np(IV) which then remains in the solvent.

Development of the experiment and hydrodynamic behaviour of the solvent

Before the begining of the experiment the mixers-settlers were washed with tertiobutyl benzene, the extraction battery was filled with respectively HNO_3 5M from stage 1 to 10 and HNO_3 2M in stages 11 and 12. Back-extraction battery was filled with a mixture of 0.1N HNO_3 and 0.2M hydroxylamine nitrate. The solvent was then introduced and when the equilibrium was reached the real feed solution was send in the batteries. Total duration of the experiment was 42h20' so 5.5 renewals of the batteries volume.
Both phases were sampled in each stage to analyse the repartition of every elements. Alpha activity of the raffinate was measured to check if the equilibrium was reached.

Hydraulic behaviour of the solvent was satisfactory during the experiment : quick disengagement, clear phases, limited amount of phases entrained, were observed.
Nominal flow rates were controlled automatically during the experiment in the extraction battery (percentage of dispersion of the values measured was roughly 5%). The back-extraction battery wich needed to be thermostated was made of stainless steel then the interface level control was more difficult. Level probes were used measuring the conductivity of the medium. Flow rates were measured with test tubes.

A deficiency in the stripping aqueous phase flow occured which has created a large increase in the acidity of the aqueous phase (2.2N). The consequence was a poor back-extraction of plutonium and neptunium which were retained in the solvent. When the neptunium and plutonium were found in the organic phase a treatment of the solvent was undertaken with 0.5M oxalic acid which had eliminated Np and Pu completely. However total duration of the experiment was not sufficient to eliminate completely this disturbance and to reach equilibrium again.

The solvent is recycled in the extraction battery. When it contains a too high concentration of nitric acid the equilibrium concentration of H^+ is more important than the expected one.

Results for each elements

Acid repartitions are given in fig (3a) and (3b), higher than the calculated ones in the extraction battery (5.8N instead of 4.9N) due to the disturbance that occured in the back-extraction battery which causes recycle of some nitric acid.
For H^+ the mean value of the distribution coefficient is 0.15 which is a little greater than the value obtained in test tubes. In the stripping battery the acidity found 0.87N is too much important, that explains partly the lack of efficiency of this battery.

Total alpha emitters are well extracted, decontamination factor is : $DF_\alpha = 1.1\ 10^4$, for the stripping battery decontamination of the solvent is not so good : $DF_\alpha = 125$. Extraction for Am^{3+} and Eu^{3+} is one of the most significant results these elements are the most difficult to extract (Europium simulates Curium). Repartitions in the extraction battery are given in fig (4a) and (5). Adjustements to calculated profiles are good (for an acidity of 6N in the feed giving the same acidity at equilibrium). Observed perturbations on the curves are due to the precision of the analytical method, in some cases counting are very weak (counting for long time : one hour have been then undertaken). For the first stages the high activity found can be attributed to a contamination by the recycled solvent.

Decontamination factors found are : $DF_{Am} = 1.26\ 10^4$; $DF_{Eu} = 6.7\ 10^3$. For the stripping battery efficiency is not so good decontamination factors are : $DF_{Am} = 800$; $DF_{Eu} = 250$.
This fact is partly due to acidity disturbances previously described (Am repartition is given in fig (4b)). Besides, the solution contained technetium and laboratory experiments have shown that there is a coextraction phenomenon between Am^{3+} and TcO_4^- in weak acid medium, these results are given in fig (6). These coextraction could explain the lack of achievement of the back-extraction battery.
Plutonium has been correctly extracted : DF > 10^5 for back-extraction a small accumulation was observed for previously described reasons. For neptunium, results are not completely significant, there was a deficiency in back-extraction (due to acidity) recycling and accumulation in the extraction battery, at the end of the experiment the accumulation peak was not completely reabsorbed : DF ex = 144 ; DF str = 400. It was then not possible to conclude on the effect of iron. Uranium is correctly extracted DF > 1000 and weakly stripped as planned.
Iron is strongly extracted DF > 1000 and correctly stripped.
Concerning technetium roughly 79% is extracted and 21% remains in the raffinate of the extraction battery ; in the stripping battery 14% is back-extracted and 86% remains in the solvent. This element needs further researches especially because it decreases the efficiency of the actinides stripping.

Solvent stability

The solvent was recycled 1.4 time in the extraction battery. No significant degradation was found (amide 0.554M, amine not found, carboxylic acid 0.008M), only a small concentration was observed due to the diluent evaporation.

Comparison with Truex

Main decontamination factors are given in TABLE 6. They show that efficiency of both extractants is roughly the same. Alpha activity is reduced from 53 mCi/l to 5 μCi/l for the waste treated with the diamide giving then a non TRU waste as for TRUEX and no problems occured due to degradation products.

TABLE 6 - Decontamination Factors obtained with a Diamide Solvent compared to the TRUEX ones (from (1) and (2))

Process decontamination factor	TRUEX	Diamide
UO_2^{++}	$> 10^5$	$> 10^5$
Pu^{4+}	$> 10^5$	$> 10^5$
Am^{3+}	$5\ 10^4$	$1.2\ 10^4$
$Eu^{3+}(Cm^{3+})$		$6.7\ 10^3$
Np	6	144

CONCLUSION

This experiment was the first realized at a bench scale size with a diamide extractant ; it has shown the good hydraulic behaviour of the solvent and its good chemical and radiolytical stability. No precipitates and emulsions created by degradation products were observed.

Extraction of actinides is satisfactory, back-extraction is easy in spite of several difficulties due to working conditions.

This process, in spite of its lack of industrial development can be compared to the TRUEX one.
Researches are now undertaken to limit iron extraction, solve the technetium problem and optimize neptunium decontamination. Another research direction is to select a molecule which could be used with an aliphatic diluent.

SYMBOLS USED

$D_{M^{3+}}$: distribution coefficient of the cation M^{3+}.

C_{H^+} : concentration of H^+.

O/A : ratio of the organic phase volume to the aqueous phase volume.

DF : decontamination factor.

REFERENCES

(1) G.F. Vandegrift, R.A. Leonard, M.J. Steindler (1984) "Transuranic decontamination of nitric acid solutions by the TRUEX solvent extraction process. Preliminary development studies" Report ANL 84-45

(2) R.A. Leonard, G.F. Vandegrift, D.G. Kalina, D.F. Fisher, R.W. Bane, L. Buris, E.P. Horwitz, R. Chiarizia and H. Diamond (1985) "The extraction and recovery of Plutonium and Americium from nitric acid waste solutions by the TRUEX process - Continuing development studies"

(3) H. Hubert (1986) "Complexation des ions trivalents des groupes 4f et 5f par les diamides et utilisation en extraction liquide-liquide" Report CEA.R.5337

(4) C. Musikas, P. Hoël and G. Thiollet (1985) "Procédé de récupération des actinides et/ou des lanthanides par extraction dans un solvant organique contenant un propane diamide" Patent FR-2 585-692

(5) C. Cuillerdier, C. Musikas and P. Hoël (1988) "Malonamides as new extractant of actinides" Poster presented at 18ème Journées des Actinides PARIS (France) CEA/CONF 9468

(6) G. Thiollet and C. Musikas (1989) Solvent Extr. Ion Exchange 3, 813

(7) C. Cuillerdier, L. Nigond, C. Musikas, X.Vitart and P. Hoël (1989) "Analytical methods and problems for the diamide type of extractants" Poster presented at "Second Karlsruhe International Conference on Analytical Chemistry in Nuclear Technology" Karlsruhe Federal Republic of Germany. CEA/CONF 9897

(8) C. Ginisty and V. Pierlas - Unpublished results

(9) P. Baron - "Newsolv" - Unpublished programm

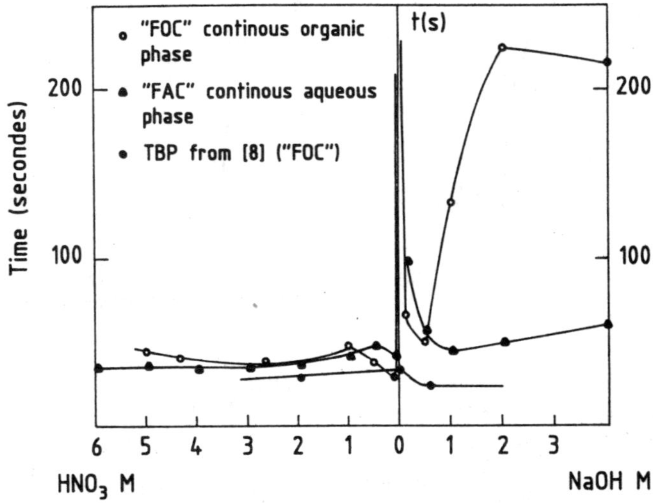

Fig (1) - Disengagement time for 75% of the volume of the dispersed phase
t = 25°C ; "foc" continuous organic phase ; "fac" continuous aqueous phase

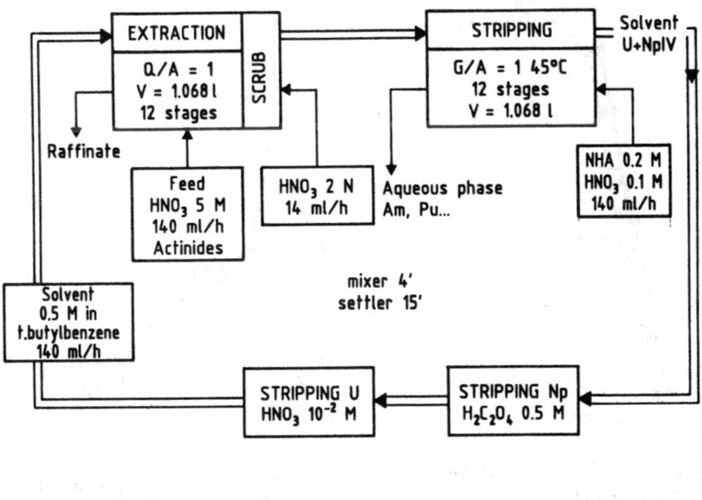

Fig (2) - Scheme of the experiment

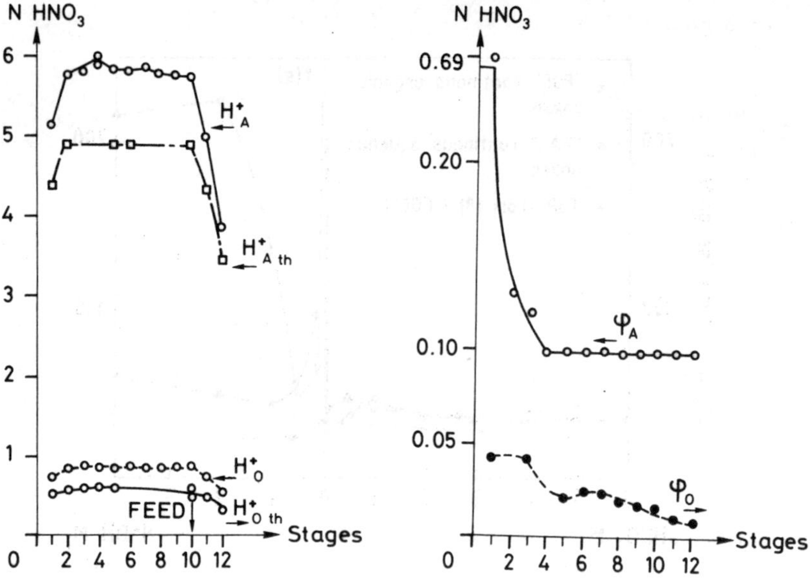

Fig (3) - Acid repartition in the extraction (3a) and stripping (3b) battery

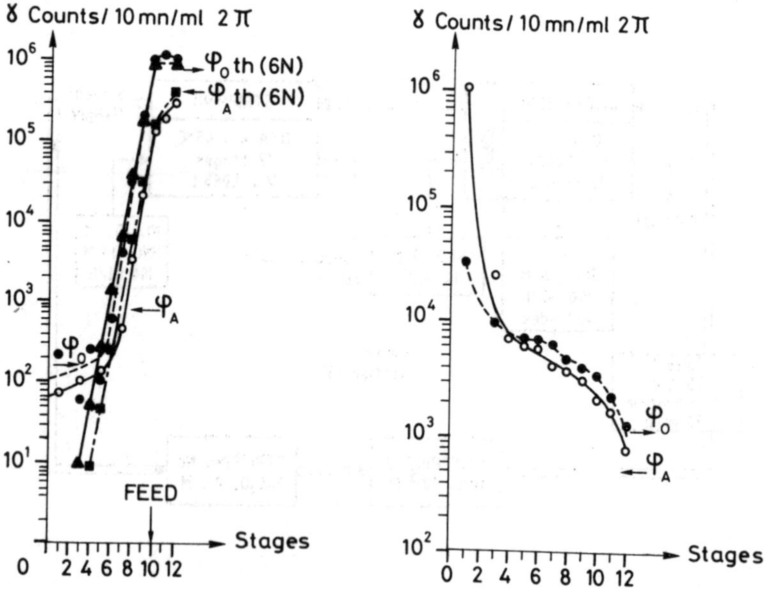

Fig (4) - Americium repartition in the extraction (4a) and stripping battery (4b)

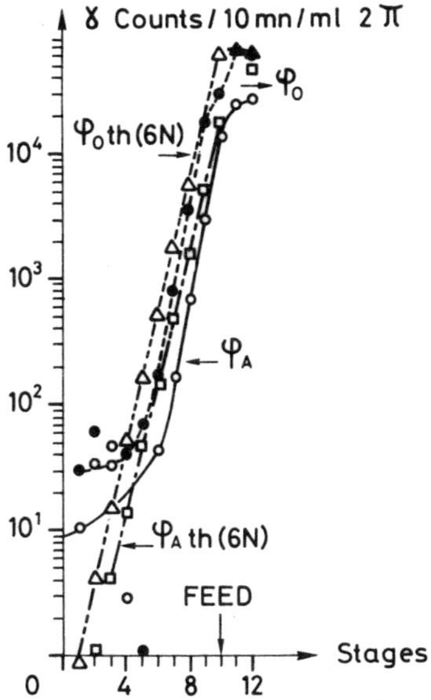

Fig (5) - Europium repartition in the extraction battery

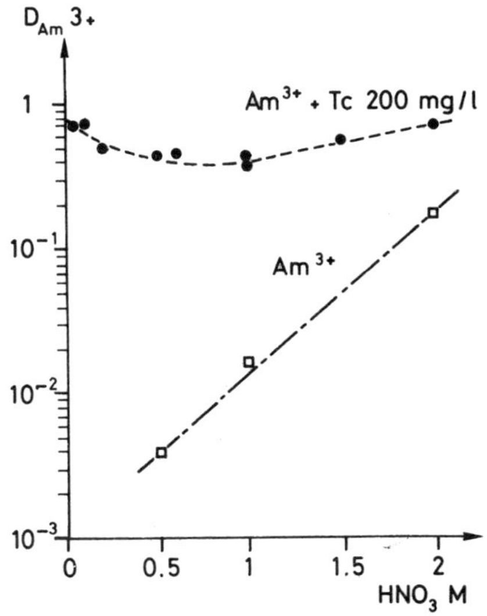

Fig (6) - Coextraction between Am and Tc

EXTRACTION OF NON FERROUS METALS FROM HYDROMETALLURGICAL WASTE WATERS. PLANNING PRODUCT RECYCLING USING SEPARATION TECHNIQUES.

F. Mijangos*, J.I. Lombrana*, F. Varona * and M. Diaz**.

This work studies possibilities to recover products from a complex hydrometallurgical effluent. Sulphate, iron as goethite and several metal ions has been considered. Metals recovery with chelating ion exchange resin is the main process. Equilibrium and kinetic results are reported and respective models has been tried to predict the system behaviour.

INTRODUCTION

Nowadays, industries processing metallic ores must treat their effluents to adjust the environmental local legislation. Usually, a first treatment in source is needed before sending it to a centralized treatment plant. In the hydrometallurgical industry, high flow of liquors with low concentrations of non ferrous ions are produced which have a high metallic value. Hence, the cost of depuration processes can be offset by metal recovery using separation techniques as Brooks reviews (1).

Sulphuric industry gives rise to a great deal of pyrite cinders from roasting. In this work, the effluent produced in the processing of these pyrite cinders, in order to obtain iron, copper, silver, gold and zinc, has been considered.

The following stages makes up the hydrometallurgical process: roasting, leaching, cementation and extraction (Figure 1). Pyrite cinders are roasting with sodium chloride to promote the solubilization of non ferrous metals. These are extracted with HCl in the leaching stage. A ferric matrix remains insoluble which is called "purple ore" and usually sent to iron industry when quality specifications are fulfilled. The leaching liquor mainly contains copper and zinc which are separated by

(*) Chemical Engineering Dpt., University of Pais Vasco. Apdo. 644 Bilbao. Spain.

(**) Chemical Engineering Dpt. University of Oviedo. Spain.

TABLE 1.- Averaged concentrations (kg/m^3) corresponding to the mean components of the zinc extraction raffinates.

PARAMETER	RAFFINATE I	RAFFINATE II
Acidity	2.7	2.5
Iron	10.6	0.005
Manganese	1.24	0.005
Cobalt	0.15	-
Nickel	0.01	0.005
Copper	0.15	0.034
Zinc	0.93	0.97
Cadmium	0.012	0.020
Lead	0.065	0.003
Chloride	32.2	10.5
Sulphate	92.0	0.80

cementation and extraction (ZINCEX process), respectively. As a result of all these stages, it is generated the effluent here considered. A general description can be found in the Habashi's well-known book (2).

The aim of this work is to plan a treatment strategy to get a maximum product reclamation in the effluent and to achieve the concentration levels allowed by legislation. The viability of the process will depend on the value of the recovered metals and the operation costs, so that ion exchange with a chelating resin has been more deeply studied.

The treatment of this effluent should consider separation or removing of several species, mainly, sulphate, iron and recovery of heavy metals with economic interest like copper and zinc among others. Finally, a general fitting to fulfill the limits allowed by legislation can be done by precipitation with carbonate.

FLOW-SHEET AND DESCRIPTION FOR THE PROPOSED EFFLUENT TREATMENT.

Since the zinc extraction process is made up of two cycles (anionic and cationic extractants), an effluent is produced from each one. Their characteristics are shown in Table 1.

The effluent from the anionic cycle (raffinate I) contains all the metals and products used or introduced into the process; i.e. sulphate, iron and metals from pyrite cinders, except for zinc and cadmium. These last metals are mainly contained in the second effluent, from the cationic cycle, which is mainly a diluted solution of $CaCl_2$.

The aim of the effluent treatment is to recover all the products with high economical value. A separated treatment for each effluent should be better since the metals recovered from require different separation methods. The proposed schedule is shown in Figure 1.

We suggest sulphate recovery as Glauber's Salt and iron precipitation as goethite, both from the first raffinate. This treatment could be introduced before the extraction stage because many problems such as interference of iron(III) and salt crystallization during zinc extraction could be avoided. Glauber's Salt ($Na_2SO_4 \cdot 10H_2O$) obtained from this liquors has been produced before at the Kuperhutte factory (DKH process). The salt was crystallized by cooling at 0 °C the liquors from cementation with a yield of 73 %, the anhydrated salt obtained by centrifugation and dehydration is often sold to paper and detergent industries. It is assumed that 80-85 kg Na_2SO_4/ton cinder (3% S) was obtained, being energy supply estimated in 4.2-4.5 tons of steam, 120 kwh and 80000 kcal from hot gases. The purity of the product was around 99.8 %. These results are reported in the Ullmann's Encyclopedia (3).

Iron(II) is the metal ion in higher concentration in the effluent. The total metal discharges decrease a 95 % by removing this metal. The Hematite, Jarosite and Goethite processes are very well-known in hydrometallurgy. All these processes could be applied to liquid effluents treatment. Nevertheless, the Hematite and Jarosite processes are not considered here because the obtained product is not adequate for the iron industry and/or the process technology is expensive for waste water treatment. The main reaction in goethite process could be carried out at 25 °C or higher temperature and atmospheric pressure. Moreover, the required reactives are cheap and the product (goethite) could be sold together with the "purple ore" to the iron industry.

TABLE 2.- Operation conditions and composition of the obtained iron precipitate.

OPERATION VARIABLES		SOLID COMPOSITION (% dry solid)	
Oxidant (% stoich.)	110	Iron	42.7
Temperature (°C)	37.5	Zinc	0.013
Slurry pH	3.12	Lead	0.11
Reaction time (h)	1.25	Cobalt	0.065
Agitation (rpm)	38	Copper	0.038
Floculant (g/t solid)	125	Na_2O+K_2O	2.26

In order to study the iron removing from this effluent, we have carried out a set of experiments using a factorial design technique. The effect of temperature, agitation , pH, oxidant, ORP, time and floculant has been considered on several variables such as solid wet, iron content and metal losses. An averaged result for the intermediate level of the independent variables are summarized in Table 2.

A good quality level of the solid can be deduced from results because of the quite low concentration of non ferrous metals. A final iron (III) solution concentration ranging 5-10 mg/l is obtained by this kind of precipitation with a yield close to 100 %. The economic viability of the process is highly dependent on reactive prices so that residual products from other process should be used.

A thermogravimetric analysis (DTA-TGA) has been made of the solid (Table II). The results have been compared with those obtained by Davey & Scott (5) for solids from different goethite processes and media; i.e. Electrolytic Zinc Co. of Australasia Ltd. (EZ) and the Vielle Montagne Co. (VM) for sulphate and chlorine media. The DTA-TGA curves for our product was found to be similar to that obtained for those authors with a sample of the EZ process (sulphate medium). This precipitate is formed by α and β-FeOOH with good characteristics for filtration.

The recovery of some heavy metals by means of ion exchange resins would play a key role in compensating for waste water treatment within the proposed scheme. As Hiskey discuss (6), other techniques like L-L extraction could be considered. Nevertheless ion exchange with solid resins presents advantages. This technique is able to produce highly enriched product from low concentration solutions using equipment with multistage fluidized bed which can work with slurries. A chelating resin has been applied and studied looking for equilibrium and kinetic behaviour in hydrometallurgical liquors.

Finally, the residual metal concentration should be reduced below legal limits by addition of sodium carbonate to precipitate basic species mainly composed by manganese which could be exploited by some industries. This step has not been studied experimentally but there is a large experience on these reaction from the chemical point of view (2).

Zinc and cadmium could be separated and easily recovered by an anionic ion exchange process from the second effluent (raffinate), which is produced in the cationic cycle of zinc extraction. Moreover, $CaCl_2$ could be concentrated and the resulting solution employed in the pelletization of the pyrite cinder powder. The remaining and purified water could be recycled to the process.

ION EXCHANGE THEORETICAL BACKGROUND

Ion exchange resins with chelating groups have been applied in different fields due to the high selectivity for heavy metals. A commercial type containing iminodiacetic groups was selected because its capacity for recovery heavy metals from hydrometallurgical effluents. As a consequence of its particular behaviour, electroselectivity, chelating and protonation reactions should be considered in its applications. In this work, kinetics and equilibrium results in a multimetallic system corresponding to an industrial practical case are presented.

Equilibrium Model

The solution pH at the equilibrium is an important variable in resin behaviour. Several authors like Marinsky (7) has found out its effect on heavy metals retention. Protonation constants of a simplified model must be calculated for a basic study using eqn. (1):

$$\alpha_h = \frac{\beta_1(H) + 2\beta_2(H)^2 + 3\beta_3(H)^3}{1 + \beta_1(H) + \beta_2(H)^2 + \beta_3(H)^3} \qquad (1)$$

No difference between the concentration inside and outside the resin was assumed. Ionic strength is high and constant in hydrometallurgical liquors during ion exchange reaction, then the thermodynamic reference state can be defined at these experimental conditions so that activity coefficients should be constants equal to 1.

Since three protons can be attached in the case of the iminodiacetic group, three different protonation constants can be defined. In order to determine their values, a potentiometric titration was carried out and data correlated to the equation (1) using an non linear regression method.

A mass balance for the chelating group gives place to equation (2):

$$Q_t = q_R\, f(H) + a\, \Sigma q_i \qquad (2)$$

f(H) means the protonated and free fraction of the functional group, according to equation (3).

$$f(H) = \left[1 + \sum_{i}^{3} \beta_i (H)^i\right]^{1/2} \qquad (3)$$

Then, an ion exchange model of pseudohomogeneous equilibrium, equation (4), can be applied for these multimetallic solutions which considers the effect of the external pH and retained metals

$$k_i = \frac{q_i^a}{q_R^b\, C_i^m} \qquad (4)$$

Definition of an apparent constant, is given by the expression:

$$k_i' = \frac{k_i}{f(H)^n} \qquad (5)$$

that could be useful in experiments carried out at constant external pH. In such a way, equation (6) is obtained from eqn. (2), (4) and (5)

$$k_i' = \frac{q_i^a}{(Q_t - a\Sigma q_i)^b\, C_i^m} \qquad (6)$$

Values of a, b and m can be obtained by adjusting

experimental results to the proposed model. Apparent constants determination was carried out by measuring q_i vs. C_i in experiments at constant pH. The exponent n could be finally calculated from considering different apparent constants and metals.

An algorithmic method was proposed to solve the equilibrium equations and compare the predicted and experimental values. The iterative sequence converges quickly to the solution by a simple computer program.

A quite important simplification in such a complex medium is to consider cations retained in significant amounts only. Thus, in this case only copper, nickel, zinc and cobalt were considered. This last one, which has the lower conditional constant, is always retained in amounts lower than 5 % of the whole capacity of the resin.

Kinetic Model

The mathematical treatment can be simplified using pseudosteady state approximation but several author like Carberry (8) think that this approximation is not accurate for liquid since diffusion coefficients are lower in few magnitude orders than for gases; it is a typical example of liquid-solid non catalytic reaction where the pseudosteady state simplification cannot considered. Nevertheless, just in case of chelating resin, where the reaction may take hours or days, this approximation could be valid. Dealing with chelating resins, a shrinking core model with pseudosteady state simplification has been applied and accepted for many authors like Yoshida and Kataoka (9) and Helfferich (10).

Those experimental systems always consisted of a monometallic solution so that the obtained results are difficult to extrapolate to hydrometallurgical solutions which usually are multimetallics. Nevertheless, considering that all species diffuse independently and depending on the controlling step, the model can be expressed as:

$$t = \tau_j F_j(X_i) \tag{7}$$

where for external diffusion control (j=1),

$$\tau_1 = \frac{q_i^* \rho R}{3 k_{Li} C_{io}} \quad \text{and} \quad F_1(X_i) = X_i$$

for internal diffusion control (j=2),

$$\tau_2 = \frac{q_i^* \rho R^2}{6 D_i C_{io}} \quad \text{and} \quad F_2(X_i) = 1 - 3(1-X_i)^{2/3} + 2(1-X_i)$$

and for kinetic reaction control (j=3),

$$\tau_3 = \frac{q_i^* \rho R}{k_{ci} C_{io}} \quad \text{and} \quad F_3(X_i) = 1 - (1 - X_i)^{1/3}$$

X_i the fractional coverage referred to equilibrium is defined according to the expression

$$X_i = \frac{q_i(t)}{q_i^*} \tag{8}$$

RESULTS AND DISCUSSION

In preliminary tests, resin capacity and density were determined for kinetic experiments, and particle size selected under microscope. Kinetics experiments were carried out in infinite volume conditions;i.e.:

$$(\Sigma C_i) L >> (\Sigma q_i) S \tag{9}$$

Experiments were developed in a discontinuous stirred tank

provided with pH and temperature controls. Firstly, by means of the equation (1) were calculated the values of log β_i from a potentiometric titration curve of the resin resulting 9.72, 13.32 and 14.27, for i=1,2 and 3, respectively.

Equilibrium parameters a,b and m in eqn (6) were determined from experiments conducted at pH = 3.0. Consequently, values of 2,2,1 were found out, respectively. The calculated apparent constants are shown in Table 3 for other different set of experiments. The effect of pH, cobalt concentration and salinity (as sodium sulphate) is shown in this table. Salinity gives rise to a reduction in the apparent constant value. Cobalt concentration ranging between 5.04 to 15.27 mol/m^3 does not affect to apparent constants of other cations, except for cobalt itself where a slight variation seems to be observed. The effect of external solution pH increases the calculated constants with a maximum for copper and nickel around pH = 4.25, probably due to the formation of basic species.

In Figure 2, the linear relation between log (k_i/k_i') vs. log f(H) is plotted for copper, nickel, zinc and cobalt. Experimental results fit accurately with a slope of n=1. That means that the supposed relation between the constants and pH was right and the influence of this last variable on metal retention can be predicted.

The average value of the conditional constants were calculated using n=1 in eqn. (5), resulting k_{Cu} = 1.05 E6, k_{Ni} = 1.58 E4, k_{Zn} = 2.24 E4 and k_{Co} = 2.82 E3 L/mol . These constants are independent of pH and , consequently, they can be used to predict its effect upon metal retention in resin for multimetallic solutions, eqns. (5) and (6). In Figure 3, the calculated and experimental results of loaded copper vs. solution concentration are shown (nickel, zinc and cobalt present similar behaviour). It can be observed that there is a maximun load of copper at pH around 4.2. That means that other reactions like the formation of basic especies could be important at higher pH values. This equilibrium equations can be used to predict q_i which is employed into the kinetic models.

TABLE 3.- Apparent constants (m^3/mol) calculated from experiments using industrial waste water without iron.

EXP. CONDITIONS				APPARENT CONSTANTS (m^3/mol)			
pH	Log f(H)	(Co)	($NaSO_4$)	COPPER	NICKEL	ZINC	COBALT
2.99	3.72	1.83	0.58	0.19	2.26E-3	3.30E-3	5.5E-4
3.87	3.02	1.83	0.58	0.50	1.41E-2	2.44E-2	2.8E-3
4.25	2.78	1.83	0.58	2.96	4.18E-2	5.16E-2	4.5E-3
4.62	2.57	1.83	0.58	0.74	2.50E-2	5.50E-2	4.8E-3
3.65	3.17	5.09	0.58	0.75	1.71E-2	1.96E-2	3.0E-3
3.65	3.17	10.86	0.58	0.62	1.10E-2	2.29E-2	4.2E-3
3.65	3.17	15.27	0.58	0.73	1.26E-2	2.26E-2	5.7E-3
3.90	3.00	1.83	0.26	0.69	1.83E-2	2.95E-2	3.0E-3
3.87	3.02	1.83	0.58	0.50	1.41E-2	2.44E-2	2.8E-3
3.87	3.02	1.83	0.98	0.24	1.41E-2	1.40E-2	1.8E-3

Firstly, kinetics experiments were carried out using a mono-metallic solution of copper. Data of q_i vs. time were obtained in a stirred tank running experiments at different reaction time. Results were treated by means of the unreacted core model according to equation (7). Only $F_2(X_i)$ (Figure 4) seems to be linear and then the internal diffusion is the mean controlling step. From the slope a value of D_{Cu} = 1.04E-10 m^2/s was calculated.

The effective diffusion coefficient for copper was also calculated from experiments conducted at different pH values. The result are shown in Figure 5. The diffusion coefficient increases with the external pH which means that the fluxes of both species are coupled. Also in Figure 5, the calculated values by Höll et al. (11) for this commercial resin during the exchange Cu/H are represented.

In Figure 6, the loaded metals vs. time as a result of the experiments developed with the hydrometallurgical effluent are shown. Cobalt and nickel, which are loaded in lower concentration, quickly achieve equilibrium, while an opposite behaviour is observed for copper and zinc.

These results do not agree with unreacted core model-internal diffusion control. It seems that reaction step could be also involved or the kinetics behaviour of the ion exchange resin changes when several cations are involved in the system.

CONCLUSIONS

It has been proposed a flow-sheet for hydrometallurgical waste water treatment which allow a maximum product recycling. The treatment involves techniques like crystallization, precipitation and ion exchange with chelating resins. Iron precipitation and ion exchange has been experimentally studied. This last technique has been studied using an equilibrium model to predict metal retention from the hydrometallurgical multimetallic solution. The conditional constants are k_{Cu} = 1.05 E6, k_{Ni} = 1.58 E4, k_{Zn} = 2.24 E4 and k_{Co} = 2.82 E3 L/mol. Ion exchange kinetic behaviour has been fitted accuratly to the unreacted core model for copper. The calculated value of the diffusion coefficient was 1.04 E-10 m^2/s at pH = 4.0. Result from the complex hydrometallurgical could not be adjusted to this model.

SYMBOLS USED

a = Stoichiometric coefficient (-).
b = Stoichiometric coefficient (-).
C_i = External concentration for metal i (mol/L).
D_i = Effective diffusion coefficient (m^2/s).
f(H)= Non chelated fraction of resin group, eqn. (3) (-).

k_i = Conditional constant of metal i (L/mol).
k_i' = Apparent constant of metal i (L/mol).
k_{ci} = Chemical reaction constant (s^{-1}).
k_{Li} = Mass transfer coefficient (m/s).
L = Total liquid volume (L).
m = Stoichiometric coefficient (-).
n = Stoichiometric coefficient (-).
Q_t = Ion exchange capacity (mol/kg dry resin).
q_i^* = Chelated group in equilibrium (mol/kg dry resin).
q(t)= Amount of loaded metal at time t (mol/kg dry resin).
q_R = Free functional group (mol/kg dry resin).
R = Wet particle radius (m).
S = Mass of dry sodium-form resin (kg).
t = Time (s).
X_i = Fractional attainment to the equilibrium (-).
α_h = Protonated fraction of the functional group (-).
β_i = Overall protonation constant (L^i/mol^i).
ρ = Wet resin density (kg/m^3).
τ_j = Defined in equation (7).

REFERENCES

1.- Brooks, C.S.,1986, J. Met., 50, Jul., 50-57.

2.- Habashi, F.,1960, Principles of Extractive Metallurgy", Gordon & Breach Sci. Pub., NY.

3.- Ullmann, F., 1984, Ullmann's Encyclopedia of Industrial Chemistry, VCH Verlagsgesellschaft, 5th Ed., Weinhein (1984)

4.- Mijangos, F., Andrés A. and Díaz M.,1987, XII Cong. Interam. Ing. Quím., Lima, Juny.

5.- Davey, P.T. and Scott T.R., 1975, Trans. Instn. Metall. (Sec. C: Mineral Process Extr. Metal.) 84, 83-86.

6.- Hiskey, J.B.,1986, J. Met., 41, July, 40-46.

7.- Marinsky, J.A., 1986, in Mass Transfer and Kinetics of Ion Exchange, NATO ASI Series E, No 71, 75-120.

8.- Carberry, I, 1976, Chemical and Catalytic Reaction Engineering Mc Graw Hill, NY, 311.

9.- Yoshida, H., Kataoka T.,1985, Can. J. Chem. Eng., 63 (3), 422-429.

10.- Helfferich, F., 1986, in Mass Transfer and Kinetics of Ion Exchange, NATO ASI Series E, No 71, 121-180.

11.- Höll W., and Sontheimer, H.,1977, Chem. Eng. Sci., 32, 755-762 .

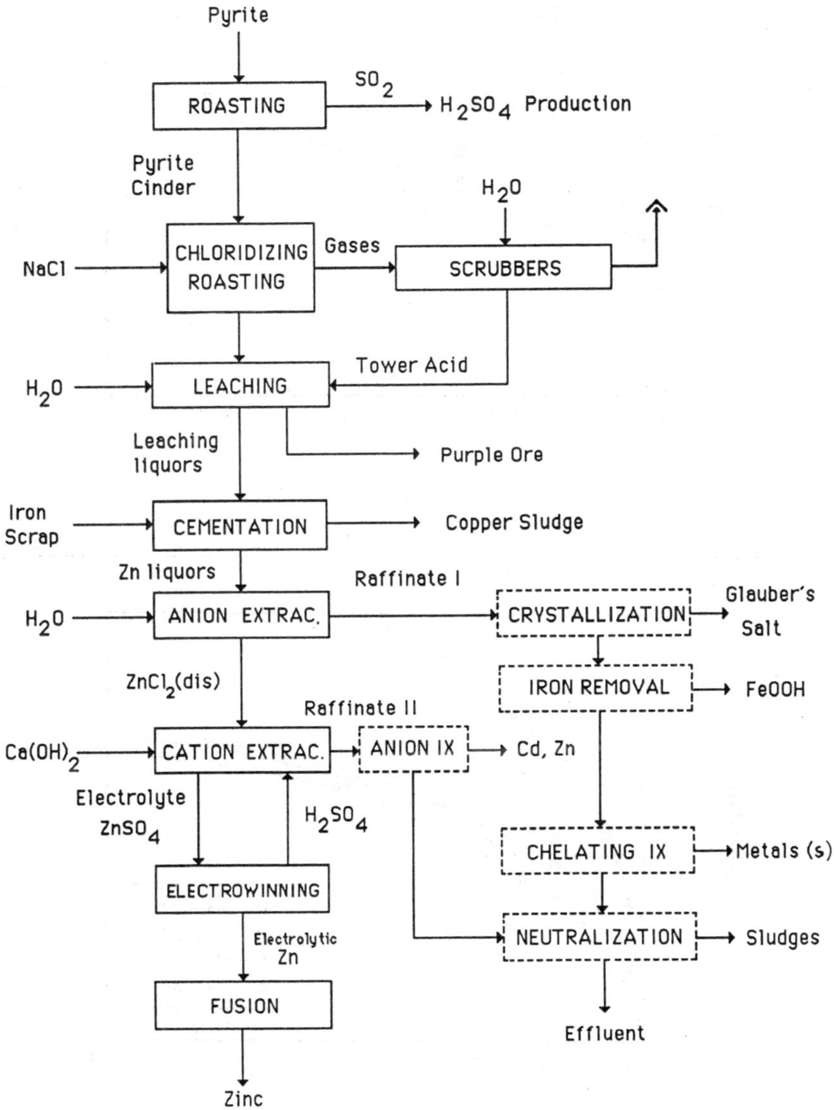

Fig 1. Flow-sheet of the industrial hydrometallurgical process. Dash lines denotes proposed operations.

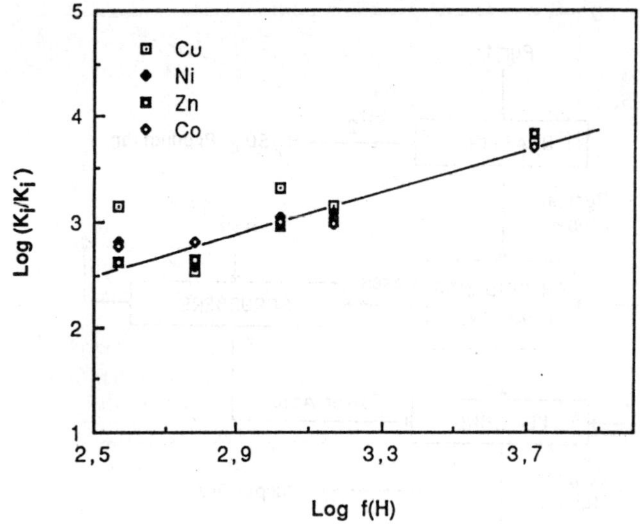

Fig 2. Correlation between conditional and apparent constants in relation with the pH function.

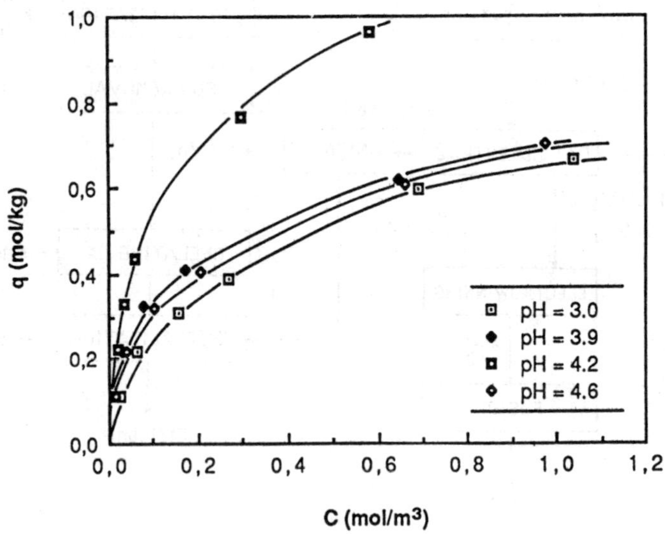

Fig 3. Experimental and calculated results for copper retention at different pH levels.

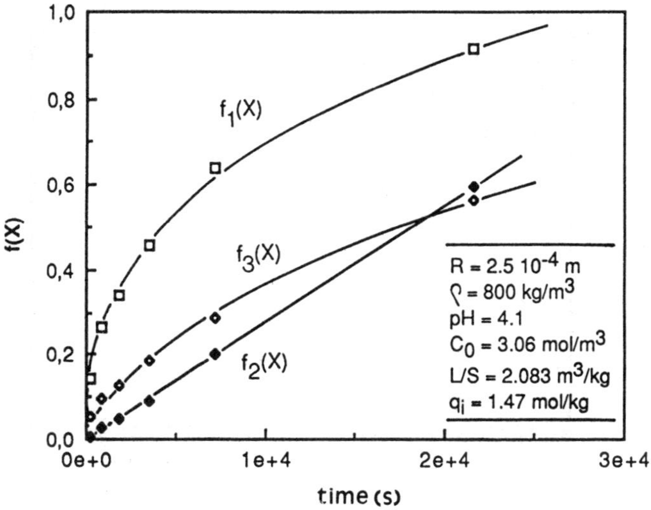

Fig 4. Kinetic results of loaded copper according the unreacted core model for different controlling step.

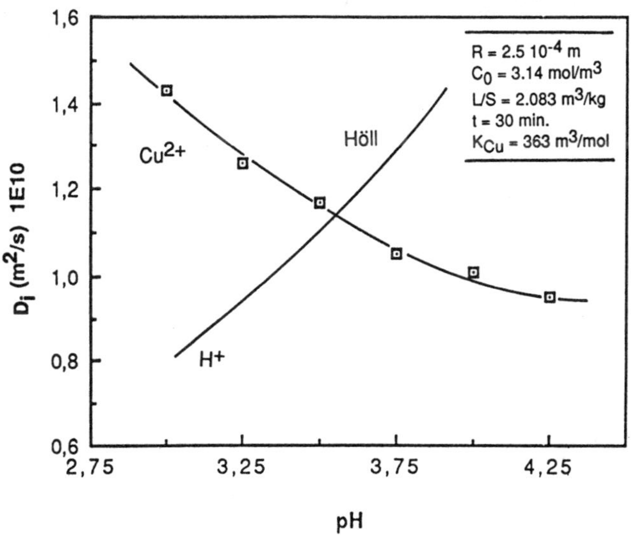

Fig.5. Effective diffusion coefficient of copper as a function of external solution pH.

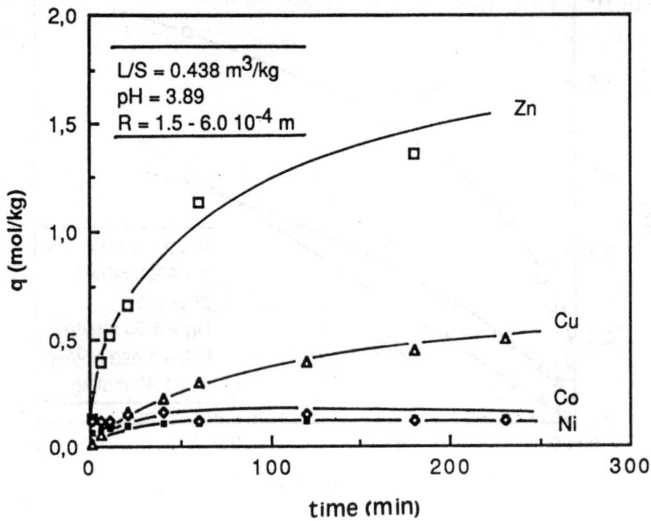

Fig 6. Metal retention kinetics onto the iminodiacetic resin from the hydrometallurgical solution.

FLOW DISTRIBUTION IN FIXED BED ION EXCHANGE UNITS

D J GUNN*

When fluid flows through a bed of particles the distribution in the bed depends upon the configuration and the arrangements for fluid entry and outlet. For a fixed bed ion exchange unit good hydrodynamic distribution is obtained only if the fluid velocity is constant over a cross section of the bed normal to the velocity vector, and if the length of the flow path is constant. If this requirement is not met, breakthrough of the ionic front will be experienced from some parts of the bed at an earlier time than others so contributing to underutilisation of resin in parts of the bed. Two other major problems in resin utilisation are due to wall effects, and the susceptibility of shallow beds to hydraulic damage.

INTRODUCTION

The obvious restrictions of experiments on large equipment dictate that proving experiments for ion exchange resins will be carried out on a small scale. The water to be treated is usually passed through a narrow column packed with resin particles, and the conductivity of the effluent is monitored for breakthrough. In contrast, the full scale ion exchange unit for large flows will often be large in diameter with a relatively shallow bed of resin. Internal piping is provided either for the regeneration of the resin in situ, or by removal of the resin to a regenerating vessel and return after regeneration. In some cases piping is immersed in a shallow bed of resin, and it appears that the functional requirements of the internal piping have dominated the design.

The volume of resin used in a treatment vessel is normally based upon experiments in narrow columns, and the design task is seen as the accommodation of a fixed volume of resin in a vessel of economical size and shape. Fluid and particle entries, exits and piping are added to the vessel as required to carry out ion exchange and regeneration. Provided that the required volume of resin can be accommodated, the design of vessel, nozzles and piping is apparently regarded as a separate exercise, without consideration of interactions with the primary function of ion exchange.

However, gross liquid maldistribution in a fixed bed of resin particles may be caused by mechanical arrangements for resin containment, and piping arrangements for water flow and resin removal. Once recognised, the requirements for good liquid distribution can be implemented at the design stage. It is much better to provide at the design stage, because it may be expensive to change a design that gives maldistribution once the ion exchange unit has been constructed.

*Department of Chemical Engineering, University College of Swansea

There are probably two principal causes of liquid maldistribution; maldistribution by design because of the configuration of fluid entries and exits, and maldistribution because of damage to the bed by jetting inlet streams displacing particles to form craters in the bed of particles.

The effects of liquid maldistribution, and longitudinal dispersion are described and related to the role of transverse dispersion in modifying interactions between longitudinal dispersion and maldistribution. The requirements for good fluid distribution and the implications for designs of fluid distribution and mixing are set out.

Flow Maldistribution by jetting damage to the bed

Linear fluid velocities in pipes may be of the order of 5 m/s, and a jetting liquid stream, if allowed to fall on the surface of the bed at this velocity, will scour channels and craters in the bed. Water to be treated will then flow preferentially through the shallower regions of the bed causing saturation of the resin followed by breakthrough in these regions. In severe cases of jetting damage to shallow beds, the scoured craters may be so deep that only a fraction of the resin in the bed is utilised.

Once the possibility of damage to the bed by this cause is recognised, the bed may be protected by deflecting and retarding jets from the entry nozzles, so that liquid is fully retarded before the bed surface is entered. In addition, the integrity of the bed surface may be protected by a wire mesh or perforated plate.

The use of protective screens or perforated plates may increase the complexity of piping if regeneration of the resin takes place externally, but a screen will filter larger solids from the inlet stream that may be cleared by back washing of the bed. In any case, protection of the bed surface from jetting damage should be mandatory for wide and relatively shallow beds by deflecting and retarding jets from the inlet nozzles. A deflection plate of some 2-3 nozzle diameters in diameter, placed about a plate diameter from the nozzle will help, but an internal nozzle atachment that diverts flow away from the bed surface into a number of smaller openings that reduce the fluid inlet velocity some twenty-fold should be effective. This requirement is not as crucial when the surface of the bed is covered by a protective mesh or perforated screen.

Flow maldistribution by design

The nature of flow through a fixed bed of resin particles may be characterised by the particle Reynolds number $du\rho/\mu$. At a typical Reynolds number of 20, flow through the bed is laminar, and although this value is a little above the accepted range of application, the relationship between the fluid velocity through the bed u and the potential of pressure and gravitation p is given by Darcy's Law,

$$u_x = \frac{K}{\mu}\frac{\partial p}{\partial x} \qquad (1)$$

in the co-ordinate direction x. For a particular bed of resin, fluid at potential p_i enters the bed and leaves at a lower potential p_o. The potential distribution through the bed is given as the solution to the partial differential equation known as Laplace's equation,

$$\nabla^2 p = \frac{\partial^2 p}{\partial x^2} + \frac{\partial^2 p}{\partial y^2} + \frac{\partial^2 p}{\partial z^2} = 0 \tag{2}$$

with the boundary condition of $p = p_1$ at the inflow surface of the bed, and $p = p_o$ at the outflow face. The direction of the velocity vector is normal to surfaces of constant potential, and the velocity is proportional to the potential gradient normal to the equipotential surface.

The most effective use of resin in the fixed bed is obtained by arranging that the fluid velocity u is constant over a cross section of the bed normal to the velocity vector, and the length of the flow path is constant. The normal way of arranging this condition is the linear fixed bed in which liquid enters the bed over the full cross section. A variation on this arrangement is the radial flow fixed bed in which fluid enters the bed at an inner cylindrical surface and flows radially outwards to leave at the outer cylindrical surface.

The radial-flow fixed bed is used for the occasion when such a bed operating at a given pressure difference, can be accommodated in a vessel of more economic size and shape than when a linear fixed bed is used. However, for both types the ionic front when determined only by the hydrodynamic distribution, moves uniformly through the bed and reaches the outflow surface at the same time. Fig 1 illustrates the lines of equipotential and stream lines through the linear bed, and Fig 2 the distribution in the radial-flow bed.

Although the linear fixed bed might be expected to be the normal arrangement for the fixed-bed ion exchange unit, vessel designers have on occasions introduced changes of bed shape, or configurations of internal piping, that cause major departures from the principle of constant velocity and constant flow path. Regeneration of resin by removal from the ion exchange vessel will usually require more complex internal piping than regeneration in situ. It is for this type that the vessel designer is most likely to depart from the principles of constant velocity and constant flow path, particularly if he is not aware of them.

Because of developments in numerical methods solutions to equation (1) may be obtained more or less readily depending upon configuration, to give fluid distribution in the bed. However, the use of the fluid distribution to improve the performance of the unit is a more complex matter, and the calculation of the fluid distribution is likely to be only of interest when considering modifications to a unit with the object of improving the distribution. Even this is likely to be an unusual requirement since a good hydrodynamic distribution is given by the simplest configurations such as those illustrated in Figs. 1 and 2.

One example of a bed configuration that departs from the principles of constant velocity and constant flow path is a dished bed that has been described by Tittle et al (1980) and illustrated in Fig.3. Although a full description of the bed, particularly the fluid outlets, has not been given, an estimate of fluid distribution may be obtained by noting that according to equation (1) the fluid velocity at a point will be inversely proportional to the bed thickness when measured along a stream line passing through the point. Based simply upon the gross hydrodynamic distribution, the rate of advance of the ionic front will also be inversely proportional to the bed thickness, and therefore when the ionic front reaches the base of the bed at the periphery, the shape of the front will be similar to that indicated in Fig.3. The figure also shows the Na^+ front reported by Tittle et al (1980), with the similarity between the estimated and measured front indicating that hydrodynamic distribution is most probably the dominant factor causing poor utilisation of the resin. Resin below the ionic front at breakthrough will be poorly utilised.

The effect of bed damage by crater formation upon hydrodynamic distributions may also be examined from the solution to equation (1). The calculation of the potential field for a postulated or measured disturbance of the bed by crater formation may be found as a solution to Laplace's equation for the postulated bed shape. The fluid velocity throughout the bed may then be found from the potential gradient and the permeability.

The fluid path through the bed may be indicated by fluid streamlines, and if a two-dimensional or axisymmetric bed is considred, a stream function ψ may be introduced that for a two-dimensional bed is related to the fluid velocities u_x and u_y by

$$u_x = \frac{\partial \psi}{\partial y} \tag{3}$$

$$u_y = -\frac{\partial \psi}{\partial x} \tag{4}$$

The equation of continuity for a two-dimensional incompressible flow is,

$$\frac{\partial u_x}{\partial x} + \frac{\partial u_y}{\partial y} = 0 \tag{5}$$

and by substitution of equations (3) and (4) in (5) it may be verified that the equation of continuity is satisfied. The use of the stream function in demonstrating distribution is useful because the difference in stream function between two streamlines is equal to the rate of fluid flow between the surfaces represented by the streamlines in cross section.

If the streamlines are shown for equal increments of stream function, when adjacent streamlines are far apart the fluid velocity in that region is low, and if close together the velocity is high. A section of a two dimensional bed with a single crater and streamlines through the bed is shown as Fig.4 where the clustering of the streamlines about the lower regions of the crater shows that resin there will be exhausted more quickly than elsewhere in the bed, with localised breakthrough from that region.

The distribution of streamlines in which the liquid effluent is collected in a common header immersed in the resin is shown as Fig.5. Resin in the approaches to the fluid outlets will be saturated before resin at the bed periphery, again showing poor resin utilisation.

At Reynolds numbers beyond the range of application of Darcy's Law, the pressure gradient is dependent upon velocity to a power greater than 1.0. The rate of advance of the ionic front is still proportional to the fluid velocity, but the velocity is less sensitive to variations in bed thickness than for lower flows. The ionic front will therefore tend to be flatter at higher velocities for bed configurations such as that shown in Fig.3. The flatter distribution is obtained at the expense of increased pressure loss, so that high velocities will not usually be attained in beds of ion exchange resins because of the small particle size.

However, the design of an ion exchange unit with protection from jet damage and with good hydrodynamic distribution is straightforward if the principles that have been set out are followed. The effect of axial and radial dispersion causing interaction between fluid regions of different velocity and particle holdup is more complicated and therefore a description and analysis of the effects is a necessary preliminary.

The interaction between dispersion processes and flow in fixed beds

When the concentration at the inlet of a fixed bed of resin is changed from one constant level to another, the concentration front travels through the bed, progressing as the resin particles become saturated. If mass transfer resistance is neglected at the particle surface and within each particle, the motion of the front may be represented by the partial differential equation

$$D \frac{\partial^2 c}{\partial x^2} - U \frac{\partial c}{\partial x} - \phi \frac{\partial c}{\partial t} = 0 \qquad (6)$$

where U is the interstitial velocity, D is the coefficient of axial dispersion and $\phi = \left[1 + K_s\left(\frac{1-e}{e}\right)\right]$ with K_s the volumetric ratio of the ionic capacity of the solid to fluid and e the bed porosity. The solution to this equation and boundary conditions when the initial concentration is zero, may be expressed (Carslaw and Jaeger, 1959),

$$\frac{c}{c_o} = 1 - \text{erf}\left[\frac{\phi x - Ut}{2\sqrt{(Dt\phi)}}\right] \qquad (7)$$

$$\text{where the error function } \text{erf}(a) = \frac{2}{\sqrt{\pi}} \int_o^a e^{-\chi^2} d\chi \qquad (8)$$

The coefficient of axial dispersion D is a consequence of velocity differences in the fluid flowing through the bed and molecular diffusion. Experiental measurements of dispersion are usually expressed as the dependence of Peclet group Ud/D upon the characteristic Reynolds number of flow for the particles. The form of the dependence of axial Peclet group upon Reynolds number for large particles during liquid flow is shown in Fig.6 (Gunn 1969). The Peclet number characterising liquid phase dispersion in beds of small diameter particles is particularly sensitive to the conditions of flow, a finding that has been attributed to disturbance of particles by liquid in flow. The disturbance takes the form of apparently creating channels for liquid flow, therefore causing a variation in liquid velocity over the cross section, so increasing dispersion.

Fig.7 shows that the effect of an increase in the coefficient of axial dispersion is to extend the concentration front. For a resin bed in ion exchange the effect of an increase in the coefficient of axial dispersion is to reduce the time at which breakthrough is observed. Thus any factors in the design of fixed bed ion exchangers that will increase axial dispersion will cause a deterioration in the performance of the ion exchange unit. The data summarised by Fig. 6 are mainly based upon experiments with impermeable particles at ratios of tube to particle diameter greater than 10, but for small particles and non-spherical particles some reported experimental results lie significantly below the correlating line of Fig.6. Hence the dispersion coefficients for these experiments are greater than would be expected from Fig.6.

One of the factors that contributes to an increase in dispersion coefficient was suggested by experiments of Schwartz and Smith(1953) who measured the distribution of velocity some 20 mms from the end of the fixed bed. They found velocity maxima in the vicinity of the wall with a trough in velocity at the centre of the bed. The velocity maxima extended 2-3 particle diameters from the wall. It has been subsequently shown (Price (1968), Vortmeyer and Schuster (1983)) that Schwartz and Smith overestimated the extent of the enhanced velocity region, because the velocity distribution develops sharply from the end of the fixed bed. Price's experimental arrangement did not permit development of the profile from the end of the fixed bed, and he found that the velocity in the wall region leaving the bed was 1.65 times the velocity in the bulk region, and the extent of the wall region was no more than half a particle diameter.

Such experiments determine the velocity in free space leaving the fixed bed, but not within the fixed bed because of the reduction in porosity due to particles. Roblee et al (1958) and Vortmeyer and Schuster (1980) have measured porosity distributions from the wall for irregular and regular particles. The main transition from the porosity of the wall of 1.0 to a bed porosity of 0.37 occurred within 0.4 particle diameters. By continuity, the velocity in the bed U_b is related to the velocity in free space measured just above the bed U_o, by $U_b e = U_o$, and in the wall region $U_{bw}\bar{e}_w = U_w$. As the porosity in the wall region lies in the range of 0.37 to 1.0, and as Price found that $U_w = 1.65\ U_o$ then

$$U_{bw}\bar{e}_w = 1.65\ U_b e \tag{9}$$

Caycik and Gunn (1988) have measured the interstitial velocities U_w and U_b by salt injection and conduction sensors, and have found that they are approximately equal within limits set by variations generated by repacking and experimental precision. However, the fraction of solid in the wall region $(1-\bar{e}_w)$ is significantly different from $(1-e)$ the fraction of solid phase in the bulk of the bed. Thus if K_s is large compared to 1 the ratio of $\bar{\phi}_w/\phi_b$ will be about

$$\frac{\bar{\phi}_w}{\phi_b} = \frac{1 + \left(\frac{1-\bar{e}_w}{\bar{e}_w}\right)K_s}{1 + \left(\frac{1-e}{e}\right)K_s} \approx \frac{(1-\bar{e}_w)e}{(1-e)\bar{e}_w} \approx 0.4 \tag{10}$$

Taking a mean value of $\bar{e}_w$ as 0.6, ϕ_b in the bulk is 2.5 times as great as ϕ_w.

To examine the effect of this disparity, we may consider the performance of the bed by describing the motion of the front in the wall and in the bulk regions by equations of the form of (6), but with provision for mass exchange between the wall and the bulk regions. As we have in equation (7) a means of interpreting the effect of changes of D upon breakthrough, if we take equation (6) to describe the motion of the front for the whole bed, then the coefficients D, U and ϕ apply to the whole bed. Hence the mean interstitial velocity U is related to the interstitial velocities for the bulk and wall regions U_b and U_w by

$$U\bar{e} = U_b\ \delta e + U_w(1-\delta)\bar{e}_w \tag{11}$$

where $(1-\delta)$ is the fraction of the bed cross section that is the wall region. Similarly,

$$\phi = \phi_b\delta + \bar{\phi}_w(1-\delta) \tag{12}$$

However, the dispersion coefficient for the bed is not given by a relationship similar to (11) and (12).

The dispersion coefficient is determined not only by dispersion in the wall and bulk regions, but also by the different velocities of the concentration front in the two regions, and mass exchange. If we consider the possibility of interchange betwen the bulk and wall regions by diffusion, two equations may be derived representing the conservation of mass for the wall and bulk regions (Gunn 1980). They are

$$\frac{D_b}{\phi_b}\frac{\partial^2 c_b}{\partial x^2} - \frac{U_b}{\phi_b}\frac{\partial c_b}{\partial x} - \frac{\partial c_b}{\partial t} - \gamma c c_b + \beta c_w = 0 \tag{13}$$

$$\frac{D_w}{\phi_w}\frac{\partial^2 c_w}{\partial x^2} - \frac{U_w}{\phi_w}\frac{\partial c_w}{\partial x} - \frac{\partial c_w}{\partial t} - \beta c_w + \gamma c_b = 0 \tag{14}$$

where $\gamma = -\dfrac{\alpha_1^{\ 2}D_R}{\phi_b R_b^{\ 2}}$, $\beta = \dfrac{\gamma}{(1-\theta)}$ and $\alpha_1^{\ 2}$ is the first zero of $J_o(u) = 0$.

The initial conditions are $c_b = \theta$ at $x = 0$, $c_w = (1-\theta)$ at $x = 0$. with both zero elsewhere. An estimate of the effect of the regions may be obtained by comparing the properties of the solution to equations (13) and (14) with the solution to equation (6) in which a tracer particle is placed at $x = 0$ at $t = 0$.

The procedure for comparison is to calculate and identify the first and second moments of the solutions in space, and it has been described by Gunn (1980) for a similar system. The two-sided Laplace transform $\int_{-\infty}^{\infty} e^{-sx}c\,dx$ is taken of both equations (13) and (14) to give two first order differential equations in time. The moments of the distribution may then be found by the following limiting processes applied to the solutions,

$$\mu_1 = \underset{s\to 0}{\mathrm{Lt}} \frac{d}{ds} \int_{-\infty}^{\infty} e^{-sx}(c_b + c_w)dx \tag{15}$$

$$\mu_2 = \underset{s\to 0}{\mathrm{Lt}} \frac{d^2}{ds^2} \int_{-\infty}^{\infty} e^{-sx}(c_b + c_w)dx \tag{16}$$

For the distribution given by the initial condition of c = 1 at x = 0, t = 0 applied to equation (6), the variance σ^2 is

$$\sigma^2 = \mu_2 - \mu_1^2 = 2Dt \tag{17}$$

so that on comparing variances found from equation (15) and (16) with 2Dt, the following expression for the overall dispersion coefficient D_L is obtained (Gunn, 1980),

$$\frac{D_L}{D_b} = \frac{\theta\phi}{\phi_b} + \frac{(1-\theta)\phi}{\phi_w}\frac{D_w}{D_b} + \left(\frac{U_w}{\phi_w} - \frac{U_b}{\phi_b}\right)^2 \frac{\theta(1-\theta)^2 R_b^2 \phi_b \phi}{\alpha_1^2 D_R D_b}$$

$$+ \left(\frac{U_w}{\phi_w} - \frac{U_b}{\phi_b}\right)^2 \frac{\theta(1-\theta)^3 R_b^4 U \phi_b^2}{\alpha_1^4 D_R^2 D_b x} \left[\exp\left(- \frac{\alpha_1^2 D_R x \phi}{U R_b^2 (1-\theta)\phi_b}\right) - 1\right] \tag{18}$$

One other condition of equation (18) is of interest. When the group of $\alpha_1^2 D_R x \phi / (U R_b^2(1-\theta)\phi_b)$ is small the exponential of equation (18) may be represented by

$$\exp\left(- \frac{\alpha_1^2 D_R x \phi}{U R_b^2(1-\theta)\phi_b}\right) \approx 1 - \frac{\alpha_1^2 D_R x \phi}{U R_b^2(1-\theta)\phi_b} + \frac{\alpha_1^4 D_R^2 x^2 \phi^2}{2 U^2 R_b^4 (1-\theta)^2 \phi_b^2}$$

+...neglected terms (19)

When equation (19) is substituted into (18), equation (20) is obtained

$$\frac{D_L}{D_b} = \frac{\theta\phi}{\phi_b} + \frac{(1-\theta)\phi D_w}{\phi_w D_b} + \left(\frac{U_w \phi_b}{U_b \phi_w} - 1\right)^2 \left(\frac{\phi^2}{\phi_b^2}\right) \frac{U_b^2 \theta(1-\theta)x}{2 U D_b} \tag{20}$$

and this equation represents the limiting condition of no exchange between the regions.

These equations have several interesting features. The ratio ϕ_w/ϕ_b is given by $(1-\bar{e}_w)e/(1-e)\bar{e}_w \sim 0.4$, so that even though we have not been able to establish a significant difference between U_w and U_b, the difference in the fractions of solid in the wall and bulk regions ensures that there is a large difference between the velocities of the fronts in the two regions.

The dependence of the dispersion coefficient ratio upon x/d_t and d_t/d may be illustrated for a particle Reynolds number of 20. For this Reynolds number the radial Peclet group is 22, and the axial is 0.76 (Gunn, 1969). Fig.9 shows the dependence of the dispersion ratio upon x/d_t. The graph is doubly logarithmic and the dependence of the dispersion ratio upon d_t/d is shown for several values of that parameter.

The effect of transverse mixing is shown by a comparison of the predictions of equation (20) with those of (18). For small values of x/d_t the differences between the predictions are small, but the differences grow with an increase in x/d_t. For laboratory scale columns and full scale operated at the same linear velocity and of similar depth, Fig.9 shows that differences between the performances of the large and small scale units may not be a source of concern, mainly because of the effect of transverse mixing.

There is one very important proviso. Experiments on axial mixing in fixed beds of small solids can show much greater dispersion coefficients than given by Fig.6, and this is thought to be related to the difficulty in the uniform packing of beds of small particles particularly if the diameter of the bed is large, and to disturbances of the particles by the flowing fluid. The susceptibility of a large diameter shallow bed may be illustrated by a calculation based upon equation (20), in which it is supposed that the particles are disturbed so that the effective thickness of the bed is reduced to one half of the initial value for 1% of the cross sectional area of the bed. Therefore over this region the fluid velocity is increased by a factor of 2, and the value of ϕ is reduced by a factor of 2. From equation (20) taking as before the value of $\frac{\theta\phi}{\phi_b} + \frac{(1-\theta)\ \phi D_w}{\phi_b D_b}$ to be 1

$$\frac{D_L}{D} = 1 + \left[\frac{U_w\phi_b}{U_b\phi_w} - 1\right]^2 \frac{Pe}{2} \left[\frac{x}{d}\right] \theta(1-\theta) \tag{21}$$

so that

$$\frac{D_L}{D} = 1 + 0.04\ x/d \tag{22}$$

For ion exchange units the value of x/d is normally very large and therefore equation (22) shows the beds are very susceptible to hydraulic damage.

When transverse mixing may be neglected as supposed for equation (20), the breakthrough curves may be found by superposition of equation (7) weighted appropriately for the heterogeneous regions. If transverse mixing is significant however a single expression of equation (7) with weighted values of ϕ and U, and the dispersion coefficient D given by equation (18) should give an acceptable representation of the breakthrough curve.

The calculations behind Fig.9 are illustrative of a single Reynolds number, and may not be fully representative of the range of conditions likely to be encountered in practical ion-exchange units. The analysis in the latter section of the paper may be used to examine the broad range of practical configurations.

However, it is clear that the primary concern in the design of an ion exchange unit is to contrive good fluid distribution in the bed using the simple principles stated here, to ensure uniform and efficient packing of the particles in the bed, and to protect the bed from hydraulic damage. In addition to provision for fluid retardation at the vessel inlets, perforated plates or substantial mesh protection for the bed will reduce the likelihood of hydraulic damage, improve fluid distribution, and if sufficiently firm, maintain uniform packing of particles in the bed. Once the prerequisites have been established, the nature of axial dispersion and the importance of transverse mixing may be examined using the relationships presented in this paper. This sequence of design priorities should be mandatory, even if the whole question of regeneration, whether in situ or not, has to be reconsidered. The question of liquid distribution and mixing is particularly important for the production of ultra pure water.

SYMBOLS USED

c = concentration, moles or equivalents per unit volume

d = particle diameter

d_t = bed diameter

D = coefficient of axial dispersion

D_L = total coefficient of axial dispersion

D_R = coefficient of radial dispersion

e = bed porosity

J_o = Bessel function of zero order

K_s = volumetric ratio of the ion capacity of solid to fluid

p = fluid potential of pressure and gravity

R = radius

t = time

U = interstitial velocity

U_0 = superficial velocity

x, y, z = Cartesian coordinates

α = root of $J_o(u) = 0$

β, γ = defined exponents

δ = area fraction of bulk region

θ = capacity fraction of bulk region

ϕ = ionic capacity

ψ = stream function defined by equations (3) and (4)

$\mu_1 \mu_2$ = first and second moments

Subscripts

b = bulk

w = wall

REFERENCES

Carslaw, H S and Jaeger, J C, 1959, "Conduction of Heat in Solids", Oxford U.P.

Caycik, M and Gunn, D J, 1988, "True mean velocities in the wall and bulk regions of fixed beds", Chem Eng Sci, 43, 1637-1646.

Gunn, D J, 1969, "Theory of radial and axial dispersion in fixed beds", Trans I Chem E, 47, 351-359.

Gunn, D J, 1971, "Axial dispersion in packed beds: the effect of the quality of packing", Trans I Chem E, 49, 109-114.

Gunn, D J, 1980, "Theory of liquid phase dispersion in packed columns", Chem Eng Sci, 35, 2405-2413.

Price, J, 1968, "The distribution of fluid velocities for randomly packed beds of spheres", Mech and Chem Eng Trans (Austria), May, p 7.

Roblee, L H S, Baird, R M and Tierney, J W, 1958, "Radial Porosity Variations in Packed Beds", A I Ch E Jnl, 4, 460.

Schuster, J and Vortmeyer, D, 1980, Chem Ing Tech, 52, 848.

Schwartz, C E and Smith, J M, 1953, "Flow distributions in packed beds", Ind Eng Chem, 45, 1209.

Tittle, K, Tyldesley, J D, Tasker, P W and Butcher, C J, 1980, "Mixed bed performance in a condensate polishing plant", Symposium on Water Chemistry II, BNES, Paper 23.

Vortmeyer, D and Schuster, J, 1983, "Evaluation of steady flow profiles in rectangular and circular packed beds by a variational method", Chem Eng Sci, 38, 1691-1699.

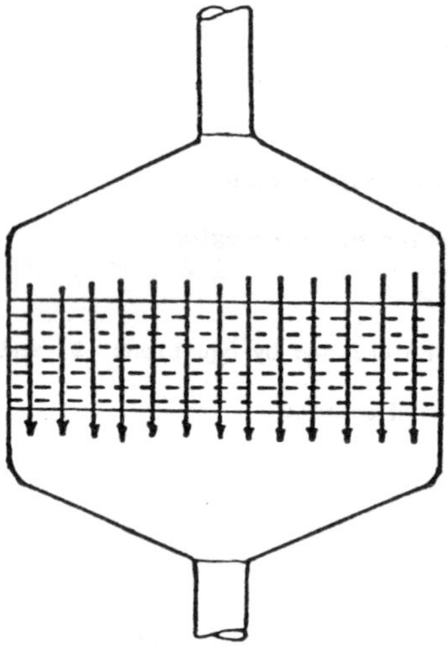

Fig. 1 Equipotential and streamlines for the linear fixed bed

---------------potential
——————streamlines

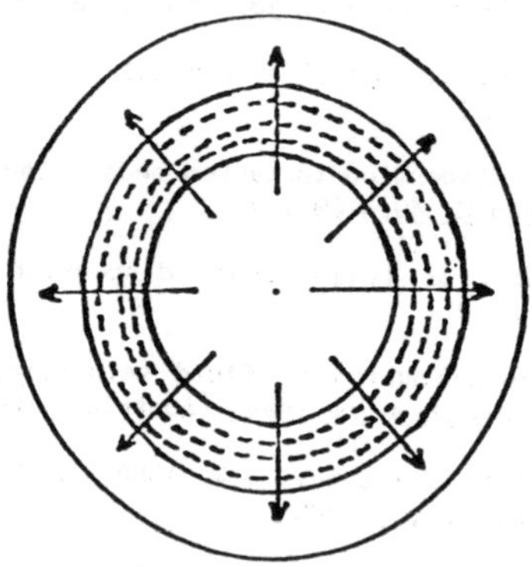

Fig. 2 Equipotential and streamlines for the radial-flow fixed bed

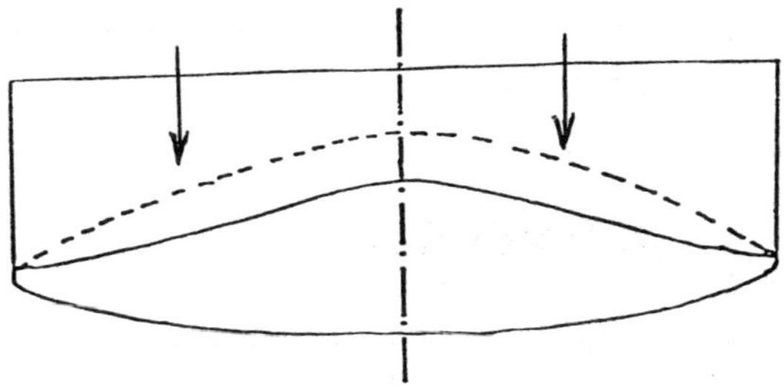

Fig.3 Dished bed ion exchange unit (Tittle et al, 1980)

----------------Na$^+$ front

————————hydrodynamic front

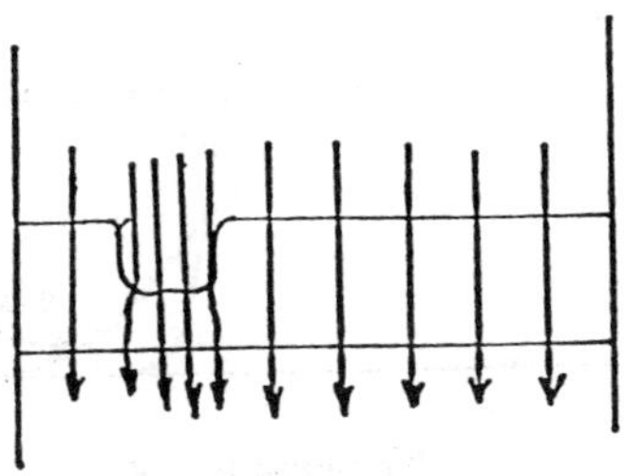

Fig.4 Estimated equal interval streamlines for a two-dimensional bed showing crater formation by hydrodynamic damage

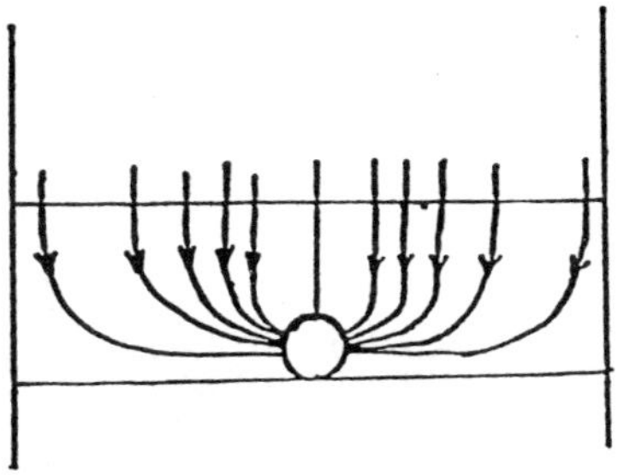

Fig.5 Estimated equal interval streamlines for common header collection of effluent

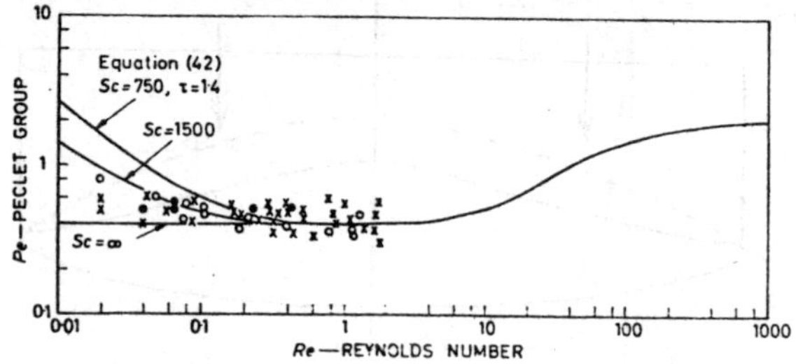

Fig.6 Dependence of Peclet group (axial) upon Reynolds number for liquid phase dispersion in beds of particles

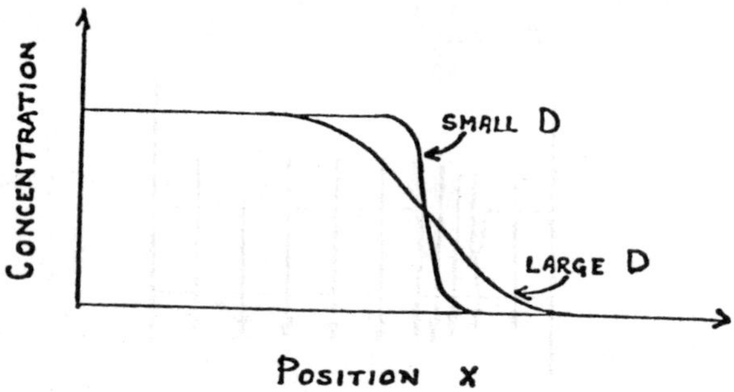

Fig.7 The effect of an increase in coefficient of axial dispersion upon breakthrough

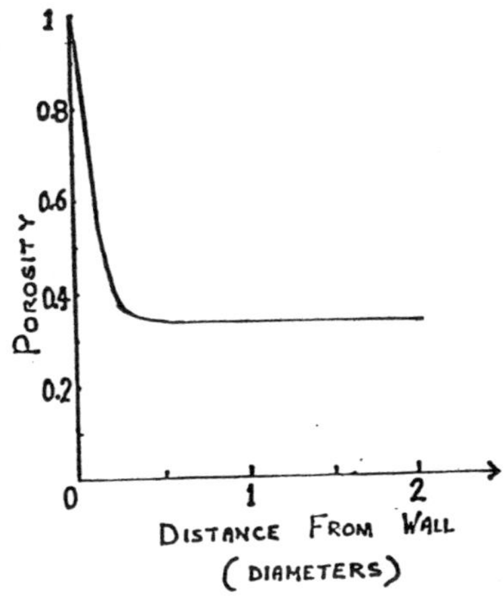

Fig. 8 Porosity distribution near a solid surface (Roblee et al (1953), Vortmeyer and Schuster (1980))

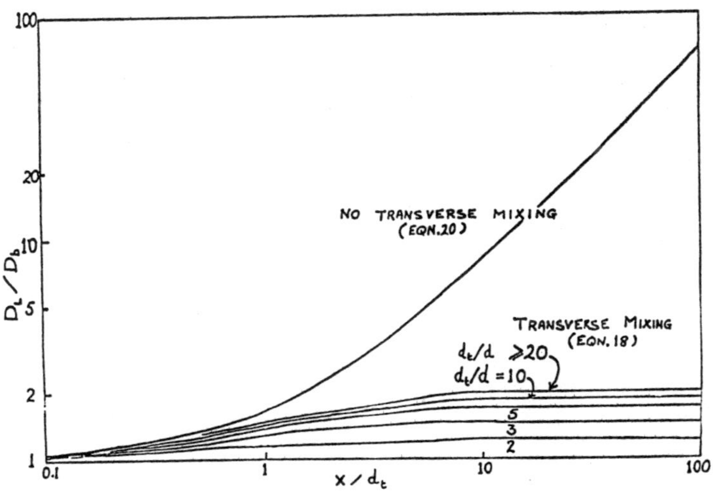

Fig. 9 The dependence of dispersion coefficient ratio upon bed configuration

DEVELOPMENT OF CMP-IMPREGNATED SOLID EXTRACTANT FOR TRU REMOVAL FROM RADIOACTIVE LIQUID WASTES

K.TAKESHITA*, S.MATSUMOTO**, M.KUMAGAI*,
Y.ENDO***, J.KOGA** and Y.TAKASHIMA*

A new solid extractant, which was supported by macroreticular copolymer of styrene and divinylbenzene (SDB), impregnated with a bidentate organophosphorus compound, DHDECMP (dihexyl-N,N-diethylcarbamoylmethylphosphonate; abbreviated as CMP), was developed for removal of transuranium elements (TRU) from radioactive liquid wastes. The solid extractants with different crosslinkage were prepared and the extraction experiments using cerium(III) as a substitutive material for TRU were carried out.

The amount of CMP impregnated in SDB particles, or the CMP content, increased with decreasing degree of crosslinkage of SDB particles. The maximum content of 2.15 g-CMP/g-resin was obtained for a resin crosslinked by 10wt% divinylbenzene. It was 1.8 times of that for a commercial resin, Amberlite-XAD4.

The distribution ratio of cerium was proportional to the 2.5 power of the CMP content and the 3 power of the concentration of nitrate ion in the aqueous phase. Thus, the stoichiometry for the formation of a complex between CMP and cerium can be expressed by

$$Ce^{3+}(aq) + 3NO_3^-(aq) + 2.5CMP(resin) \rightleftharpoons Ce(NO_3)_3 \cdot 2.5CMP(resin)$$

The distribution ratios obtained for the solid extractants were higher than those for some liquid-liquid extraction processes, in which CMP was diluted with organic solvents such as n-heptane, o-xylene and carbon tetrachloride.

The extraction rates of cerium of the solid extractant with smaller diameter were higher with the smaller diameter of the solid extractant, suggesting the rate-controlling step to be the inter-particle diffusion of cerium. The time to reach 50% of equilibrium was only 10 seconds for that of 0.2 mm in diameter.

INTRODUCTION

A small amount of long-lived alpha emitters such as transuranium elements (TRU) exists in high-level or intermediate-level radioactive liquid wastes arising from a spent nuclear fuel reprocessing plant. At present, such alpha bearing wastes, often called TRU wastes, are processed directly as glass or bitumen products(1), of which the isolation time will be required for several tens of thousands years against the long-term potential hazard of TRU(5). However, if TRU elements can be separated from other fission products, the storage period of the remaining wastes will be reduced and the stripped TRU elements can be burnt up by neutron bombardment in FBR or irradiation of proton beam

* Institute of Research and Innovation,
1201 Takada, Kashiwa, Chiba 277 Japan
** Dept. of Environmental Chem., Faculty of Eng., Saitama Univ.,
255 shimo-okubo, Urawa 338 Japan
*** Ishikawajima-Harima Heavy Industries Co.,LTD.
1 Shin-nakahara, Isogo, Yokohama 235 Japan

arising from particle accelerator(5).

The separation of TRU elements from radioactive liquid wastes has been achieved by solvent extraction processes, in which organophosphorus compounds such as TBP (tributylphosphate) and HDEHP (di(2-ethylhexyl) phosphoric acid) are used(1,3,10). Recently, bidentate organophosphorus compounds such as CMP (carbamoylmethylphosphonate) and CMPO(carbamoylmethylphosphine oxide) have been studied, since these compounds may be more effective for the removal of TRU than the conventional ones(7,8). They have been used in the TRUEX process, ICPP process and so on(4). Thompson reviewed various processes developed for TRU separation and pointed out that these processes will need additional large scale equipment, comparable to the main partitioning process of a reprocessing plant and will produce a large amount of secondary waste solutions(10).

In the Rocky Flats Plant, the americium recovery process was developed, which used CMP-impregnated copolymer of styrene and divinylbenzene(2). Faubel et al. tried to remove TRU in the intermediate-level liquid wastes using CMPO-impregnated acrylic resin(3). The results of these studies suggest that by the use of such solid extractants, the TRU separation process will be made very simple and the volume of the secondary wastes will become smaller than those from solvent extraction processes.

We are developing a CMP-impregnated solid extractant for the selective removal of TRU from radioactive liquid wastes. For this purpose, CMP was impregnated in porous SDB particles varying in crosslinkage, and batchwise experiments to extract cerium(III) as a substitutive material for TRU, were carried out. The effects of the size and crosslinkage of the solid extractants on the distribution ratio and the extraction rate of cerium are discussed.

EXPERIMENTAL PROCEDURE

Preparation of CMP-impregnated solid extractants

Macroporous SDB particles were used as a support for the solid extractant. The SDB particles were produced by the capsular polymerization method(6,9). This method consists of several steps, such as the formation of single-sized and concentric two phase droplets, the solidification of the outer phase, the polymerization of monomer in the inner phase and the removal of shell materials.

The mixture of styrene, divinylbenzene(DVB) and n-heptane was used as the monomer solution. The SDB particles with different crosslinkage were prepared by changing the contents of DVB in the monomer solution. The content of DVB represents the degree of crosslinkage of polymer, which is described by the weight fraction of DVB in the monomer mixture of styrene and DVB. N-heptane contributes to the formation of macropores by causing phase separation during the polymerization step. The porosity of the particles increases with the content of n-heptane, which is defined by the volume fraction of n-heptane in the monomer solution. It was kept constant at 50 vol% in this study. SDB particles crosslinked by 5, 10 and 20wt% of DVB were prepared, which hereafter we call 5%-resin, 10%-resin and 20%-resin, respectively.

A bidentate organophosphorus compound, DHDECMP (dihexyl-N,N-diethylcarbamoylmethylphosphonate; abbreviated as CMP) was used as an impregnating agent. The purity of CMP was 92%. The impregnation of CMP in SDB particles was achieved by suspending the SDB particles in a solution of CMP with a diluent, dichloromethane for 10 hours. Excess CMP in the particles was removed by washing with water several times. The diluent was removed in vacuo at 50 °C. The amount of CMP impregnated into SDB particles was estimated from the difference of particle weight before and after CMP impregnation. The CMP content, f, was defined as follows;

$$f\ [\text{g-CMP/g-resin}] = \frac{[\text{Weight of impregnated CMP}]}{[\text{Weight of SDB particles}]} \quad (1)$$

Measurement of distribution ratio of cerium

The distribution ratio of cerium was measured by contacting 1 gram of solid extractant with 10 ml of a test solution in a stirring cell for 10 hours. The test solution contains 240 mg/l of cerium(III) and 0.33 to 2.0 mol/l aluminum nitrate. The temperature in the cell was kept constant at 25 °C and the agitating speed was 600 rpm. The concentration of cerium in the test solution was measured by ICP spectrometry.

The distribution ratio D of cerium is defined as

$$D = \frac{[\text{moles of cerium extracted in 1 gram of support resin}]}{[\text{moles of cerium remaining in 1 ml of aqueous phase}]} \quad (2)$$

In preliminary tests, it was confirmed that the concentration of cerium in the test solution became unchanged with time in 10 hours. The concentration of cerium in the solid extractant was calculated from that in the test solution by the mass balance.

Measurement of extraction rate of cerium

The extraction rate of cerium was measured in a stirred cell, in which 1 gram of solid extractant was added to 100 ml of the test solution. The composition of the test solution, the temperature in the cell and the agitating speed were the same as described previously. In preliminary experiments, we confirmed that the extraction rate did not change at the agitating speed of faster than 600 rpm, and the mass transfer around the solid extractant was fast enough at these agitating speed. Therefore, the observed extraction rate of cerium should be controlled by either the diffusion rate of cerium into the particles or the reaction rate between cerium and CMP.

RESULTS AND DISCUSSION

Impregnation of CMP into SDB particles

Table 1 shows the amount of CMP impregnated in the SDB particles which are different in diameter, specific surface area, average pore diameter and crosslinkage. The CMP content increased by using a low-crosslinked resin. The maximum value of 2.15 g-CMP/g-resin was obtained for the 10%-resin. This value was 1.8 times of that in the solid extractant produced from a commercial resin, Amberlite XAD-4, used by Alford

Table 1 The effects of the diameter of SDB resin, the degree of crosslinkage by DVB, the specific surface area, the average pore diameter and the temperature of CMP impregnation on the CMP content.

RunNo.	dia. (mm)	Properties of SDB resin DVB conc. (wt%)	Ave.pore dia. (A)	Surface area (m^2/g-resin)	Temp.of CMP Impregnation (°C)	CMP content, f (g-CMP/g-resin)
1a	1.4	10	1500	128	25	2.15
1b	2.0	10			25	2.14
1c	1.4	10			70	2.10
2	1.4	20	1500	140	25	1.20
3	1.4	45	—	—	25	0.35
Amber-lite	0.5	–	50	784	25	1.24

and Navratil(2). The higher CMP content in the lower crosslinked resin may be explained by EPMA. Figure 1 shows the line profiles of CMP content across the cross sections of two different extractants obtained from the 10%-resin and 20%-resin, respectively. CMP was impregnated over the whole cross section of the low-crosslinked resin, whereas it was not impregnated near the center of the high-crosslinked one.

Two photographs in Fig. 2 show the SEM images of the interior of the SDB resin before and after impregnation of CMP. It seems that the SDB particle consists of microglobes and that spaces between the microglobes form communicating pores. The microglobes swelled and grew up in size after the impregnation of CMP. The specific surface area of the SDB particles decreased from 128 to 10 m^2/g as a result of the CMP impregnation. These results indicate that many pores in the particles are lost by the growth of the microglobes.

Distribution ratio of cerium

Figure 3 shows the distribution ratio of cerium as a function of the CMP content. Two solid extractants produced from the 5%-resin and the Amberlite-XAD4, were tested. The distribution ratios for both the extractants increased in proportion to the 2.5 power of the CMP content. The ratios for the extractant developed in this study, however, tend to saturate at CMP contents over 1.0 g-CMP/g-resin. This would be because the impregnated CMP becomes less effectively utilized with increasing CMP content as a result of the disappearance of pores caused by the growth of microglobes as shown in Fig. 2.

Figure 4 shows the distribution ratios of cerium as a function of the concentration of nitrate ion. The CMP contents of the extractants were 1.95 g-CMP/g-resin for the 5%-resin and 1.0 g-CMP/g-resin for the Amberlite-XAD4, respectively. The distribution ratio for each extractant increased in proportion to the third power of the concentration of nitrate ion.

The stoichiometry for the formation of a complex between CMP and cerium is hence represented as

$$Ce^{3+}(aq) + 3NO_3^-(aq) + 2.5CMP(resin) \rightleftharpoons Ce(NO_3)_3 \cdot 2.5CMP(resin), \quad (3)$$

Comparison with solvent extraction processes

The solvent extraction experiments of cerium were carried out by using three CMP solutions of different organic solvents, n-heptane, o-xylene and carbon tetrachloride. The ratio of the volume of organic phase to that of aqueous phase was kept constant at 0.2 ml-org.phase/ml-aq.phase and the extraction temperature was 25 °C in all experiments. Figure 5 compares the distribution ratios measured by the solvent extraction experiments with those obtained by the solid extractants which were supported by 5%-resin. Abscissa in this figure means the amount of CMP in 1ml of the organic phase or 1ml of the resin. Vertical axis denotes the distribution ratio of cerium, which is defined as

$$D_{solvent} = \frac{[\text{moles of cerium extracted in 1ml of org.phase}]}{[\text{moles of cerium remaining in 1ml of aqueous phase}]} \qquad (4)$$

for the solvent extraction, or

$$D_{solid} = \frac{[\text{moles of cerium extracted in 1ml of solid extractant}]}{[\text{moles of cerium remaining in 1ml of aqueous phase}]} \qquad (5)$$

for the solid extraction.

The distribution ratios obtained by the solvent extraction experiments depended on the organic solvents used. The distribution ratios for the CMP/o-xylene and the CMP/CCl_4 systems were lower than that for the solid extraction. The values for the CMP/n-heptane system were comparable to those for the solid extraction. But the formation of secondary organic phase was observed, and so this solvent extraction system may be inapplicable to continuous processes such as mixer-settler and pulse column. On the other hand, the solid extractants could extract cerium selectively and stably even at high contents of CMP.

Extraction rate of cerium

Figure 6 shows the changes with time in the amount of extracted cerium relative to total extraction capacity of solid extractant. The solid extractants of 0.1, 0.2, 1.4 and 2.0 mm in diameter were used. The initial extraction rate of cerium increased with decreasing diameter of the extractant. This result suggests that the rate controlling step is the diffusion process of cerium into the extractants. The extraction rate, however, was independent of the size of the extractant at diameters smaller than 0.2 mm. This means that the rate-controlling step changed from the inter-particle diffusion of cerium to the reaction between CMP and cerium. The time to reach 50% of equilibrium was only 10 seconds for the extractant of smaller than 0.2 mm in diameter.

CONCLUSIONS

1. The CMP content increases with decreasing degree of crosslinkage. The maximum content, 2.15 g-CMP/g-resin, which have been obtained for the resin crosslinked by 10wt% DVB, is 1.8 times of that for the Amberlite XAD-4.

2. The formation of complex between cerium and CMP is represented by

$$Ce^{3+}(aq) + 3NO_3^-(aq) + 2.5CMP(resin) \rightleftharpoons Ce(NO_3)_3 \cdot 2.5CMP(resin)$$

The use of low-crosslinked resin is effective to increase the distribution ratio of cerium. The distribution ratio is higher than those for the liquid-liquid extraction.

3. The rate-controlling step is the inter-particle diffusion of cerium for particles larger than 0.2mm in diameter and the reaction between cerium and CMP for the particles smaller than 0.2 mm in diameter. The time to reach 50% of equilibrium is only 10 seconds for the particles of smaller than 0.2 mm in diameter.

ACKNOWLEDGMENT

The authors are grateful to Mrs. Mami Yuki and Mr. Tsuyoshi Yoshioka for their assistance in the experimental work.

LITERATURE CITED

1) Benedict,M., T.H.Pigford and H.W.Levi: Nuclear Chemical Engineering, Chap.10, McGrow-Hill,1981
2) Alford,C.E. and J.D.Navratil: RFP-2625 (1977)
3) Faubel,W. and S.A.Ali: Nuclear Technology, 86, 60 (1989)
4) Horwitz,E.P., H.Diamond and D.G.Kalina: ANL-84-45 (1984)
5) Kubota,M.: J.Atomic Energy Soc. Japan (Japanese), 29, 775 (1987)
6) Matsumoto,S., K.Takeshita, J.Koga and Y.Takashima: J.Chem.Eng.Japan, 32, 691 (1989)
7) Schulz,W.and J.D.Navratil:RFP-2625 (1977)
8) Schulz,W.and L.D.McIsaac: ISEC, 21, 619 (1977)
9) Takeshita,K.,S.Matsumoto,M.Kumagai,J.Koga, M.Sazarashi,Y.Takashima and T.Tamura:Proc. of 20th DOE/NRC Nuclear Air Cleaning Conf.,vol.1, 205 (1988)
10) Thompson,R. : AERE-R-11966 (1986)

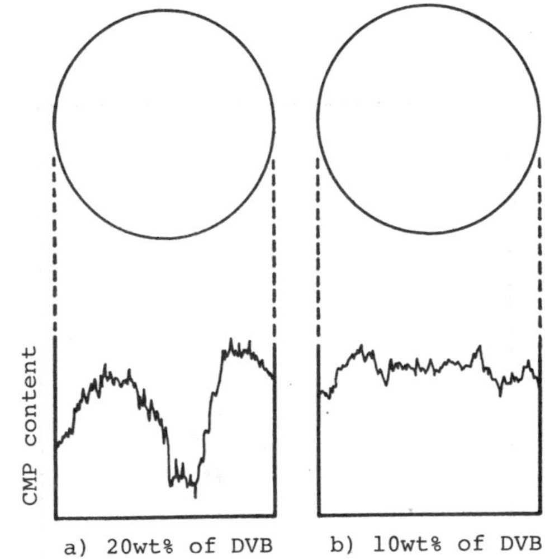

Fig.1 The line profiles of CMP content across the cross-sections of two different solid extractants obtained from SDB particles crosslinked by 10wt% and 20wt%, respectively.

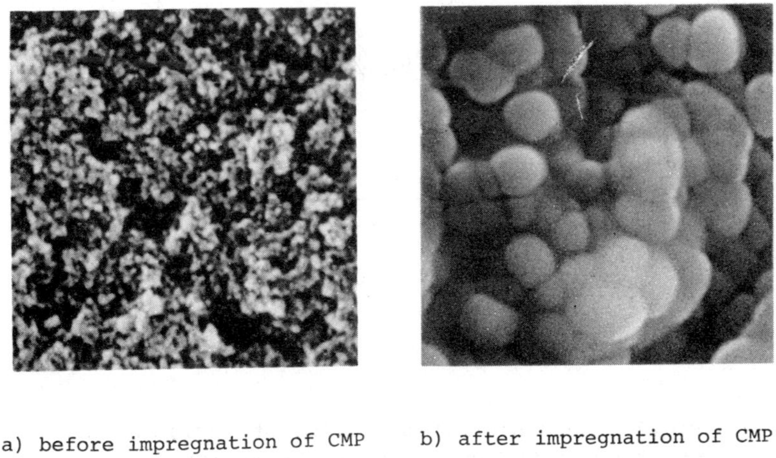

Fig.2 The SEM images of the interior of the SDB resin before and after the impregnation of CMP.

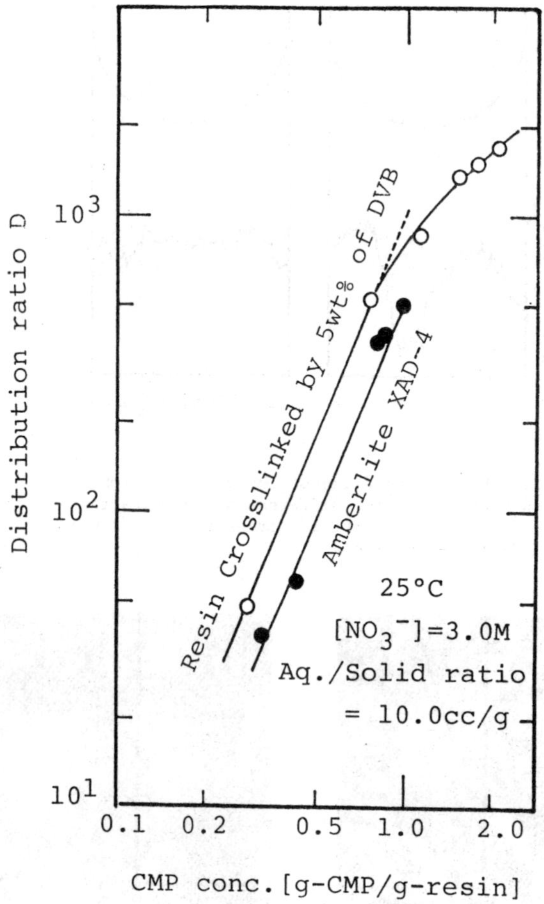

Fig.3 The distribution ratios of cerium as a function of CMP contents of SDB resins.

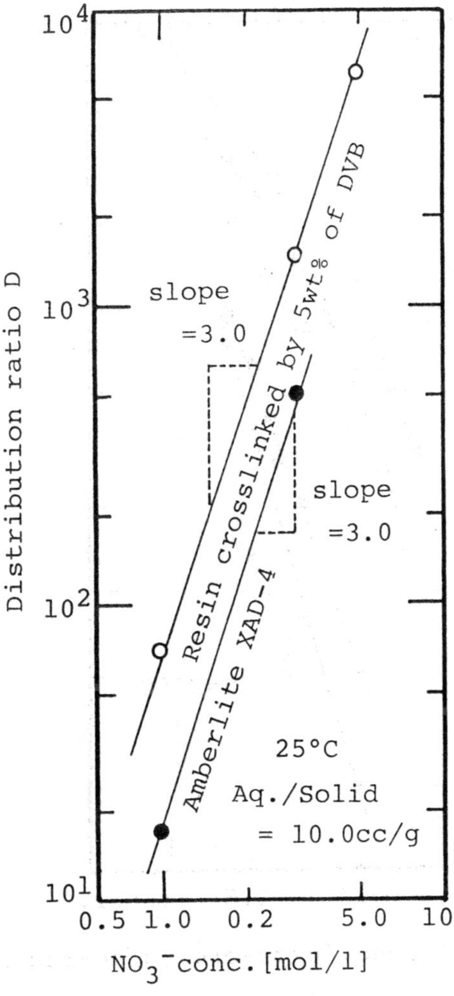

Fig.4 The distribution ratios of cerium as a function of the concentration of nitrate ion in the aqueous phase.

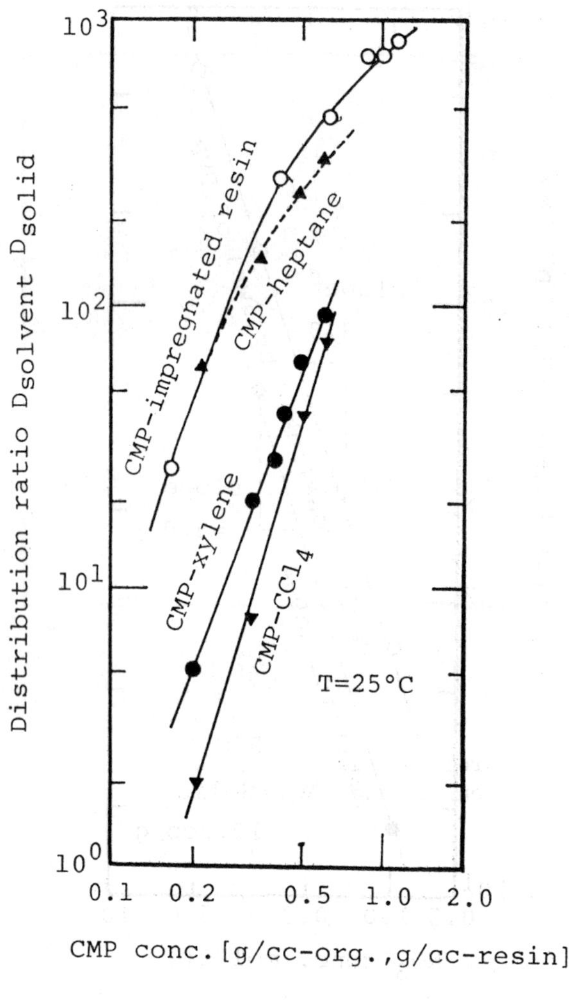

Fig.5 Comparison between the distribution ratios of cerium which were measured by the solid extraction experiments and the solvent extraction experiments.

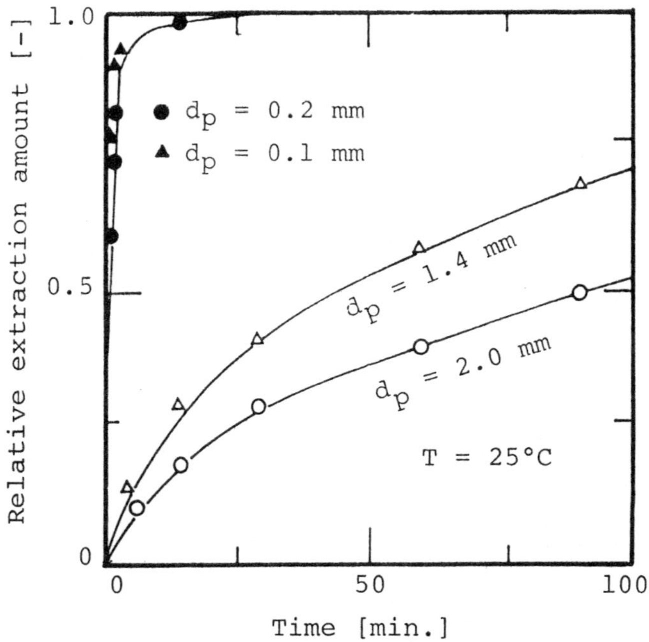

Fig.6 The relative extraction amount of cerium as a function of time.

SORPTION OF CAESIUM BY SODIUM NICKEL HEXACYANOFERRATE(II): COLUMN PERFORMANCE AND PREDICTION OF BREAKTHROUGH CURVES

D.L.Jacobi* and M.Streat**

The column performance of sodium nickel hexacyanoferrate(II) (NaNiHCF) for the removal of caesium from aqueous solution was investigated. Experimental breakthrough curves were determined using prepared samples of NaNiHCF in small glass columns with continuous analysis of the effluent solution. The adsorption isotherm was also determined experimentally and found to be extremely favourable, fitting a Freundlich type expression.
The experimental breakthrough curves were fitted to a mathematical model which was solved numerically using the Freundlich isotherm and also analytically when the isotherm was assumed to be rectangular. It was found that both models allowed a good fit of the experimental data and therefore enabled the effect of flowrate, bed height and caesium concentration on the breakthrough volume to be predicted with some precision.

INTRODUCTION

The use of complex hexacyanoferrates (HCFs) for the removal of caesium from radioactive waste streams is now a well established procedure in the nuclear industry. These complexes are highly selective for caesium ions in high salt content solutions and possess excellent thermal and radiation stability. A wide variety of HCFs have been prepared using different transition metals as reviewed by Lowenschuss (1), Hooper *et al.* (2) and Loos-Neskovic *et al.* (3). In general HCFs are made by mixing a salt of the transition metal with a ferrocyanide solution to give a finely divided precipitate which has to be granulated for packed column use. In the above reviews, the nickel complex in particular is noted for its exceptionally high affinity for caesium in comparison to other transition metals such as copper, cobalt, zinc and manganese.

The column performance of sodium nickel hexacyanoferrate(II) for the removal of caesium from aqueous solution, using samples prepared at Imperial College, is described in this paper. The adsorption isotherm was determined from a series of batch tests and experimental breakthrough curves obtained using small glass columns.

The curves were fitted to a mathematical model which involved the solution of the material balance and kinetic equations. A comparison was made between the numerical solution obtained using a Freundlich type isotherm and the analytical solution which is possible when the isotherm is assumed to be rectangular.

* British Gas, LRS, Michael Road, Fulham SW6 2AD

** Department of Chemical Engineering, Loughborough University of Technology, Leicestershire, LE11 3TU

EXPERIMENTAL METHODS

Preparation of Materials

Samples of NaNiHCF in a granular form were prepared by the authors at Imperial College using a freeze-thaw method which was based on earlier work by Vol'khin *et al.* (4,5). $NiSO_4$ and $Na_4Fe(CN)_6$ were mixed with a ratio of Ni:Fe = 1.3 to give a finely divided precipitate which was then washed. By freezing and thawing the precipitate under controlled conditions a granular material was formed which could be easily filtered. The product was then dried and the fraction with a particle size of 250-420 microns was used in the column tests.

Column Performance

The sorption of caesium in a packed bed of NaNiHCF was investigated using the experimental rig shown in Fig. 1. In each test 1 or 2 g of the exchanger was placed in a glass column of 1 cm diameter where it was supported by a sintered glass filter. The adsorber was gently fluidised by back washing with 1 M $NaNO_3$ which served to remove trapped air bubbles and also to fully wet the adsorber.

Feed solutions were prepared using 500 or 1000 ppm caesium as inactive $CsNO_3$ in 1M $NaNO_3$. Tracer caesium-134 (γ emitter - 0.605 MeV) was used which allowed efficient and convenient analysis of caesium in the samples by gamma scintillation counting. It was calculated that the tracer caesium concentration in the solutions was negligible. A constant speed peristaltic pump was used for the feed which was found to give a constant flow over the long operating periods that were required.

From an initial calculation based on the batch capacity of the adsorber and the concentration of caesium in the feed, an estimate was made of the volume of feed required and the time taken to reach breakthrough. With the flowrates used (0.075 and 0.21 $m^3\ m^{-2}h^{-1}$) it was found that periods in excess of 24 hours would be required. In order to run the tests over these long periods a method of automatically counting the column effluent was designed as shown in Fig.2. The effluent was passed through a coil which was attached to an intrinsic germanium detector that was connected to a Canberra Series 35 multichannel analyser (MCA). The analyser was programmed to repeatedly count the effluent over a fixed period of time and provide a printout of the data.

In order to obtain an accurate count of the feed solution, the sample coil was filled through an empty column. To check for the build up of caesium on the surface of the sample coil or the carrying over of fines, independent samples of the feed and column effluent were collected in phials and counted separately to check the ratio of the caesium concentrations.

Calculation of Working Capacity

The working capacity q_o(mg/g), of the adsorber was calculated as the average caesium loading per dry mass of the adsorber at the end of a column test. The capacity was calculated from the area above the breakthrough curve, and was also checked by mass balance from the caesium concentration and total volume of the column effluent.

Determination of Adsorption Isotherm from Batch Tests

The adsorption isotherm for NaNiHCF was determined in a series of batch contact tests. In each test a known mass of the adsorber was contacted with a volume of feed solution. The samples were shaken over a period of 72 hours which was assumed sufficient for equilibrium to be reached. The caesium concentration of the solution (c) was determined by counting and the solid phase concentration (q*) determined by mass balance. The solid phase concentrations were expressed as mg Cs per g of dry mass of exchanger.

Determination of bed voidage

The voidage of the packed bed was measured directly using 1 g of the adsorber which was wetted and then loaded as a slurry into a small glass column which had a sintered glass bottom. The bed was allowed to settle and then the solution level was lowered carefully to that of the adsorber. The column was centrifuged to remove the interstitial solution and the value of the bed voidage (e) calculated from the weight of fluid removed.

PREDICTION OF COLUMN PERFORMANCE

The prediction of the onset of breakthrough and the shape of the breakthrough curve are of fundamental importance in the design and scale-up of an adsorption column. A considerable amount of literature is available on the prediction of the breakthrough curves and the effect of different mass transfer mechanisms and equilibrium relationships. An extensive review of the different cases is given by Ruthven (6). In general, a solution of the mass and heat balances, coupled with an adsorption rate expression, is required to obtain the breakthrough curve.

For the sorption of trace quantities, it can be assumed that the column is isothermal and the breakthrough curve can be obtained by the simultaneous solution of a differential mass balance and a sorption rate equation. Assuming plug flow through the column and that there is negligible axial dispersion, the usual form of the differential mass balance is given by:-

$$u\frac{\partial c}{\partial z} + \frac{\partial c}{\partial t} + \left(\frac{1-e}{e}\right)\rho_p\frac{\partial q}{\partial t} = 0 \qquad (1)$$

The form of the sorption rate equation and definition of a suitable mass transfer coefficient are dependent on the mechanism of mass transfer and a representative driving force. From kinetic tests carried out by the authors (7), it is proposed that the sorption of caesium by NaNiHCF is controlled by particle diffusion within the range of conditions used. Using a pseudo mass transfer coefficient, k, and applying the Glueckauf and Coates (8) approximation of a linear driving force across a solid film, the rate of sorption may be represented by :-

$$\frac{\partial q}{\partial t} = k(q^* - q) \qquad (2)$$

The value of k is a function of the particle diameter, diffusivity and a correction factor for the linear driving force. The value of q*, the solid phase concentration in equilibrium is obtained from the isotherm which, for the sorption of caesium by NaNiHCF, was found to be a Freundlich type expression:-

$$q^* = a\,c^b \qquad (3)$$

A limitation of using a Freundlich expression is that the isotherm is open ended, with no account being made for the saturation of the adsorber. This may be avoided by restricting the expression to a specified concentration range and expressing the isotherm in dimensionless form:-

$$y^* = x^b \tag{4}$$

Equations (1) and (2) may also be transformed into dimensionless form by defining the usual time and bed parameters as follows:-

$$\frac{\partial x}{\partial \xi} + \frac{\partial y}{\partial \tau} = 0 \tag{5}$$

$$\frac{\partial y}{\partial \tau} = (y^* - y) \tag{6}$$

For an initially unreacted bed, to which a feed of concentration c_o is introduced at a time $t = 0$, the initial and boundary conditions are:-

$$I.C. \quad t=0: \ \tau = -k\frac{z}{u} \quad x = y = 0 \quad all \quad z$$

$$B.C. \quad t>0: \ x = 1 \quad at \quad z = 0$$

The solution of equations (4), (5) and (6) requires a numerical solution due to the non-linearity of the Freundlich equation. An implicit finite difference method was used in the present work, which was based on the algorithm described by Kyte (9).

The non-linearity may be removed, if the isotherm is extremely favourable by assuming that the adsorption is irreversible, which would be described by a rectangular isotherm (i.e. b=0). Equation (4) may then be replaced by:-

$$\begin{aligned} y^* &= 0 \quad for \quad x = 0 \\ y^* &= 1 \quad for \quad x > 0 \end{aligned} \tag{7}$$

The applicability of using a rectangular isotherm to represent a favourable ion exchange system has been discussed by Vermeulen (10). Prediction of the breakthrough curve is obtained by an analytical solution of equations (5), (6) and (7) which was developed by Cooper (11). The solution falls into three cases:-

$1 + \tau < \xi$	$y = x = 0$	*(unreacted bed)*
$\xi < 1$	$y = 1 - e^{-\tau} \quad x = 1 - \xi e^{-\tau}$	*(initial formation of pattern)*
$1 < \xi < 1 + \tau$	$y = x = 1 - e^{\xi - \tau - 1}$	*(constant pattern)*

More recently, Ruthven and Yoshida (12-14) have used this analytical solution and have compared the form of this solution with that obtained for the case of liquid film controlling. The use of the analytical solution allows the model to be fitted directly to the experimental breakthrough curve, and also leads to a greater understanding of the shape of the curve. The accuracy of the numerical method was also checked with the analytical solution at the limit of $b = 0$.

RESULTS

The experimentally determined isotherm for sorption of caesium from 1 M $NaNO_3$ onto NaNiHCF is shown in dimensionless form in Fig. 3. The isotherm was fitted with a Freundlich type expression, $y = x^{0.248}$ which was valid for concentrations at which the adsorber was not saturated.

The experimental details of two of the column tests carried out are listed in Table 1.

The differences between the two tests are the solution flowrate and packed bed height. The experimental breakthrough curves are shown in Figs. 4 and 6 with the fitted analytical solutions and also in Figs. 5 and 7 with fitted numerical solutions. In these breakthrough curves the dimensionless solution concentration at the column exit is plotted against the number of void bed volumes of solution passed through the column.

In order to fit the models to the experimental curves, the working capacity of the adsorber was first calculated as described above. A value for the pseudo mass transfer coefficient was initially obtained by directly fitting the analytical model and then this value varied slightly to achieve the best fit with the numerical model. The values of q_o and k which were used in the solutions are listed in Table 2.

DISCUSSION

The general shape of the breakthrough curves obtained from the experimental column tests was good, with no caesium detected in the effluent up to the breakthrough point which was characterised by a sharp rise in the caesium concentration. The effect of the solution flowrate and bed height is clearly seen, with the breakthrough point occurring after about 100 bed volumes in Test A and only after about 250 bed volumes in Test B. The general shape of the breakthrough curve is also much sharper in the case of Test B, which may account for the slight increase in the working capacity.

The plots of the breakthrough curves shown in Fig 4 - 7 show that, in general, both the numerical solution with the Freundlich isotherm and the analytical solution with assumed rectangular isotherm provide adequate fits of the experimental data. The values of the pseudo mass transfer coefficients that were used were higher in Test A when the flowrate was higher, and in both tests a slightly lower value was used for the rectangular isotherm.

The accuracy of the models needs to be assessed by consideration of the fit along the shape of the curve and also by the occurrence of breakthrough. It is observed that both models seem to predict the shape of the curves adequately although it is found that with the use of the Freundlich isotherm the occurrence of breakthrough is more accurately predicted. It is generally not clear if the slightly premature breakthrough observed is due to the form of the isotherm or to axial dispersion. Axial dispersion was not taken into account in the models. A change in the mass transfer mechanism at low caesium concentrations may also occur and this could affect the occurrence of breakthrough.

CONCLUSIONS

It has been shown that the use of NaNiHCF in a packed column is very effective for the selective sorption of caesium from salt solutions. The adsorption was found to be very favourable and can be fitted with a Freundlich type expression.

The experimental breakthrough curves show that no caesium was detected in the effluent up to the breakthrough point, which was observed by a sharp rise in the caesium concentration.

The curves were fitted to a model which assumed that the sorption was controlled by a linear driving force across a solid film. The model was solved numerically using the Freundlich isotherm, and also by an analytical solution using an assumed rectangular isotherm.

It was found that both models could be fitted to the experimental data using the measured parameters and an appropriate value for the pseudo mass transfer coefficient. The use of these models will enable the effect of the solution flowrate, feed concentration, and bed height on the column performance to be predicted with some precision.

SYMBOLS USED

a	=	constant in Freundlich expression (ml/g)
b	=	Freundlich coefficient
c	=	solution concentration (mg/ml)
c_o	=	feed solution concentration (mg/ml)
k	=	mass transfer coefficient (solid film) (h^{-1})
q	=	solid phase concentration (mg/g[dry])
q_o	=	solid phase concentration in equilibrium with c_o (mg/g[dry])
q*	=	solid phase concentration in equilibrium with c (mg/g[dry])
u	=	interstitial velocity (cm/h)
z	=	distance through bed (cm)
e	=	bed voidage
ρ_p	=	particle density (dry mass/volume of adsorber) (g/cm^3)

Dimensionless terms

x	=	solution concentration (c/c_o)
y	=	solid phase concentration (q/qo)

y^* = solid phase concentration in equilibrium with x (q^*/q_o)

τ = time parameter $\left[k\left(t-\frac{z}{u}\right)\right]$

ξ = bed length parameter $\left[\frac{k\rho_p q_o z}{c_o u}\left(\frac{1-e}{e}\right)\right]$

REFERENCES

1. Lowenschuss, H., 1982, Radioactive Waste Management, 9(4), 327-341

2. Hooper, E.W., Phillips, B.A., Dagnall, S.P., Monkton, N.P., 1984, AERE-R11088

3. Loos-Neskovic, C., Federoff, M., Garnier, E., Gravereau, P., 1984, Talanta, 31(12), 1133-1147

4. Vol'khin, V.V., Kolesova, S.A., 1967, Zh. Prikl. Khim., 40, 342-345

5. Vol'kin, V.V., Koshcheeva, E.A., 1968, Inorganic Materials, 4(6), 803-807

6. Ruthven, D.M., 1984, Principles of Adsorption and Adsorption Processes, Wiley, New York

7. Jacobi, D.L., Unpublished work.

8. Glueckauf, E., Coates, J.E., 1947, J. Chem. Soc., 1315

9. Kyte, W.S., 1973, Chem. Eng. Sci.,, 28, 1853-1856

10. Vermeulen, T., 1953, Ind. Eng. Chem., 45, 1664

11. Cooper, R.S., 1965, Ind. Eng. Fund., 4, 308

12. Yoshida, H., Ruthven, D.M., 1983, Chem. Eng. Sci., 38(6), 877-884

13. Yoshida, H., Kataoka, T., Ruthven, D.M., Chem. Eng. Sci., 39(10), 1489-1497

14. Yoshida, H., Kataoka, T. Ruthven, D.M., 1988, Chem. Eng. Sci., 43(7), 1647-1655

ACKNOWLEDGEMENT

One of us (DLJ) wishes to acknowledge the award of a research studentship from the UKAEA. The work described in this paper was carried out as part of a research contract with the UKAEA, Harwell Laboratory, and was sponsored by the Department of the Environment.

TABLE 1 - Experimental Details of Column Tests

	A	B	
Column height	1.8	3.4	cm
Column diameter	1.0	1.0	cm
Bed voidage	0.42	0.42	-
Feed caesium concentration	1.0	1.0	mg/ml
Flowrate	16.8	5.9	ml/h
Mass of adsorber	1.0	2.0	g(dry)
Particle density of adsorber	1.22	1.29	g(dry)/ml
Bed volume (void)	0.59	1.12	ml

TABLE 2 - Parameters used to fit Models

	A	B	
Working capacity of adsorber			
From area above curve	160	174	mgCs/g(dry)
From bulk effluent	165	172	mgCs/g(dry)
Fitted values of pseudo mass transfer coefficient			
Analytical solution	0.20	0.127	h^{-1}
Numerical solution	0.24	0.15	h^{-1}

Figure 1. Column Test Rig

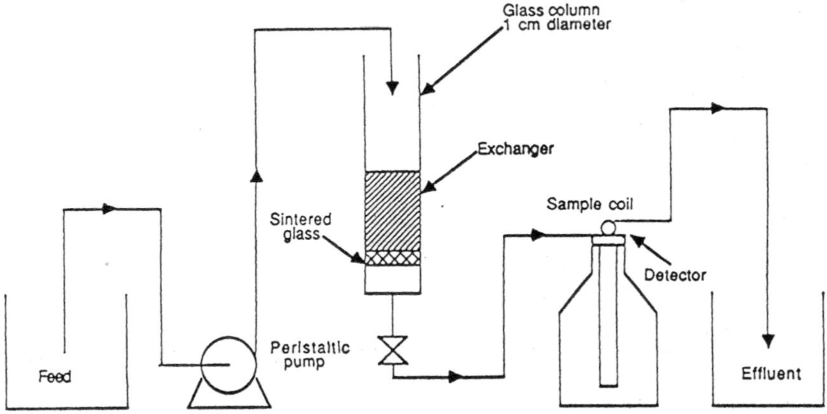

Figure 2. Arrangement of Counting Equipment for Column Tests

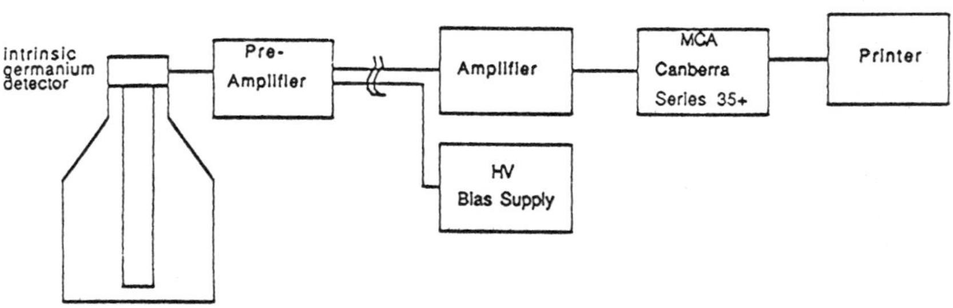

Figure 3. Adsorption Isotherm for NaNiHCF

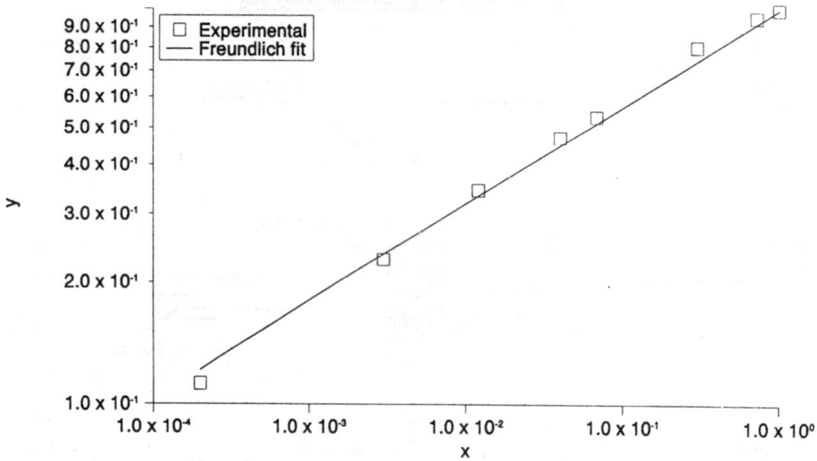

Freundlich fit with b=0.248

Figure 4. Breakthrough curve for Test A Fitted with Analytical Solution

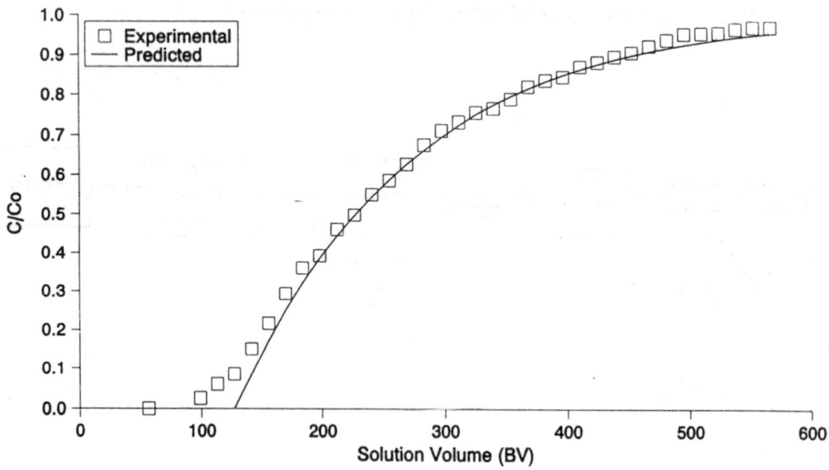

b = 0.0 k = 0.20 h-1

Figure 5. Breakthrough Curve for Test A
Fitted with Numerical solution

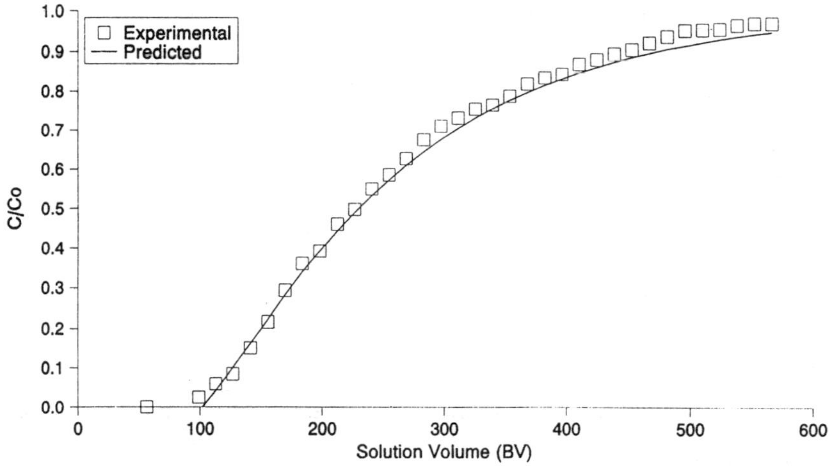

b = 0.248 k = 0.24 h-1

Figure 6. Breakthrough Curve for Test B
Fitted with Analytical Solution

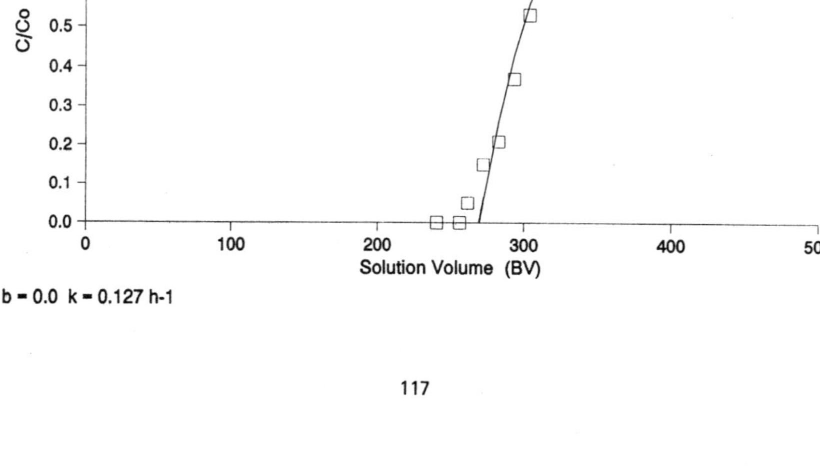

b = 0.0 k = 0.127 h-1

Figure 7. Breakthrough Curve for Test B Fitted with Numerical Solution

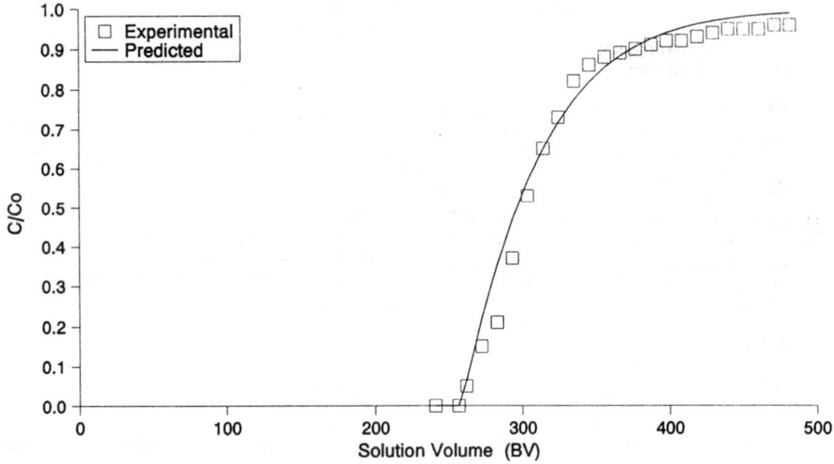

b = 0.248 k = 0.15 h-1

ELECTROKINETIC PROCESSES FOR BIOCHEMICAL EXTRACTIONS

L R Weatherley

The basic principles of a range of electrokinetic processes of current interest in the field of biochemical separation are reviewed. The importance of the use of electric fields for separation of both low molecular weight and high molecular weight compounds from complex mixtures is emphasised. Two categories of process are considered and include separations based upon differences in electrical mobility, for example electrophoresis and processes in which electric fields are used to improve kinetics, for example in liquid-liquid extraction. The status of the engineering development of electrokinetic processes in this context is briefly reviewed.

*Department of Chemical and Process Engineering
Heriot Watt University , Riccarton , Edinburgh EH14 4AS

INTRODUCTION

Examination of recent literature would indicate that the potential application of electrokinetic phenomena for separation of biochemically derived products is wide ranging. There are three broad categories of process which can be defined involving the use of electrical fields for separation.

i) Separation processes which are based upon differences in electrical mobility of the species comprising the mixture . Such differences are used to achieve selective separation of species which are difficult to separate by other means.

ii) Processes which use electrostatic forces to enhance the kinetics of a process. This type of process involves the induction of an electrostatic charge upon a macro species , for example a particle or a droplet, which is present in the system. In this category, processes usually would work in the absence of an electric field but at a rate which is unacceptably slow or which is limited in terms of total degree of separation achievable. Also in this category would fall processes which are augmented by other means, for example by operation at high temperature and pressure, or by injection of high intensity mechanical energy.

iii) Electrochemical conversion processes which rely upon electrode processes to achieve a Redox type transformation (1). This application of electric fields to biochemical transformation is particulary relevant to enzyme catalysed reaction systems involving cofactors and their regeneration.

Within the above framework of applications the use of electrokinetic phenomena is relevant to a wide range of species, which could range in size from ionic species through macromolecules, colloidal species, up to and including particles of visual size. Documented applications and potential applications include: -

Macromolecule / Colloid separations
Solid / Liquid separations
Separation and removal of ionic species
Enhancement of membrane and other filtration processes
Enhancement of rate processes in liquid liquid systems
Cofactor regeneration in continuous biotransformation processes

THEORETICAL BACKGROUND

I Bulk Electrolytes

With reference to figure 1 the transport of ionic species under the influence of an electric field may be quantified by equation 1 which comprises three terms: -

$$N_i = -z_i.u_i.F.c_i.\nabla E - D_i.\nabla c_i + c_i.v \qquad (1)$$

N_i	- Flux of species i	c_i	- concentration
z_i	- valence	∇E	- field strength
u_i	- transport number	D_i	- diffusivity
F	- Faraday constant	v	- velocity

i) Electrokinetic transport due the electromotive force acting upon the charged ionic species. The magnitude of this term is influenced by the magnitude of the local field strength acting on the ion, and by the magnitude of the charge on the ion.

ii) Transport by molecular diffusion due to the existence of a concentration gradient. The magnitude of this term is affected by the magnitude of the concentration gradient and of the diffusion coefficient.

iii) Transport by convection due to the action of viscous and momentum, and pressure forces acting upon the continuum .

Each of the three terms can be of significance in terms of the total rate of ionic transport, but only terms i) and ii) have any substantial effect on the selectivity of separation.

The constraints upon the solution of equation 1 are as follows:-

i) Electroneutrality;

$$\Sigma z_i . c_i = 0 \qquad (2)$$

ii) Conservation of species

$$\frac{\delta c_i}{\delta t} = - \nabla N_i \qquad (3)$$

The total flux of charged species may be also expressed in terms of the total electrical flux or the current density, I, thus

$$I = F . \Sigma z_i . N_i \qquad (4)$$

Solution of equations 1 - 4 is possible in certain selected cases (2) and enables prediction of the flux of individual species for given conditions of convection, diffusion and applied electric field. The use of electrokinetic effects for separation of ionic species for normal electrolyte solutions is therefore based upon differences in molecular diffusivity and ionic mobility.

II Charged Particles (2)

In the case of discrete particles dispersed in an ionic liquid continuum, departure from electroneutrality occurs in the vicinity of the solid liquid interface according to the Helmoltz double layer theory, the magnitude of the effective potential around the particle being given by the Zeta potential (3). If an electric field is externally imposed upon the system, see figure 2, the movement of the particulate species is influenced primarily by a combination of viscous forces and electrostatic forces.

The balance of electrostatic forces and viscous forces is given

by:-

$$\mu . \frac{\delta^2 v}{\delta y^2} + p_e . \frac{\delta E}{\delta y} = 0 \qquad (5)$$

μ - viscosity

p_e - charge density

E - potential

Also the relationship between field strength and charge density

is given by :-

$$\frac{\delta^2 E}{\delta y^2} = - \left[\frac{p_e}{\epsilon} \right] \qquad (6)$$

ϵ - permittivity

Elimination of the field strength term between equations 5 and 6 yields an expression for particle velocity (in the direction of the field) in terms of the permittivity of the solution, the field strength, the zeta potential of the particle surface and the solution viscosity thus:-

$$v_o = -\epsilon . \frac{\nabla E . \S}{\mu} \qquad (7)$$

§ - zeta potential

▼ E - nominal field strength

The term $\frac{\epsilon . \S}{\mu}$ is the electrophoretic mobility.

In general the magnitude of the electrophoretic mobility depends upon: -

The net charge on the species

The size and shape of the species

The temperature

The nature and concentration of the solution

III Electrostatically Enhanced droplet/particle formation . (4), (5).

The essential principle involved in enhancing rates of mass transfer in separation systems is the electrokinetic acceleration of liquid droplets through a second immiscible liquid phase. **Figure 3** shows in outline a liquid liquid extraction process in a spray column in which a dispersion of droplets of one phase is created in a second continuous phase. Enhancement of the process is achieved by inducing an electrostatic charge on the droplets as they are formed. Providing the charge is maintained for a significant time after formation the droplet (or particle) may be accelerated through the second liquid by imposition of an electric field across the continuum formed by the second phase.

If droplet formation at a single nozzle is considered, see **figure 4**, the additional accelerative force acting upon the droplet due to electrostatic forces depends upon the magnitude of the charge on the droplet,(ie the charge density), and the local field strength experienced by the droplet both during formation and subsequently.

The charge held by the droplet is acquired upon formation and may be expressed in terms of the initial charge Q_{do} thus,

$$Q_d = Q_{do} . \exp\left[- \frac{t}{\tau_d} \right] \qquad (8)$$

Where τ_d is the relaxation time for the dispersed phase:

$$\tau_d = \epsilon_d . \epsilon_o / K_d \tag{9}$$

K_d - conductivity of the dispersed phase

ϵ_d - dielectric constant of dispersed phase

ϵ_o - permittivity of free space

A short relaxation time is desirable and thus the dispersed phase should be of high conductivity and a low dielectric constant.

The charge leakage to the continuous phase is given by an analogous relationship in terms of the initial charge Q_{co},

$$Q_c = Q_{co} . \exp\left[-\frac{t}{\tau_c}\right] \tag{10}$$

τ_c, the relaxation time based on the continuous phase

is expressed: $$\tau_c = \epsilon_c . \epsilon_o / K_c \tag{11}$$

ϵ_c - dielectric constant of the continuous phase

K_c - conductivity of the continuous phase

For low charge leakage a high relaxation time is desirable and thus a fluid of low conductivity and high dielectric constant is required.

The degree of enhancement in transfer rate across the liquid liquid interface is influenced by three factors

i) Droplet acceleration and improvement in interfacial shear rates

ii) Droplet oscillation , increase of which leads to enhancement in rates of internal circulation

iii) Droplet size at breakoff from the formation nozzle.

Beneficial manipulation of these parameters can be achieved by conducting liquid liquid extraction operations in the presence of electric fields. In each case, production of charged droplets moving under the influence of an electric field are subjected to additional accelerative force leading to improvements in overall transfer rates.

APPLICATIONS

I Electrophoretic Processes - Definitions

Electrophoresis offers a powerful and versatile method for the separation of biochemicals and a range of developments are reviewed by Mat (6) . Exploitation of differences in electrophoretic mobility under the influence of an electric field can be used for separation of a number of species which could include ions, colloids, cellular materials, organelles, and whole cells. The two fundamental advantages of the process are high resolution, and the ability to retain the bioactivity of the products.

Disadvantages of the process generally accrue during scale-up when the problems of Joule heating, electroosmosis, and dispersive mixing of the separated products can reduce separation paration efficiency.

The basic principles of several important types of electrophoretic are now briefly discussed.

Moving Boundary Electrophoresis: (7). This version of electrophoresis is based upon the original demonstration. With reference to **Figure** 5 a long band of sample is placed between layers of buffer solution in an extended U tube apparatus. Application of the electric field results in the migration of components in accordance with their electrophoretic mobilities. Complete separation is not achievable and only the components having the highest mobilities are partially purified, with substantial overlap of all other components.

Zone Electrophoresis: (8). This is a variation of moving boundary electrophoresis. The sample for separation is surrounded by buffer solution, see **Figure** 6, and separation into a number of discrete zones is achieved. The resolution effect, based upon differences in electrophoretic mobility can be used in a continuous zone electrophoresis process, see **Figure** 7. In this process the mixture is continuously injected as a narrow source into a body of carrier fluid flowing between two electrodes. The solute is resolved according to electrophoretic mobility and a number of fractions may be taken off at some suitable point downstream of the inlet.

Efficient resolution depends heavily upon the minimisation of convective mixing in the direction of migration and thus the avoidance of substantial heating effects is important. Ionic strength of background electrolytes should be kept as low as possible to minimise Joule heating effects.

Isotachophoresis: (9). This is another variation of zone electrophoresis. The principle of the technique is illustrated in **Figure** 8. The sample mixture is located between two electrolyte solutions, a leading electrolyte, and a terminating electrolyte. The technique is only feasible for separation of species of the same charge either all positive or all negative. Cationic separation, for example, would involve location of leading electrolyte, L+, adjacent to the cathode where the mobility of the ions, L+, is greater than that of any of the species to be separated and location of terminating electrolyte, T+, adjacent to the anode. The mobility of ion T+ is less than that of any of the species in the sample. Upon application of the electric field a potential gradient results. The profile of the potential gradient is in accordance with the mobilities of the ions present in the mixture and a selective separation according to mobility results. Species of low mobility concentrate in regions of high field strength and species of high mobility in regions of low field strength . A steady state situation is attained in which a series of pure bands containing each charged species migrate at the same velocity along the potential gradient. The gradient of potential is constant within each band but changes stepwise at the boundary between each band. Isotachophoresis is particularly useful as an analytical technique since the field strength in any particular band can be used to determine the presence of a known species, and the length of a band used to determine the amount of the species present in relation to the total.

<u>Isoelectric Focusing: (10), (11)</u>. This technique uses a combination of an electric field and a pH gradient to achieve selective separation of amphoteric species. In chemical terms an ampholyte is a species which exhibits both acidic and basic behaviour and ampholyte behaviour in the presence of hydrogen ions is summarised below.

$$Am^- + H^+ = Am^o$$

$$Am^o + H^+ = Am^+$$

Thus the charge held by the amphoteric species Am is dependent upon the pH of the environment. Isoelectric focusing uses this pH dependent charge behaviour as the basis for the separation technique. Thus if a an ampholyte is placed in a continuum across which a pH gradient is imposed and an electric field applied the species will migrate according to its charge. Thus if the species is negatively charged at the pH of initial location it will migrate towards the anode. With reference to **Figure 9**, as it migrates it will encounter a progressively lower pH until, according to equation (a) above, it will attain a status of neutral charge. This is known as the **isoelectric point** and in a uniform electric field the species will cease to move as the effective electrokinetic force acting upon it tends to zero. Each species in the mixture will thus migrate until it reaches the pH corresponding to its own isoelectric point.

The main advantage of isoelectric focusing is that complete separation of a mixture is possible . Also, diffusional and convective transport do not interfere with the efficiency of the separation while the field is maintained.

The basic requirement of the technique is the ability to create and to maintain a pH gradient in the equipment. The most effective method uses a mixture of synthetic amphoteric molecules with isoelectric points spanning the desired range of pH. Application of an electric field to a mixture, suitably configured, results in migration of the ampholytes to their respective isoelectric points and the formation of a stable pH gradient . Gel polymerisation and incorporation of a buffer into the system results in a suitable medium for isoelectric focusing of protein mixtures with differences in isoelectric point as low as 0.001 pH units.

Electrodecantation: (12),(13). Electrodecantation involves the stratification of colloidal components at a membrane/fluid interface across which an electric field is applied. With reference to **Figure** 10, the mixture to be separated is contained in a compartment formed by two uncharged permeable membranes behind which are located anode and cathode respectively. The resulting electric field promotes the electrokinetic migration of the species according to their charge towards the membrane. The stratified layer of colloid which collects at the membrane surface will eventually move up or down according to its density relative to that of the surrounding medium. Thus in principle, the stratified layers can be separated.

Further refinement of the technique involves the use of electro- osmosis to augment particle coagulation at the membrane/fluid interface producing a compacted layer which can be periodically released from the surface by reversal of electrode polarity. At this point the frictional resistance between the deposit and the membrane is greatly reduced leading to immediate stratification of the deposit.

Examples of applications of the technique include:-

Separation of proteins and biological sera from non migrating globulins. (14)

Separation of tetanus and diphtheria toxins from protein fractions. (15)

Isolation of the active immunoglobulin hepatitis - associated antigen. (16)

Sedimentation of albumin by electrodecantion of serum proteins. (17)

Enzyme purification from pig kidney extract and from sheep testicular extract. (18)

Latex rubber concentration.

Dielectrophoresis (19), (20), (21):

Uncharged particles can be selectively separated using a non uniform electric field. With reference to **Figure** 11, uncharged particles placed in an electric field will be subjected to induced charge on each side according to orientation. If the field is non uniform such as is present in the case of a pin and plate electrode pair shown, the species will migrate to the point of maximum field strength, i e to where the lines of electrostatic force are of the maximum intensity. Under these conditions the field polarity is inconsequential and the species will migrate in the same direction regardless . Thus an alternating field can be employed to achieve the separation effect.

Exploitation of the technique at commercial scale is limited to recovery of inorganic particulates from low conductivity liquids and the performance of dielectrophoresis is comparable to mechanical methods. In principle, separation of biological colloids appear to be very suitable for dielectrophoretic separation to the presence of an electrical double layer at the surface of such species. Laboratory scale separations for a number of systems have been demonstrated (21):

Separation of Euglena cells

Separation of Human erythrocytes

Yeast separation

Liver mitochondria

Amoeba separation

E Coli separation

Mycoplasma separation

Flavobacteria separation

A major obstacle to the application of the technique to biological systems is the high conductivity of the liquid phases involved. Subject to further developments to modify the separation environment the technique could be promising for the selective recovery of biocolloids from cell debris suspensions and for the separation of different cell types.

II Electrophoretic Processes - Equipment

The engineering challenge of scaling up electrophoretic processes focuses on the need to reconcile the apparently conflicting requirements of high throughput with the laminar flow conditions necessary to achieve a sharp separation between the product rich bands. A further factor which can adversely affect separation efficiency is the presence of free-convection instabilities which arise from temperature gradients in the flow field due to Joule heating effects.

Design methods for a range of electrophoretic processes are well established (22,23,24). The key design variables include electric field strength, flow path length and width, flow orientation, control of temperature, and physical & electrical properties of the carrier phase.

The earliest continuous unit was the continuous free-flow zone electrophoresis unit described by Philpot (25) and further developed by Stricker (26) and by Hannig (27). The basic principle of operation is already referred to in Figure 7.

A further improvement is described by Gobie et al (28) who introduced product recycle into the continuous flow electrophoresis process. The use of recycle has also been successfully incorporated into equipment for isoelectric focusing and zone electrophoresis (29).

Martin & Hampson (30,31) describe a technique for large scale preparative isoelectric focusing in which the use of expensive amphoteric buffers is replaced by the use of amphoteric membranes whose isoelectric points vary in accordance with their position in the equipment. The substantial reduction in flow channel dimensions minimises joule heat generation, and further temperature stability is achieved by supplementary heat exchange.

The BIOSTREAM separator developed at the Harwell Laboratory (32) is one of the most significant equipment developments in this field. The separator operates in the zone electrophoresis mode and provides a high throughput capability for protein separation. The novel aspect of the design is the use of a vertical rotating cylinder at the outer edge of an annular flowpath to control hydrodynamics.

Larger scale electrophoretic separation operations at the kg/h scale are contemplated in two further designs. Sova (33) describes equipment for industrial scale autofocusing comprising a large rectangular box filled with a solution of the feed material in distilled water which is separated batchwise in the absence of ampholytes and in the absence of significant convection. A continuous rotating annular bed electrophoresis column (CRAE) is described by Datta et al (34) and uses an adsorptive packing to reduce convective mixing in the separation zone. The migration zone thickness is small thus reducing heating effects, and the design proposed allows incorporation of a large number of feed and product points thus giving the potential for high degrees of separation of complex mixtures.

III Solid Liquid Separations (2), (35), (36)

For separations involving removal of larger solid particles from a supernatant liquid stream one of the major applications of electrokinetic effects is in the maintenance of a cake free filter medium. This is typically achieved in a process such as cross flow electrofiltration, see **Figure 12.** Improvement in filtration efficiency exploits two electrokinetic effects. Electrophoretic migration of particulate material away from the filter medium (which may be a membrane) is used to reduce solids build up at the interface between the filter medium and the fluid thus reducing the pressure driving force required. Secondly electrooosmosis of liquid through the filter medium is exploited to augment the permeate flux due to the pressure driving force. Typically electric fields in the range 0 - 60 v/cm are used in aqueous systems involving solids in the size range 0 - 10 microns.

Other effects which influence filtration efficiency include

radial migration of particles, which occurs under the influence of inertial forces acting upon the particles.

The solids concentration profile in the vicinity of the filter medium/fluid interface is substantially modified in the presence of electrophoretic particle migration.

i) At zero and low applied electric fields the concentration of particle will be a maximum at the interface.

ii) As the effects of the electric field upon particle motion increase, the concentration profile near to the interface flattens and is equal to zero when the sum of the combined particle fluxes due to electrokinetic migration, radial migration and diffusion are equal and opposite to that due to convective transport. The applied voltage at which this occurs is known as the **CRITICAL** voltage.

iii) Above the critical voltage, the electrokinetic effect becomes predominant and there is a net movement of particles away from the interface.

Electro-osmosis of permeate through the filter medium further enhances the solid liquid separation process.

Applications of cross flow electrofiltration in the biological field have been limited again due to the high conductivities of the process fluids. In principle the small particle sizes frequently present, together with the well suited charge properties of biological particulates, should raise the possibility of successful use of the technique in the future.

IV Other Electrokinetic Processes

Electrodialysis. Electrodialysis has been developed widely for the treatment of brackish and saline waters as a route to potable water. The principle of the technique is shown in **Figure** 13 and relies upon the ion selectivity of anionic and cationic ion exchange membranes as a basis for selective separation. The driving force for the process is a current-driven ionic flux across a stack of compartments formed by alternate cationic and anionic ion exchange membranes. Application of an electric field across the compartments results in electrokinetic transport of positively charged species through the cationic membranes and negatively charged species through the anionic membranes. Thus for a continuous flow system, at steady state, the effluents from alternate compartments will comprise concentrated and depleted streams with respect to ionic species. Selective separation of mixtures of ionic species is possible and relies upon the differing ionic mobilities within the ion exchange membranes.

The maximum ion removal rate is mass transfer limited by a phenomenon analogous to electrode polarisation which occurs in electrolysis. With reference to **Figure** 13, at high current densities, concentration gradients form adjacent to each membrane until the ionic concentration at the interface falls to zero. Under these conditions water splitting may occur resulting in pH changes in both concentrated and dilute effluent streams which may cause membrane fouling.

Other variations of conventional electrodialysis are also worthy of mention :-

Ion Replacement (37), (38). Here membranes of only one charge are used, see **Figure 14**. Application of the electric field across the anionic membrane configuration shown results in the migration of anions B-, and F-, along the potential gradient. the cations B+, and F+ remain in the compartments into which they were fed. The net effect is that the eluent stream discharging for example from compartment M contains anions B- in addition to F-, thus a degree of ion exchange is achieved.

Transport Depletion (37). In this mode, the anionic membranes in the conventional stack are replaced by neutral membranes which allow transport of both cations and anions as shown in **Figure 15**. Application of the electric field results in transport of cations towards the cathode through both sets of membranes. The anions, by contrast, only move through the neutral membranes and their flux is due to charge compensation in addition to electrokinetic transport. The main advantage of this technique for biochemical applications is that the use of anionic membranes which are particularly prone to fouling is avoided.

Ion Substitution (37). Three different types of compartment are created in the stack in this configuration which is shown in **Figure 16**. The net result in terms of performance is the replacement of anion P- by anion D- and in this respect is similar to the Ion replacement technique although theoretically 100 % substitution is possible ion the case of Ion Substitution.

The general advantage of conventional electrodialysis and the variations described, with respect to biological processing is the high degree of separation achievable under relatively mild conditions.

Areas of Applications of ED in the context of Biochemical Engineering include (39):-

i) Separation of low molecular weight non ionic products

ii) Separation of charged low molecular weight products

iii) Recovery and purification of reagents.

Recovery of non ionic low molecular weight products frequently involves removal of ionic species. For example desalting is important in processes such as recovery of proteins from whey where the removal of minerals is essential without despoilage of the product (40). The conversion of cheese whey involves the separation of milk proteins and uncharged lactose from salts. The salts are separated by electrodialysis into a concentrated waste stream through membranes with low water transport properties. Whey is essentially serum remaining from the cheese production process and contains lactose, lactalbumin, lactoglobulin, and minerals. The protein is a potential , high value product providing that separation from the salt species is achieved. Compared to ion exchange which may also be used for whey desalting, electrodialysis offers a number of advantages:-

i) Lower protein denaturation

ii) Continuous processing

iii) Reduced losses of biodegradable material and lower effluent discharge.

Separation of charged low molecular weight species includes conventional deionisation processes around which the majority of commercial scale electrodialysis processes have developed. Conventional deionisation processes include desalting, adjustment of pH, electrolytic conversion and waste treatment, and pretreatment processes (prior to electrophoresis for protein separation it is desirable to reduce the conductivity of the solution by removal of ionic salt species for example.) Electrodialysis can be used to concentrate salts and small charged species (of MW less than 300) to very significant concentrations and achieve efficient separation of larger species and uncharged species. Ion exchange membranes will normally reject products having molecular weights in excess of 500.

A number of recent applications of interest to the biotechnologist are listed as follows (41), (42): -

i) pH adjustment of process streams can be achieved without increase in overall salt concentration. This can be done by introduction of an acid stream into the process stream followed by removal of other cations and anions by electrodialysis (39).

ii) Ion modification of biological solutions without change of pH can be achieved. This would have an application where a sodium free dietry stream was required and sodium ions can be replaced by potassium or calcium without product denaturation. (39)

iii) There are a variety of food processing applications including deionisation of fruit juices, wine, milk, sugar molasses and whey (43). The demineralisation of skimmed milk for example is inhibited by membrane fouling (particularly the anionic membranes) by proteins although this problem can be avoided by using acetyl cellulose membranes in conjunction with the ion exchange membranes. The molecular weight of the acetyl cellulose is lower than that of the fouling proteins (40).

iv) There increasing interest in the use of electrodialysis as part of extractive fermentation processes. Here the overall conversion of substrate to product is enhanced, by continuous removal of the product thus offsetting the effects of product inhibition on the microorganisms. The mild processing conditions and the absence of any extractive reagents in the electrodialysis process are attractive in the context of fermentation systems. Examples of processes which could benefit from this approach include Production of Lactic acid by glucose fermentation (41) and acetic acid production (42).

The desalting capability of electrodialysis can be used for the fractionation and separation of higher molecular weight species. The sensitivity of protein solubility to salt concentration is one major effect which may be exploited. For example in the case of plasma proteins, reduction in the salt concentration results in rapid precipitation of the non albumin proteins.

Another application exploits the differing isoelectric points of amino acids (43) to effect separation. The principle of the process is pH adjustment of a mixture of amino acids to ensure a significant difference in the charge held by the desired species compared to others present. Selective migration of the desired species towards the electrode of opposite polarity is achieved in the electrodialysis stack.

Electrically Enhanced Liquid Liquid Extraction:

Liquid liquid extraction offers a highly selective means of product separation for biochemical products. Operation at ambient temperature and a high degree of selectivity control offer considerable advantages when compared to other methods of separation, particularly in the context of high volume products occurring in dilute solutions from fermentation. Conventional liquid liquid extraction processes of untreated fermentation broths is however relatively inefficient due to general reductions in interfacial mass transfer rates which occur due to the accumulation of cell debris and other biological surfactant material at the liquid liquid interface. A second difficulty is the high propensity for the liquid liquid mixtures involved in the extraction to form poorly separating emulsions leading to poor extractor performance.

Prosecution of liquid liquid extraction operations in the presence of electric fields results in several fundamental advantages which to some extent have the potential to ameliorate some of the worst problems of whole broth processing. With reference to **Figure 4** production of a dispersion of aqueous broth at a charged nozzle in contact with a continuum of low conductivity extractant can result in the production of charged droplets. An electric field is readily imposed across the continuum of solvent through which the charged droplets are then accelerated by the resulting electrostatic force acting upon them.

The mass transfer rate is enhanced in three ways :-

i) Increased droplet slip velocity results in greater interfacial shear and reduction in the continuous phase mass transfer resistance

ii) Increased droplet oscillation results in reduction in the dispersed phase mass transfer resistance

iii) Increased interfacial area for transfer results due to the smaller droplet sizes achieved at the inlet nozzle.

The potential of electrically enhanced extraction for the improvement of existing biochemical extraction has been demonstrated at laboratory scale. The major benefits accrue during the extraction of untreated fermentation broths (yeast based and mycelial culture broths) which normally require high energy extraction processes to facilitate efficient dispersion and coalescence. Creation of whole broth droplet dispersions through fine diameter nozzles is frequently not feasible due to problems of blockage. Electrically enhanced extraction enables very fine dispersions of broth to be created through nozzles of relatively large diameter by application of sufficiently high nozzle voltages to result in formation of unstable droplets which break to form an electrostatic dispersion.

Laboratory scale demonstration experiments indicate enhancements (44) in mass transfer rate of an order of magnitude for yeast based broth and mycelial culture broth extractions involving recovery of products such as Ethanol and Penicillin.

DISCUSSION

The general factors governing separation processes for biological products include the complexity of the feed mixtures, the sensitivity of the product molecule to process conditions and the low concentration of the system. The major recent developments in using electrokinetic processes for separation in biological systems appear to have centred on the recovery and separation of high molecular weight, low volume high value products such as therapeutic and diagnostic enzymes and proteins, by some form of electrophoretic process. In many of these cases the commercial product throughput is so small that scale-up from laboratory scale is relatively trivial and carries few process engineering implications. Clearly the anticipated growth in the volume of many biologically produced agents in the high molecular weight category has added impetus to the scale -up and development of novel electrophoretic processes. The priority for medium term application of electrokinetic processes for high volume, high molecular weight product separations may lie in the food industry area for example in dairy product processing for protein recovery.

The use of membrane based electrokinetic processes such as electrodialysis is also likely to be important in this area for protein treatment and desalting. Processes such electrodialysis are already well developed in the context of water treatment and therefore further engineering development is likely to focus on the development of novel, fouling resistant membranes, sterilisation of membrane systems, and integration of the separation process with fermentation and other upstream operations. This latter aspect is very relevant to the development of low cost fermentation routes to medium volume products such as lactic acid and citric acid where product dilution and product inhibition are important limiting factors.

The use of electrokinetic separations at commercial scale for solid-liquid separation in biotechnological systems is probably more distant except for small scale applications.

This is also the case of electrically enhanced solvent extraction where scale-up development is required at a basic level for this technique to be exploited for example in commercial scale pharmaceutical recovery. The successful demonstration of the ability to electrostatically spray untreated viscous fermentation solutions is important not only in the context of solvent extraction and recovery but also has the potential to enhance the kinetics of biotransformation processes in liquid-liquid systems. This technique typically involves contact of a cell rich aqueous solution with an organic phase containing substrate for conversion. The inherent problem of slow kinetics may potentially be overcome by electrostatic spraying techniques.

NOMENCLATURE

c_i - concentration of species i

D_i - diffusivity of species i

∇E - field strength

E - potential

ϵ - permittivity

ϵ_o - permittivity of free space

ϵ_c - dielectric constant (continuous phase)

ϵ_d - dielectric constant (dispersed phase)

F - Faraday constant

I - current density

K_c - electrical conductivity (continuous phase)

K_d - electrical conductivity (dispersed phase)

N_i - flux of species i

p_e - charge density

Q_d, Q_{do} - surface charge

t - time

τ_c, τ_d - relaxation time

μ - viscosity

u_i - transport number

v, v_o - velocity

y - distance

z_i - valence

§ - zeta potential

REFERENCES

1. Comtat M., Bergel A., Mahenc J., 1984, Proceedings of ISCRE 8, EFCE publication Noergamon, London, pp667-674.

2. Perry J., 1988, Chemical Engineers Handbook, 6th Edition McGraw-Hill, New York.

3. Moore W.J., 1966, Physical Chemistry, 4th Edition, Longmans, London.

4. Thornton J.D., 1976, Electrically Enhanced Liquid/Liquid Extraction, Birmingham University Chemical Engineering Journal, pp6-13.

5. Thornton J.D., Stewart G, 1967, Charge and Velocity Characteristics of Electrically Charged Droplets in Liquid Liquid Extraction, IChemE Symposium Series No. 26, pp 29-42.

6. Mat H.B., September 1987, Theory and Applications of Electrophoresis as an Industrial Separation Process, MSc Dissertation, University of Bradford.

7. Melvin, M (ed), 1987, Electrophoresis, John Wiley & Sons, London.

8. Deyl, Z., 1979, Electrophoresis, Elsevier, Amsterdam.

9. Everaets, F.M., Becker, J.L., Verheggan T.P.E. (eds), 1976, Isotachophoresis, Elsevier, Amsterdam.

10. Gaal, O., Medgyesi, G.A., Kereczkey, L., 1980, Electrophoresis in the Separation of Biological Molecules, John Wiley & Sons, New York.

11. Allen, R.C. (ed), 1984, Electrophoresis and Isoelectric Focusing of Proteins, Walter de Gruyter & Co., Berlin.

12. Pauli, W., 1933, Transactions of the Faraday Society Vol. 31 p19.

13. Stamberger P., 1946, Journal of Colloid Science Vol 1 p93.

14. Gutfreund, H., 1943, The Biochemical Journal, Vol 37 p186.

15. Largier, J.F., 1955, Biochemical and Biophysics Acta, Vol 16 p291.

16. Burlev, V.A., Kiryukhin, I.F., Golosova, T.V., 1975, Chemical Abstracts Vol 82 no. 71401

17. Lappe, F., 1975, Hoppe Seylers Z., Physiol Chem. vol 10 p353.

18. Winchester, B.G., Caffrey, H., Robinson, D., 1971, The Biochemical Journal, Vol.121 p161.

19. Pohl, H.A., 1973, Electrostatics and its Applications, John Wiley, New York.

20. Pohl, H.A., 1978, Dielectrophoresis : The Behaviour of Matter in Non Uniform Electric Fields, Cambridge University Press, New York.

21. Bowden, C.P., Leaver, G., Melling J., Norton M.G., Whittington, P.N., 1987, Separations for Biotechnology, eds M.S. Verrall & M.J. Hudson, Ellis Horwood, Chichester, U.K.

22. Andrews, G.F., 1985 Electrophoresis, vol 6, p483.

23. Dobry, R & Finn R.K., 1958, vol 54 (4), p59.

24. Hybarger, R.M., Tobias C.W., & Vermeulen, T, 1963, I & E C Proc Des & Dev, vol 2 (1) p93.

25. Philpot, J St L., 1940, Trans Faraday Soc. vol 36 p38.

26. Stricker, A., 1967, sepn. Sci. & Techn. col 2, p335.

27. Hannig, K., 1961, Z.Anal. Chem. vol. 181, p244.

28. Gobie, W.A., Beckwith D.B. & Ivory, C.F., 1985, Biotechnol. Prog., vol1(1), p60.

29. Bier, M, Egen, N.B., Twitty, G.E., Mosher, R.A., Thorman W, 1986, Proc. 1st Int. Conf. Sep. Sci. & Technol., eds C.J. King & J Navratil, New York, p331.

30. Martin, A.J.P., & Hampson, F., 1978, UK Patent GB 2036081B.

31. Martin, A.J.P., & Hampson, F., 1981, US Patent 4 243 507.

32. Mattock, P., Aitchinson, G.F. & Thomson, A.R., 1980 Separation and Purification Methods, vol 9(1), p1.

33. Sova, O., 1985, J. Chromatography vol 320, p15.

34. Datta, R., Yoshisato, R.A. & Carmichael, 1986, A I Ch E Symp Ser vol 82, p176.

35. Tarleton, E.S., 1988, Filtration and Separation, Vol.30. p402.

36. Wakeman, R.J., 1986, Chemical Engineering Research and Design, vol 64.

37. Lopez Leiva, M.H., 1988, Le Bensmittel - Wissenschaft und Technologie, vol 3 p119.

38. Young, B., 1974, Journal of the Society of Dairy Technology, vol 27 (3), p141.

39. Reed, P. 1985, Genetic Engineering News, vol.9 p10.

40. Min, C.H., Nomura, D., 1987, Membrane, vol.2, p106.

41 Czytko, M., Ishii, K., Kawai, K, 1987, Chem - Ing - Tech, vol 59, p952.

42. Nomura, Y., Iwahara, M., Hongo, M., 1988, Applied and Environmental Microbiology, vol. 54 (1), p137.

43. Strathmann, H., 1985, Separation and Purification Methods vol 1, p41.

44. Millar, M.K., Laughland, G.J., Weatherley, L.R., 1987, Extraction '87', EFCE publication no. 347, p263 Institution of Chemical Engineers, London.

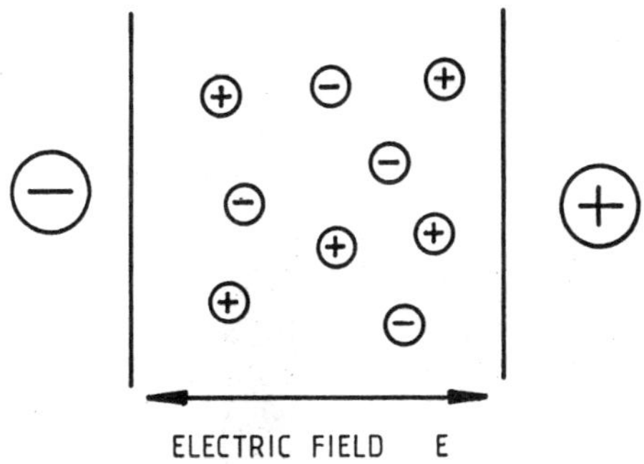

Figure 1 Bulk Electrolytes

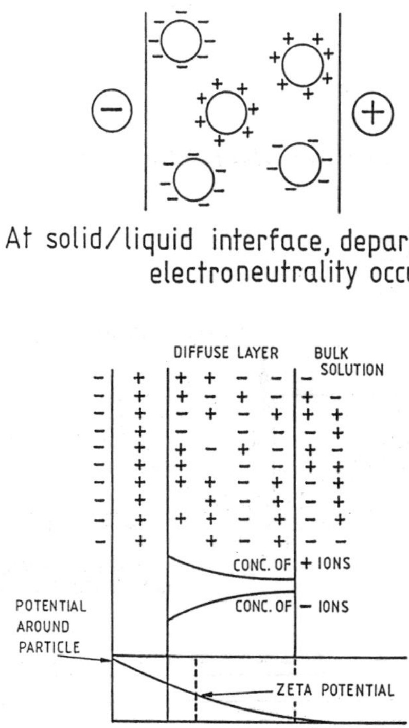

Figure 2 Charged Particles

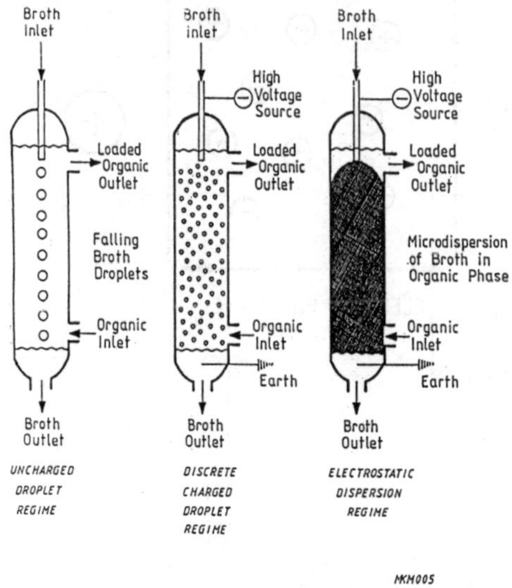

Figure 3 Electrically Enhanced Liquid/Liquid Extraction

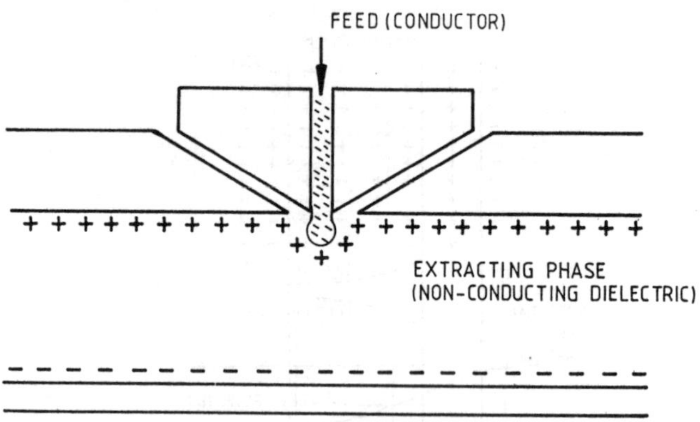

Figure 4 Charged Droplet Formation

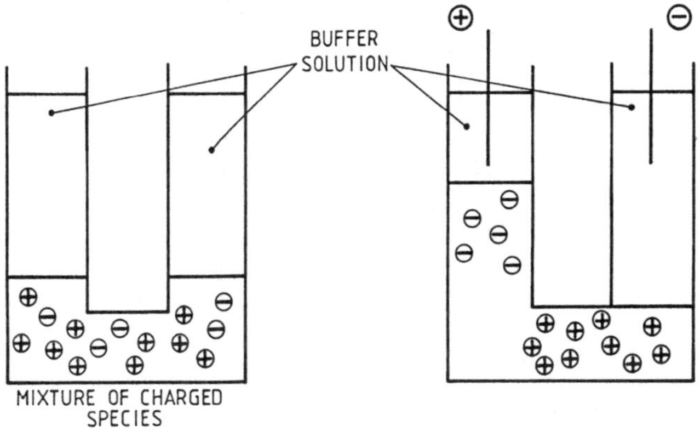

Figure 5 Moving Boundary Electrophoresis

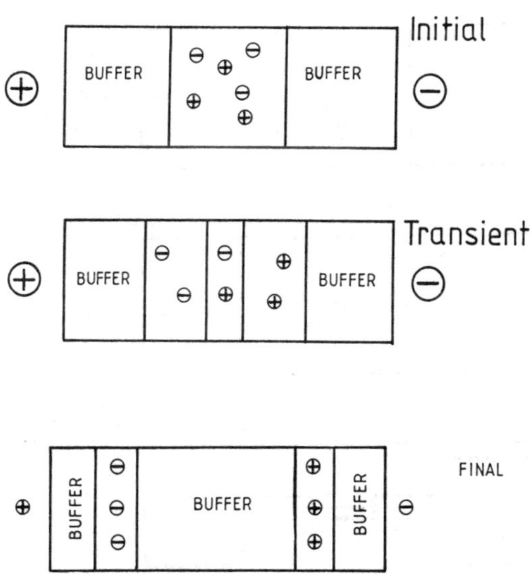

Figure 6 Zone Electrophoresis

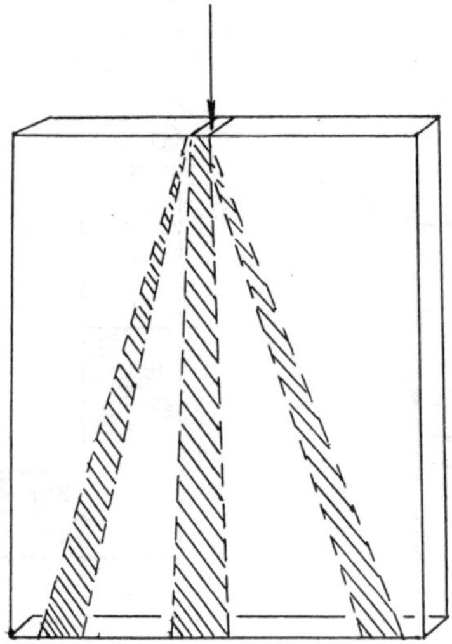

Figure 7 Continuous Zone Electrophoresis

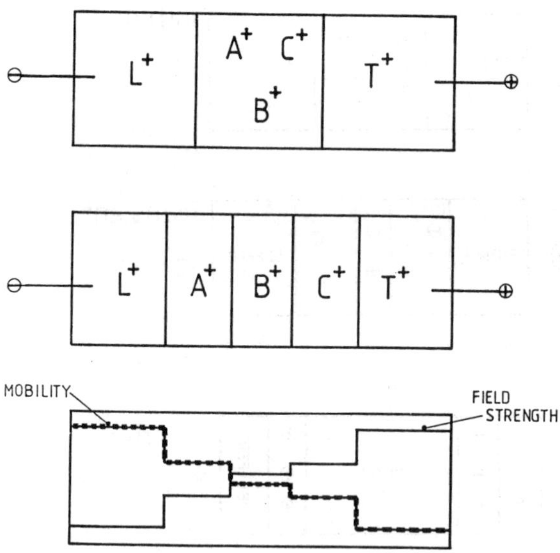

Figure 8 Isotachophoresis

For separation of Amphoteric species:

$$Am^{\circ} \rightleftharpoons H^{+} + Am^{-}$$

$$Am^{\circ} + H^{+} \rightleftharpoons Am^{+}$$

pH gradient and electric field used to separate ampholytes according to their isoelectric point.

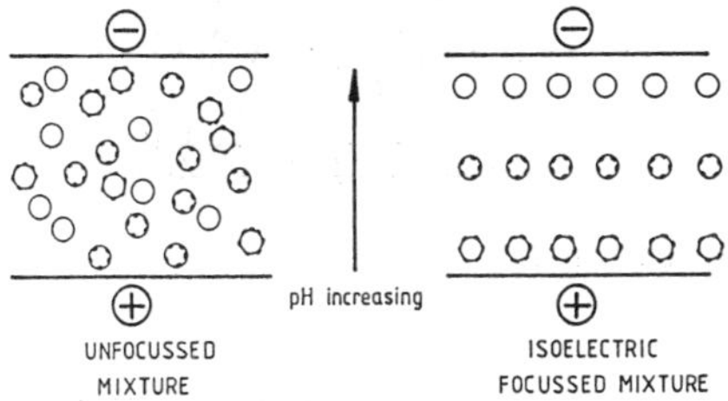

Figure 9 Isoelectric Focussing

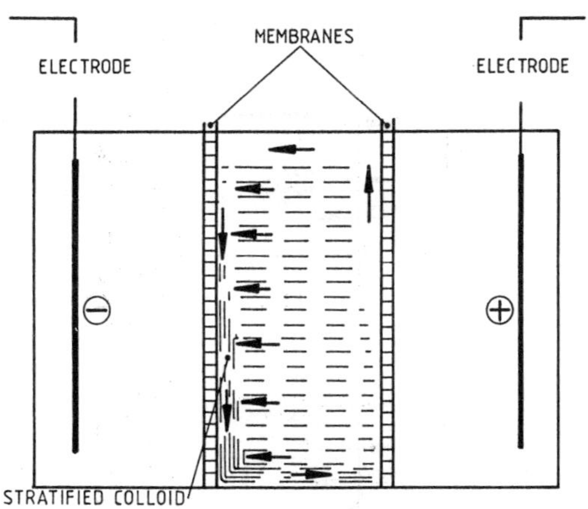

Figure 10 Electrodecantation

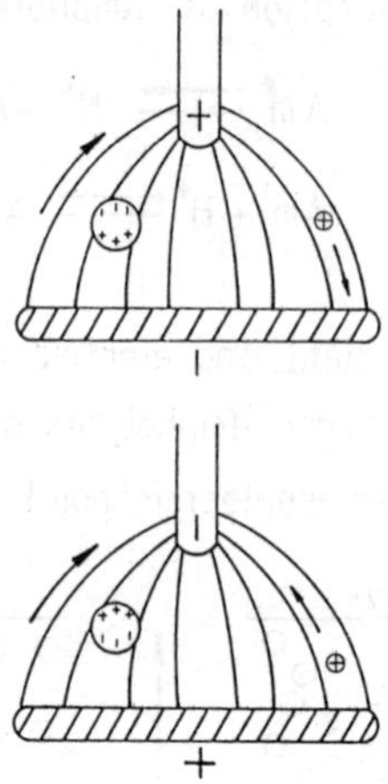

Figure 11 Dielectrophoresis

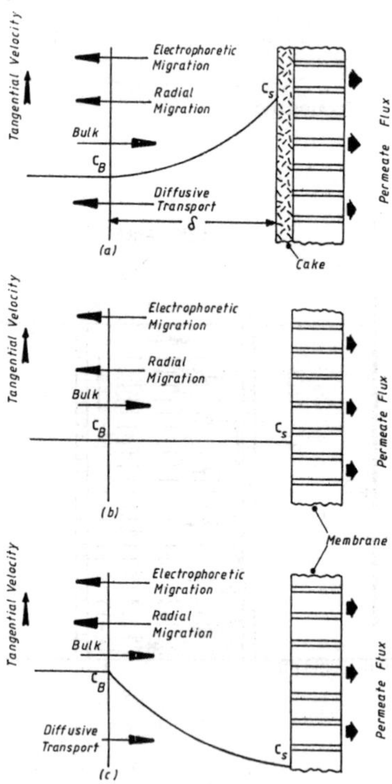

Figure 12 Cross-Flow-Electrofiltration

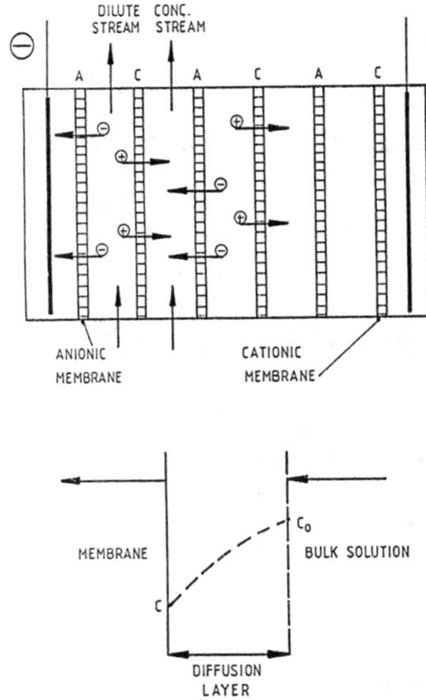

Figure 13 Electrodialysis

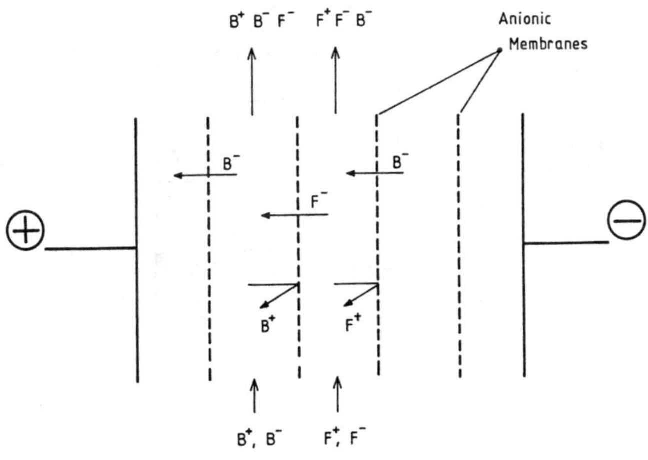

Figure 14 Ion Replacement

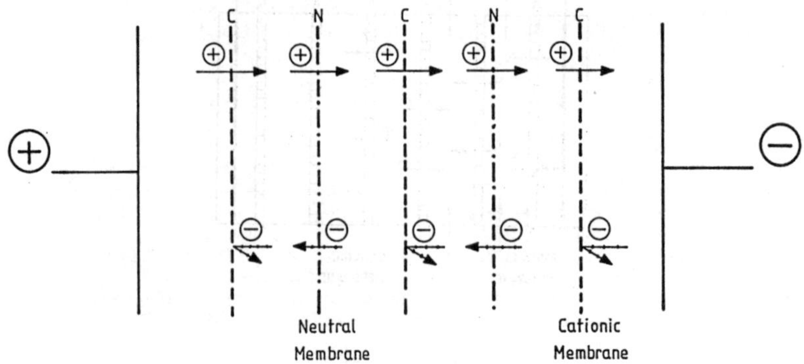

MKM 011

Figure 15 Transport Depletion

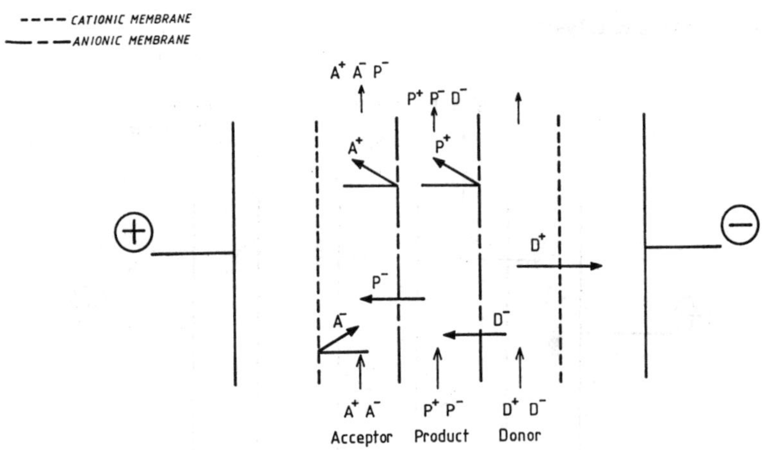

MKM 010

Figure 16 Ion Substitution

SOLVENT EXTRACTION OF PENICILLIN - A STUDY OF ELECTROSTATIC SPRAYING BEHAVIOUR DURING ENHANCED EXTRACTION IN A SPRAY COLUMN

L R Weatherley[ϕ*], I Arnott[*], J C Slaughter[@], I Campbell[@]

The conditions required for generation of electrostatic sprays in a liquid-liquid extraction column have been experimentally studied. In particular the time dependent variation of column conductance due to space charge migration, mass transfer, and due to large changes in dispersed phase hold-up were examined. It is shown that the minimum voltage required for electrostatic spray generation was significantly affected by the electrical history of the column in the absence of dispersed phase flow, and that pre-charging of the extraction column allows spray generation at reduced voltages. It was also shown that the major factor influencing changes in column conductance during the extraction of penicillin from a 25 g /l aqueous solution, as dispersed phase into dichloromethane, is solute trasfer into the organic phase. At high degrees of extraction it was shown that the effect of dispersed phase hold-up and the transfer of buffer components and water make relatively minor contributions to changes in column conductance. Preliminary drop size measurements made in the extraction column indicate the significant presence of droplets of less than 1μm in diameter.

Heriot Watt University, Riccarton, Edinburgh EH14 4AS

* Department of Chemical & Process Engineering

@ Department of Biological Sciences

ϕ To whom all correspondence be addressed

INTRODUCTION

The use of electric fields to enhance the kinetics of solvent extraction processes is well known and studies conducted since the early 1970's have confirmed the effectiveness of the technique at laboratory scale for a wide range of extraction systems. Much of the early work examined (1,2,3,4) the enhancements obtained in the discrete droplet regime in which the surface charge acquired by the droplets was sufficient to result in a significant electrostatic force upon the droplet when exposed to a field, but was insufficient to result in droplet destabilisation, otherwise known as electrostatic dispersion. Bailes (5) describes briefly some extraction studies in the electrostatic dispersion regime for the extraction of nickel from aqueous solution into di-2-ethylhexyl phosphoric acid which indicated the significant potential of operation in this regime for substantial enhancements in mass transfer rate. Other work by Scott & Wham (6) studied extraction of acetic acid in the electrostatic dispersion regime from water into methyl iso-butyl ketone in an Emulsion-Phase Contactor and droplet size reductions to within the range 1 - 10 microns were reported. Mass transfer enhancements were reported relative to the equivalent extraction performed in a York-Scheibel column and a Podbielnak contactor and enhancement factors of 17 and 10 respectively were claimed. Such improvements are consistent with other data reported by the present authors, who examined the application of electrostatic extraction in the dispersion regime to untreated fermentation broth systems (7, 8).

This work demonstrated conclusively that gross mass transfer rates for extraction of ethanol and penicillin G into TBP and methylene chloride solvents respectively could be enhanced by up to an order of magnitude by electrostatically spraying the fermentation solutions into the solvent. Significantly, it was qualitatively observed in the latter study involving penicillin extraction that the voltage required to produce electrostatic dispersion was strongly dependent upon the operational history of the continuum into which the dispersion was sprayed. Experimental data gathered for semi batch extraction of penicillin G from buffer solution into methylene chloride at pH 4 in a simple spray column with the methylene chloride as the continuous phase, showed that for a column of clean, unused solvent, the voltage required to achieve electrostatic dispersion of the aqueous phase was up to fourfold that required to achieve the same condition in the column containing a significant hold-up of dispersion, created for example in a previous experiment. Thus the relationship between mass transfer product and applied voltage determined during a succession of experiments gave rise to the hysteresis type relationship reproduced in FIGURE 1. This provided clear evidence of a change in the potential distribution profile due the build up of conducting droplets in the column, and it was postulated that the resulting attenuation of field to the immediate vicinity of the dispersed phase inlet nozzle resulted in reduction in the minimum applied voltage required for spraying.

It is clear however from other data that non-linearities in the potential distribution also arise from space charge effects caused by the migration of small quantities of charged species, for example impurities present in the continuous phase, under the influence of the applied field. Space charge build up itself is a time dependent process and was considered to be a potential contributor to the hysteresis effect observed during the earlier pen G extractions.

The principal objective of the work described in this paper was to examine in detail the time dependent behaviour of an spray extraction column operating in the electrostatic dispersion regime with respect to conductivity profile and droplet size, and to identify those factors which influence the time dependent behaviour of the minimum spraying voltage.

THEORETICAL

There is a substantial literature concerning the phenomena of electrostatic spraying, and an excellent review was recently published by Bailey (9) covering spraying of conducting and non-conducting liquids, and metallic spraying phenomena. A rigourous appraisal of droplet instability is only possible if the charge density and the local field strength are known with accuracy.

Electrostatic breakup of droplets is assumed to occur when the level of surface charge produces an outward electrostatic pressure equal in magnitude and opposite in direction to the binding interfacial tension forces which maintain the integrity of the surface. For isolated charged droplets Rayleigh (10) defined the limiting droplet charge by the expression

$$q = 8 \pi \left[\sigma \epsilon r^{3} \right]^{\frac{1}{2}} \qquad (1)$$

Zeleny (11) proposed a criterion for breakup of a charged liquid emerging from a capillary which was based on the Rayleigh criterion expressing the potential required for breakup in terms of the drop radius and the interfacial tension thus:

$$V^{2} / r_{o} \sigma = \text{const.} \qquad (2)$$

This criterion was later criticised by Taylor (12) on the grounds that it was developed on the assumption that the pressure inside the growing spheroid is the same as that on the outside. Taylor developed a revised criterion for instability which showed that instability of an droplet, elongated by an electric field, occurs when its length is 1.9 times its equatorial diameter. Hendricks et al (13) expressed this minimum spraying potential criterion, in the absence of space charge by:

$$\frac{V^2}{20.\pi.(300)^2.r} > \sigma \qquad (3)$$

In the case of charged droplets the magnitude of the charge is central to any comparison with the instability criterion. After Stewart & Thornton ,(2), under perfect dielectric conditions the induced charge density, Q/A, on a conducting droplet forming at a nozzle at high dc potential is given by

$$Q/A = \epsilon . \epsilon_o . E \qquad (4)$$

where ϵ = dielectric constant
ϵ_o = permittivity of free space
E = field strength

The mechanism of charge acquisition by drops forming at a nozzle is highly dependent upon the electrical properties of both liquid phases, and in particular the relaxation time appropriate to each phase, where relaxation time is expressed thus

$$\tau = \epsilon \epsilon_o / K \qquad (5)$$

K = conductivity

Thus the rate of charge acquisition is given in terms of the relaxation time for the dispersed phase , τ_d :

$$Q = Q_o \exp (- t /\tau_d) \qquad (6)$$

The charge retention of the drop is also reflected by the relaxation time of the continuous phase , τ_c , :

$$Q = Q_o \exp (-t / \tau_c) \qquad (7)$$

The latter term is especially important in the context of liquid-liquid extraction in the electrostatic dispersion regime since the conductivity and dielectric constant of the continuum into which the electrostatic spray is discharging will be a function of gross dispersed phase hold up, the drop size distribution, and space charge build up. Each of these parameters is likely to change considerably during the early part of an extraction process unless the continuous phase is continuously removed and replaced by pure solvent. In practice, for example in differential contact equipment, this is unlikely to be the case.

The difficulties due to space charge effects in the context of electrostatic drop breakup in a liquid-liquid environment are emphasised by Stewart and Thornton (2). Their data show that such effects may be small at nominal field strengths less than 2.5 KV/cm and reasonably accurate predictions of droplet charge density in the induced charge regime were made for single droplets of water dispersing into heptane. This finding was later confirmed by other work on the prediction of drop size in the induced charge regime (ie the discrete drop regime) by Takamatsu et al (14,15) who were able to predict drop size as a function of physical properties and nominal field strength using a semi-empirical approach. It is considered unlikely that electrostatic breakup could be predicted using the same assumption, on account of the significantly higher fields involved, and due to the substantial inhomogeneities which exist in the continuous phase following the establishment of electrostatic dispersion conditions.

EXPERIMENTAL

Objectives:

The experimental objectives were as follows:

(i) To study the gross extraction rate in an electrically enhanced spray column as a function of the applied electric field.

(ii) To study the column conductivity with respect to time during the course of an extraction and to determine the relative effects of dipersed phase hold-up and penicillin uptake upon the conductivity of the continuous phase.

(iii) To establish the effect of dispersed phase flowrate and electrical history upon the applied voltage required for the creation of an electrostatic dispersion.

(iv) To make preliminary measurements of drop size in the column under conditions of electrostatic dispersion.

The experimental system chosen was the extraction of benzyl penicillin from a sodium phosphate buffer solution at pH 4.0 into methylene chloride.

Equipment

The experimental arrangement used is shown diagrammatically in FIGURE 2 and comprised a 50mm diameter x 750mm active length spray column equipped with a single 1.98mm id stainless steel inlet nozzle located at the lower end . The column was located within a metal framed, earthed containment cabinet equipped with PVC sides and top and with a perspex door fitted with an electrical interlock to prevent access during operation at high voltage. The column was suspended from the internal framework using non-conducting string in order to minimise distortions in the electric field within the column . The column was fitted with three pairs of conductivity probes fitted at approximately equal intervals along the length of the column . These were connected to a Wayne Kerr Impedance Bridge (Wayne Kerr B905 A) thus enabling the conductivity profile along the length of the column to be monitored during extraction. Stable electric fields were introduced into the column by means of a single tungsten electrode which was fused into the glass midway along the length of the column. This was charged from a stabilised Brandenberg DC high voltage source (Type 851 -ve polarity) which allowed application of voltages in the range 0 - 50 KV and up to a maximum current of 1mA. The dispersed phase inlet nozzle and the top end of the column from which the raffinate stream was withdrawn were maintained at earth potential. The voltage was measured by means of an electrostatic voltmeter incorporated into the high voltage power supply.

The column was operated in semi batch mode by charging with a fixed inventory of methylene chloride (800 ml). The aqueous penicillin solutions were metered to the column through a peristaltic metering pump (Watson Marlowe 101 U), the raffinate being collected at the upper end of the column.

Mass Transfer Measurements

A series of detailed mass transfer experiments using the electrically enhanced spray column for the extraction of penicillin into methylene chloride are described elsewhere (8). For the purpose of the present study a straightforward comparison was made between extent of mass transfer under zero field conditions and under electrostatic spray conditions. The procedure involved passing a 100ml portion of penicillin solution (25 g lt-1 at pH 4.0) through a 1.98mm diameter nozzle at a rate of 2.59 ml min-1 into the spray column containing 800 ml of methylene chloride. At the end of the first contact the penicillin raffinate solution was sampled and assayed (by HPLC), and passed through the column for a second contact. The procedure was repeated a third time. Thus the transfer of penicillin during each pass was determined. The entire procedure was conducted under zero field conditions and then repeated with a 12kV positive voltage applied to the column, this voltage being sufficient to produce an electrostatic spray of penicillin solution.

Conductance Measurements

Previous extraction experiments under electrically enhanced conditions in the system penicillin G/ water/ methylene chloride (8), showed that the applied voltage required to achieve electrostatic dispersion of the pen G phase varied according to the history of the extraction column, see figure 1. It appeared from these data that the voltage required for achievement of the spray was much reduced as the extraction experiments, which were conducted in semi-batch mode, proceeded. It was proposed that this time-dependent effect was due to changes in the local field strength in the vicinity of the dispersed phase inlet nozzle. In the present work it was decided to examine changes in column conductance with repect to time during the course of an extraction experiment. Three causes of change in conductance were considered possible: i) accumulation of space charge due to migration of charged species present within the continuous phase; ii) changes due extraction of the penicillin and ionic components present in the buffered penicillin solutions ; iii) changes due to build up of swarms of very fine droplets of conducting dispersed phase in the continuous phase.

A series of four experiments were mounted. Three experiments used dispersed phase feeds of, penicillin solution in buffer at pH 4, pure citrate buffer solution at pH 4, and deionised water respectively. The solutions were pumped into the column containing methylene chloride at a rate of 2.59 ml min-1 through a single nozzle, 1.98mm id. A steady negative voltage of 15 KV was applied to the column electrode. In each case the conductance indicated at the probe located midway along the length of the column was recorded at intervals for a total of 30 minutes. A fourth experiment was conducted in which the column was charged with methylene chloride, the voltage applied, and the conductance at the centre of the column recorded as a function of time in the absence of any dispersed phase flow.

A further experiment was conducted to establish the relative effect of extraction of solute and the presence of dispersed phase upon conductance . This involved electrostatically spraying the penicillin solution, buffered to pH 4, into the column at a rate of 3.6 ml min-1 , with a 15 kVolt field applied to the column . The dispersion was maintained in the column for 5 minutes after which a 50 ml sample of the column contents was withdrawn and its conductance measured with respect to time. The measured reduction in conductance thus reflected the coalesence of dispersed phase, and the final conductance compared with the initial pure solvent conductance reflected the change due to uptake of penicillin.

Two further experiments were conducted in which the column was operated in the electrostatic dispersion regime i) for 10 minutes at 10 KV and ii) for 30 minutes at 12 KV. The conductance - time relationship at the centre conductance probe was recorded until no further change was detectable.

Dispersion Experiments

Earlier work (8) showed that there was a clear relationship between the voltage required to achieve electrostatic dispersion and the flowrate of the dispersed phase. In the light of the earlier findings with respect to the electrical history of the column and the apparent effect upon the conditions required for electrostatic dispersion further measurements were made. The voltage required for dispersion was determined as a function of flowrate for two cases of column history. The first case involved determining the minimum applied votage required to achieve electrostatic dispersion of the penicillin solution at a number of discrete flowrates, where a fresh inventory of solvent was loaded into the column prior to each determination. The procedure involved charging the column with solvent, commencing the dispersed phase flow, and finally, rapidly increasing the applied voltage until dispersion was achieved.

The second case involved repeating the above procedure, except a fixed inventory of solvent was used. Prior to feeding the dispersed phase, the column was subjected to an applied voltage of 7KV for a five minute period. The relationship between the applied voltage (required for spraying) and flowrate was thus determined.

A further set of experiments was conducted to examine the relationship between the internal diameter of the nozzle and the voltage required to promote an electrostatic spray. The set of experiments was conducted at a standard dispersed phase flowrate of 2.59 ml/ min of penicillin G solution (25 g lt-1) in buffer at pH 4.0. The procedure involved charging the column with a fresh amount of methylene chloride prior to each determination. No time was allowed for the column to charge or for space charge migration to occur and the voltage was increased rapidly following commencment of flow until a dispersion was achieved. The voltage at which this was achieved was recorded in each case. Nozzle diameters in the range 0.25mm - 2.50mm were studied.

Drop Size Measurements

Earlier measurements of mass transfer rates during the electrically enhanced extraction of penicillin into methylene chloride indicated the important role of enhancement of interfacial area in the observed increase in mass transfer rate. Ideally the droplet size distributions should be determined during the extraction experiments using in-line techniques (16) In the current work and as a preliminary excercise a batch sampling procedure was adopted in which a fixed sample of the contents of the spray column was transfered to a laser diffraction particle size analyser (Malvern Instruments Mastersizer MS20) and the droplet size distribution measured. By examining the coalescence dynamics of each sample it was possible to obtain an indication of the actual drop size distribution and hold-up in the column during electrostatic spraying.

The procedure involved filling the column with a standard amount of methylene chloride (800 ml) and electrostatically spraying a standard feed solution of penicillin (25 g lt-1 buffered to pH 4.0) into the column for a controlled injection period of 10 minutes at a rate of 2.59 ml /min and at an applied voltage of 15 KV. At the end of the injection period a 100 ml sample of the contents of the column was carefully drained from a specific section of the column and transfered to the particle size analyser. The analyser was used to determine the total volume fraction of dispersed phase and the droplet size distribution as a function of time. Both sets of variables were recorded over a sixty minute period in order that an indication of the coalescence behaviour of the sample be obtained.

RESULTS & DISCUSSION

The results of the mass transfer experiments are shown in TABLE 1 and show the percentage uptake of penicillin for each complete pass of the aqueous solution batch through the column. The capacity of the solvent for penicillin at pH 4.0 is relatively low and the approach to equilibrium is reflected in the data for the electrically enhanced extraction. Calculations of the average rate of mass transfer during each pass were consistent with earlier data measured under very similiar conditions and published elsewhere (8). The approximate doubling of the mass transfer rate observed with feed solutions of 25 g/lt penicillin in the presence of the electric field is modest when compared with the enhancements observed with feed solutions of penicillin at a concentration of 1 g/lt, see FIGURE 1. Operation in the electrostatic dispersion regime would be expected to yield better enhancments and it is possible that the benefits of increased interfacial area may be offset by interfacial resistances caused by complex formation at the liquid liquid interface. This was first observed by Bailes et al during nickel extraction in the electrostatic dispersion regime (5), although we have no visual evidence in this case.

The variations of conductance of the contents of the spray column with respect to time are shown in FIGURE 3. These data show comparisons between the conductance changes exhibited in the presence of dispersed phases comprising : penicillin solution in buffer at pH 4.0 ; pure buffer solution at pH 4.0 ; deionised water. The data for the penicillin solution show that the conductance of the column rises by a factor of almost 10 during the 30 minute period of the experiment. This may be compared with the much lower rise recorded in the case of the buffer solution run and the reduction in conductance recorded in the case of the deionised water run. The conductivity of the penicillin solution and the pure buffer solutions were very similiar and the difference in the observed column conductance change is probably explained by extraction of penicillin and associated water. It was concluded from these data that the presence of the dispersion in the column under these conditions of high penicillin concentration may only make a minor contribution to increase in conductance . The rise in conductance observed in the case of the buffer solution run could also be explained by the extraction of water and other components of relatively high conductivity into the organic phase. Further consideration of the earlier extractions conducted with penicllin concentrations of only 1 g l-1 (see FIGURE 1) would suggest that the effect of solute and associated water uptake upon column conductance would be much less since the concentration driving force was much lower. Thus in relative terms the influence of dispersed phase hold-up upon column conductance would be much greater. This is consistent with the hysteresis effect evident in FIGURE 1.

The results of the experiments conducted to investigate the relationship between column conductance and spray history are shown in FIGURE 4 . The conductance of the mixed phase in the column is plotted with respect to time after termination of spraying. The initial conductance of the continuous phase was 8.8 nS, thus in each of the three experiments the rise in conductance is substantial. In the case of the spray generated at 15 KV the conductance appears to fall from an initial value of 100 nS plateauing at a value in the region of 70 nS . The observed fall could be explained entirely by colaescence of dispersed phase during the measurement period. Thus the conductance of the column in this case could be explained largely by solute and water uptake due to mass transfer during spraying.

The post-spraying conductance behaviour of the column samples generated at 10 KV and 12 KV was measured over a longer time scale, see FIGURE 4, and show that an initial decrease in conductance is followed by a small increase. It is possible that the data for the 15 KV experiment would have exhibited similiar behaviour had sufficient measurements been taken. The rise in conductance may be an indication of a reversal of space charge migration of conducting species initially present in the solvent which occurs during the early stages of field application.

FIGURES 5 & 6 show the results of experiments performed to study the minimum voltages required for promotion of electrostatic dispersion (Minimum dispersion voltage). FIGURE 5 shows the effect of dispersed phase flowrate upon the minimum dispersion voltage. Curve 1 shows the relationship for the case in which the column was subjected to an initial field application (at zero flow) for a 5 minute period prior to measurement of the minimum dipsersion voltage. Curve 2 shows the relationship for the column maintained at earth potential immediately up to the start of the measurement. These data are further evidence of the importance of column history upon the creation of electrostatic dispersion. The lower values of minimum dispersion voltage observed in curve 1 suggest again the role of space charge effects in column behaviour. The initial charging period experienced by the column may promote migration of charged species present in the solvent with consequent local increases in field strength in the vicinity of the dispersed phase inlet. Beneficial increases in local field strength would also accrue from local changes in column conductance due to mass transfer and the presence of microscopic droplets of dispersed phase.

FIGURE 6 shows the effect of internal nozzle diameter upon the minimum dispersion voltage at constant aqueous flowrate. The linear velocities of the dispersed phase were below the jetting velocity and for this particular system the required voltage increases with respect to diameter. Other data (8) show that this may not be always the case and that the point of droplet destabilisation is a function not only of drop size but also of the residence time of the drop at the nozzle prior to detachment.

The results of the preliminary drop size measurements are shown in FIGURES 7, 8 & 9. The variation of volumetric concentration of droplet phase in a sample of the column contents after a ten minute period of electrostatic dispersion conditions (at 15 KV) is shown with respect to time in FIGURE 7. Experimental values in the time range 0 - 2 minutes were not available due to the time lag between sampling and measurement. However curve fitting of points beyond this time and back extrapolation to time zero are shown on the Figure. The value of 0.45 vol % dispersed phase is very similiar to values obtained in earlier work (7) for countercurrent extraction of ethanol into tri-n-butyl phosphate in an electrically enhanced spray column. The observed rapid decrease in dispersed phase volume is surprising for the size range of droplets but is perhaps evidence of coalescence enhancement, characteristic of small droplets having a relative low surface charge.

FIGURE 8 features further detail of the volume fraction variation of some of the very small droplets at sizes less than 0.53 µm. The first part of the plot (see curve(1)) shows the decay in the volume percentage of all droplets less than 0.53µm diameter. Curves (2) and (3) break down the coalescence behaviour further into size bands 0.53 - 0.28 µm (curve(2)) and less than 0.28µm (curve (3)). Each curve indicates a general decrease of volume with respect to time indicating relatively rapid coalsecence into larger droplets but with a levelling out after approximately 25 minutes. Curve fitting and extrapolation to time zero indicates approximately the percentages of dispersed volume in each size range : >0.53µm : 3.5 % , 0.53µm < > 0.28µm : 2.75 % , < 0.28µm : 1%.

The behaviour of the larger droplets in the size range 5 - 80µm is shown in FIGURE 9 and the percentage of dispersed phase within each size range is plotted against time. Overall it is clear that the majority of the dispersed phase comprises of drops within this range although the actual number of droplets will be significantly less than those in the smaller size ranges shown in FIGURE 8. Curves (1) , (2), & (3) refer to drops below 5.29µm, below 2.83 µm , and in the range 64.9 - 80 µm respectively. In each case a rise in the volume contribution is evident and is explained by coalescence of smaller drops. The troughs in curves (1) & (2) observed at 18 minutes correspond to a peak in curve (3) which is consistent with a rapid coalescence to the higher size range, although it is also clear that the existence of drops within the 64.9 - 80 µm range is largely transitional.

Quantitative interpretation of these drop size data is very limited on account of the method of measurement and future work is designed to measure drop size "on-line" in the spray column either using high speed photographic techniques or by laser scattering techniques in which the spray column is located in-situ.

ACKNOWLEDGEMENT

The authors wish to acknowledge the financial support of the Science and Engineering Research Council for this work.

NOMENCLATURE

A	surface area of drop	m^2
E	field strength	volts m^{-1}
K	conductivity	S m^{-1}
q, Q_o, Q	charge on drop	coulomb
r_o, r	radius of drop	m
t	time	s
V	potential	volts
ϵ	dielectric constant	
ϵ_o	permittivity of free space	s Ω^{-1} m^{-1}
σ	interfacial tension	N m^{-1}
τ_c	relaxation time, continuous phase	s
τ_d	relaxation time, dispersed phase	s

REFERENCES

1. Bailes, P.J., Thornton, J.D., 1971 Proc. Int. Solv. Extn. Conf., vol. 2, pp1431-1439, SCI, London.

2. Stewart, G., Thornton, J.D., 1967, I.Chem. Symposium. Ser. No. 26, ed J.M. Pirie, pp29-42.

3. Martin, L., Vignet, P., Fombarlet, C., Lancelot, F., 1983, Sepn. Sci. & Techn., vol 18, pp1455-1471.

4. Kowalaski, W., Ziolkowski, Z., 1981, International Chemical Engineering, vol 21 (2), pp323-327.

5. Bailes, P.J., 1977., Proc. Int. Solv. Extn. Conf., ed G Ritcey, Canadian Institute of Mining and Metals, pp233-241.

6. Scott, T.C., Wham, R.M., 1989, Ind. Eng. Chem. Res., vol 28 pp94-97.

7. Laughland, G J., Millar M.K., Weatherley, L.R., 1987, I.Chem.E. Symposium Series no. 103, pp263-279.

8. Weatherley, L.R., Campbell, I., Kirton, D., Slaughter, J.C., 1990, J. Chem. Tech. & Biotech. in press.

9. Bailey, A.G., 1986, Atomisation and Spray Technology, vol 2 pp 95-134.

10. Rayleigh Lord, 1882, Phil. Mag. (series 5), vol 14, pp184-186.

11. Zeleny, J, 1917, Phys. Rev. (series 2), vol 10, pp1-6.

12. Taylor, G.I., 1964, Proc. Royal. Soc. (London), vol A280 pp383-397.

13. Hendricks, C.D., Carson, R.S., Hogan, J.J., Schneider, J.M., 1964, A.I.A.A., vol 2 (4), pp733-737.

14. Takamatsu, T., Hashimoto Y., Yamaguchi, M., Katayama, T., 1981, J. Chem. Eng., Japan, vol. 14 (3), pp178-182.

15. Takamatsu, T., Yamaguchi, M., Katayama T, 1982, J. Chem. Eng., Japan, vol 15 (5), pp349-355.

16. Scott, T.C., Sisson, W.G., 1987, Proc. 5th Symp. on Sepn. Sci. & Technol. for Energy Applications, Knoxville, Tenn. USA., John Wiley, pp1541-1550.

TABLE 1 - Results of Preliminary Mass Transfer Experiments

Pass no.	percentage mass transfer (cumulative)	
	Zero field	12kV
1	18.75	39
2	35.50	56
3	46.11	55

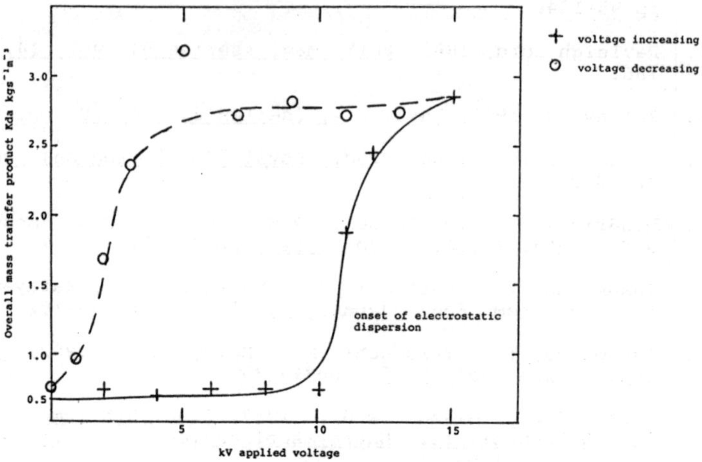

Figure 1 Electrostatic Extraction of Penicillin into Methylene Chloride (at 1.0 G L-1)

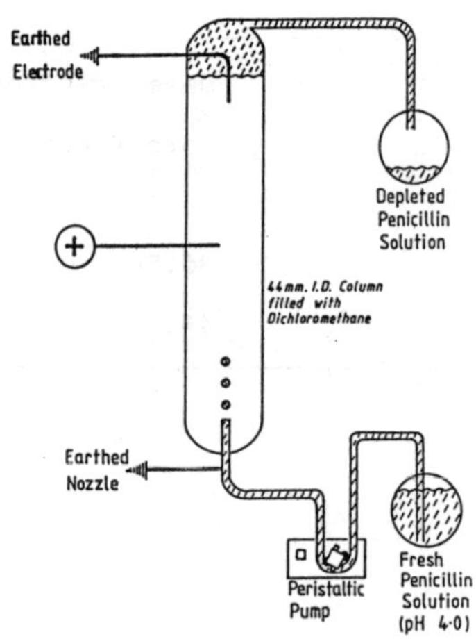

Figure 2 Electrostatic Extraction Column System

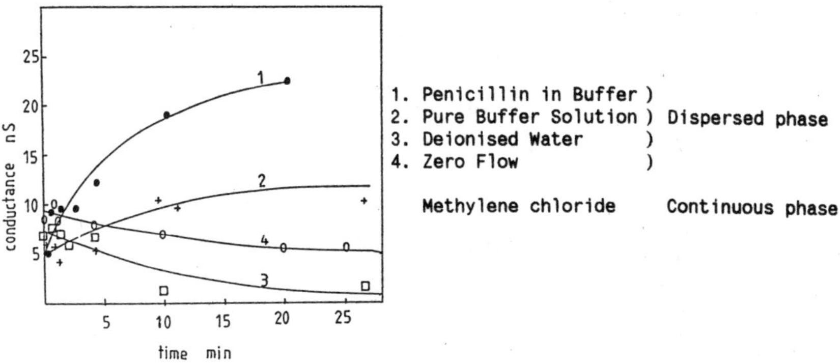

Figure 3 Variation of Column Conductance with Time

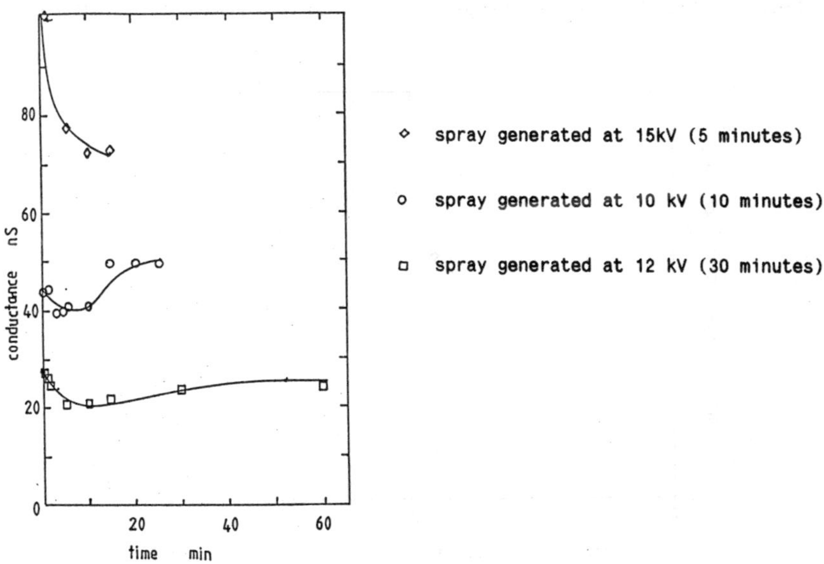

Figure 4 Variation of Conductance with Time for Coalescing Mixed Phase Samples

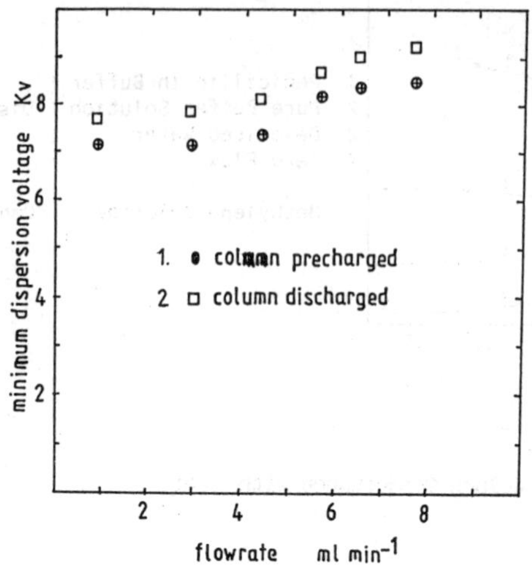

Figure 5 Effect of Column Charge History upon Minimum Dispersion Voltage

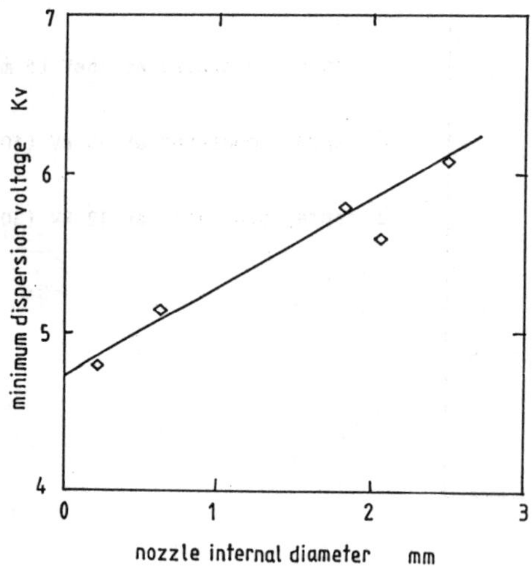

Figure 6 Effect of Internal Nozzle Diameter upon Minimum Dispersion Voltage at Constant Flowrate

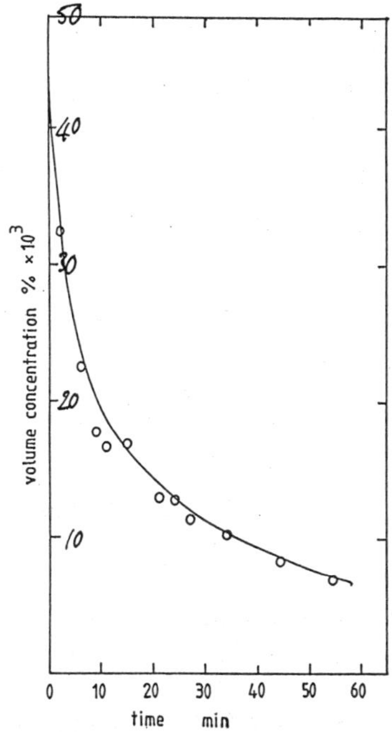

Figure 7 Variation of Dispersed Phase Hold-Up with Time Post Spraying

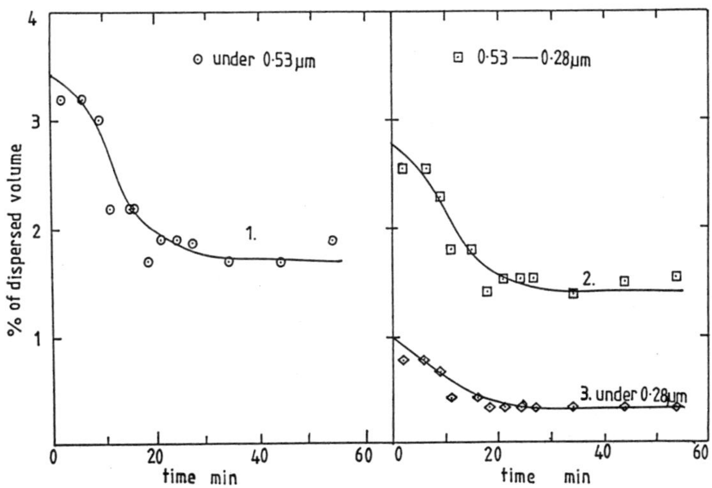

Figure 8 Variation of Dispersed Phase Hold-Up with Time Post Spraying - By Drop Size Range (Small sizes)

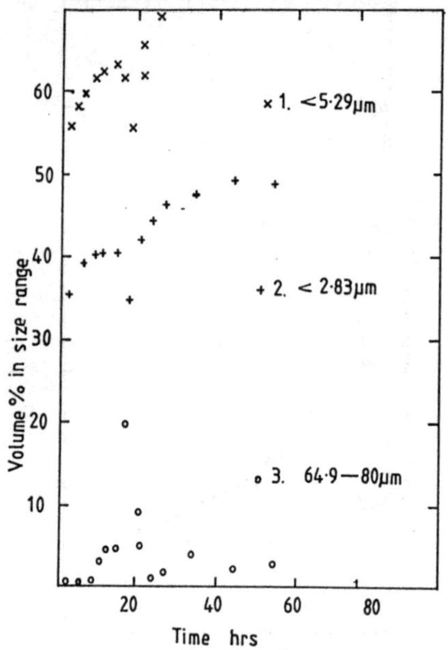

Figure 9 Variation of Dispersed Phase Hold-Up with Time Post Spraying - By Drop Size Range (Large sizes)

ENERGY AND ECONOMIC ASSESSMENT OF NEAR-CRITICAL EXTRACTION PROCESSES

M.B. King*, O.J. Catchpole and T.R. Bott

Energy consumption and capital costs for marginally sub-critical and supercritical extraction processes are discussed and the important role of solute loading in determining both these factors is emphasised. Careful attention to the solute/solvent separator conditions is found to be required if energy and capital costs are to be minimised. The scale of operation is important in determining capital costs. The discussion is illustrated by a worked example in which the separation cost for a hypothetical process for extracting oil from crushed rape seed using marginally subcritical liquid carbon dioxide is evaluated. Areas where near-critical extraction may well be shown to be cost-effective are suggested.

1. INTRODUCTION

In a typical near-critical extraction process the compressed solvent, after contacting the material (or stream) to be extracted, passes through the product recovery unit, from which the bulk of the solvent stream is recycled back to the contactor. When the extraction temperature and pressure are above the critical values for the pure solvent, the extraction is said to be "supercritical". At temperatures below the critical the fluid is "subcritical" and only the liquid phase exerts appreciable solvent power. The exact range of solvent reduced temperatures considered as "near-critical" is rather arbitrary (King et al. (1)) but is usually taken to lie between 0.9 and 1.2. Many of the processes currently in use (such as the decaffeination of coffee, the extraction of α-acids and hop oils from hops and the extraction of aroma and flavouring essences with compressed carbon dioxide) are semi-continuous with batch feed of solids. However there is no conceptual reason why the continuous extraction of suitable liquid streams using a near-critical solvent should not be possible. Indeed this is the case in propane de-asphalting, a large throughput near-critical extraction process which has been in use for many years now (Nelson (2)).

Much of the energy requirement of a conventional liquid extraction process lies in the separation of the product from the solvent. Where the product is to be used for human consumption it may be necessary to meet stringent food standards. These may well reduce the allowable amount of residual solvent to a level well below that required from an economic consideration of solvent losses. Under these conditions, near-critical

*To whom all correspondence should be addressed. School of Chemical Engineering, The University of Birmingham, Edgbaston, Birmingham B15 2TT, U.K.

extraction with a solvent which is gaseous under normal conditions has an evident advantage since nearly all the solvent will automatically be expelled when the product comes to ambient pressure. The removal of the last detectable traces from liquid product streams then presents no difficulty. (If the comparatively innocuous solvent carbon dioxide is used, the removal of the last traces, while not necessary to satisfy food regulations, may be required to give an acceptable taste.)

The energy requirements of a near critical extraction process depend on the method of separating the solute from the solvent and the degree to which solvent recovery from the product (and, where appropriate, raffinate) streams is practiced. However the process design will be based on a minimisation of total process cost per kg of product, to which energy costs is only one contribution. There will be interplay between capital costs, energy costs, solvent replacement costs and other costs, so the design giving the lowest energy cost per kg of product will not necessarily correspond to the lowest total cost requirement.

A discussion of energy costs and capital costs for possible extraction processes using a marginally subcritical (liquid) solvent and for a supercritical solvent is given in the following text.

2. EXTRACTION WITH A MARGINALLY SUB-CRITICAL SOLVENT

2.1 Process Layout

Idealised flow diagrams of possible extraction processes are given in Figs. 1 and 2. Fig. 1 is for the semi-continuous extraction of material of low volatility from a bed of solids while Fig. 2 is for the continuous extraction of more volatile material from a liquid stream. The process shown in Fig. 1 is very similar to hop extraction plants which have been operating under licence from Distillers MG Ltd. (originally Distillers (Carbon Dioxide) Ltd.) in the U.K. and Australia since about 1980 (Gardner (4)). The process shown in Fig. 2 is similar to one patented by de Filippi and Vivian (24) in 1981 for the ethanol/water separation. Both processes use liquid CO_2 as the extractant and solute/solvent separation involves a reduction in pressure with vaporization of the solvent. Having disengauged the solute, the solvent is recompressed. The solute/solvent separation recovery systems are similar in that the heat of condensation from the recompressed CO_2 stream is used to boil off liquid CO_2 from the stream entering the separator from the expansion valve and extractor. The solubilities of useful components from natural products in liquid carbon dioxide are usually quite low (typically in the range 0.01 to 1.0 weight %) and for this reason it is essential to recycle the CO_2 with a minimum of energy loss.

2.2 Energy Consumption

In the processes shown in Figures 1 and 2 most of the energy for the process is consumed by the compressor. Figure 3 shows the state of the solvent on a T-S diagram for an extraction at 25C, 65 bar with expansion either to 60 bar or 50 bar in the separator. The expressions (King et al. (3)) for the energy consumption of the compressor per unit mass of solvent passing through it (ε_c) and for its energy consumption per unit mass of product (E_c) are

$$\varepsilon_c = (\Delta H/\eta) \qquad (1)$$

and $$E_c = \left(\frac{\Delta H}{\eta Y} \right) (1 - Y\sigma_\pi) \quad (2)$$

where

ΔH = specific enthalpy difference across compressor, kJ kg^{-1}
η = compressor efficiency
Y = product loading in solvent, g product/g solvent
σ_π = solvent loading in product, g solvent/g product

(For example ΔH for extraction at 25C, 65 bar and vaporization at 50 bar is given by height CD in Figure 3.)

Two important conclusions may be drawn from Equation (2):

(1) Energy consumption increases as Y, the loading of product in the stream leaving the extractor, decreases. This is due to the greater mass of solvent which must be recirculated for a given amount of product.

(2) The term ΔH may in principle be reduced (with a corresponding reduction in E_c) by reducing the pressure drop ΔP required for product recovery.

The magnitude of ΔH for given ΔP may be determined from thermodynamic tables or charts or calculated from an equation of state. An ultimate limit to the extent to which ΔP (and hence ΔH) can be reduced is set by the elevation of the boiling point ΔT_l of the solvent boiling on the shell-side of the product recovery unit produced by the presence of solute. The heat required to boil off this solvent is provided by the condensation of virtually pure carbon dioxide at the extraction pressure on the tube-side of the recovery unit. In order to maintain a positive temperature driving force across these tubes the pressure on the condensing side must exceed that on the boiling side by an amount ΔP_l. If ΔT_l is small, ΔP_l is given by

$$\Delta P_l = (10^{-5})(\Delta H_v \Delta T_l)/(T(V^G - V^L)) \quad (3)$$

where

ΔH_v = enthalpy of vaporization of CO_2, J kg^{-1}
V^G = specific volume of saturated vapour, m^3 kg^{-1}
V^L = specific volume of saturated liquid, m^3 kg^{-1}

and ΔP_l is expressed bar.

In practice, values of ΔP_l substantially in excess of those given by Equation (3) will be required to give a realistic temperature driving force across the condenser/reboiler tubes and hence an acceptable heat exchanger area.

As ΔP is reduced the temperature driving force falls and the area required increases leading to an increase in capital cost. The actual value of ΔP used must take account not only of energy costs, which are favoured by a reduction in ΔP, but also amortization costs on the heat exchanger.

In addition to the energy required to operate the recirculation compressor, energy is also consumed when recovering solvent from the product stream and from the extracted materials. The energy consumption involved in these operations is much less than that consumed by the main compressor (3).

In some cases provision will be required for excluding air from packed bed extractors. Gahrs (6) reports that small amounts of N_2 can produce a substantial reduction in the solvent power of carbon dioxide in some systems. As noted above, (Equation 2) reductions in solubility result in increases in energy consumption.

2.3 Mass Transfer Considerations

To minimise the energy consumption of the compressor the loading of solute (Y in Equation 2) should approach its equilibrium value. To design an extraction unit in which this aim is achieved, the mass transfer behaviour of the system should be known. If, for example, a packed bed is to be extracted, the bed length should at least be sufficient for saturation at the outlet to be approached.

This and other factors influencing the sizing of the extraction towers are examined in the worked example in Section 5.

In this example a mass transfer model developed from the experimental work of Barr (5), Chami (1) and Bernardo-Gil (7) on the extraction of oil from ground rape-seed is used.

2.4 Capital versus Energy Costs

The capital and energy costs of the plant depend on the required production rate of product, the pressure and temperature of operation of the extractors and separator and the equilibrium and mass transfer characteristics of the solute/solvent system. The main units contributing to the capital cost of the plant are the main compressor (which is also the principal user of energy), the extraction towers and the product recovery unit incorporating the heat exchanger discussed in section 2.4.3. As discussed earlier a reduction in ΔP (Figures 1 and 2) results in a decrease in energy costs (and also in the capital costs of the main compressor) but leads to an increase in the capital cost of the product recovery unit.

An increase in solvent loading Y is beneficial in reducing both energy costs (Equation 2) and capital costs since it reduces the amount of solvent which must be recirculated for a given output of product. The major units will now be considered in turn.

2.4.1 Main compressor. A reciprocating compressor would be used for low tonnage operations and a centrifugal compressor for high tonnage applications. The installed cost of both types of compressor (including drive) in the range of conditions considered (viz compressible vapour not immediately adjacent to the critical point) is usually taken to vary with an exponent between 0.8 and 0.9 with respect to power output (Happel and Jordan (8), Ulrich (9)). For example, graphical data in (9) for stainless steel centrifugal compressors with electric drive can satisfactorily be represented by the equation

$$I'_{cc} = 5630(W)^{0.94}$$

where W = power delivered by compressor to fluid stream in kW. I'_{cc} is installed cost of compressor (\$ US, Mid 1982) $50 < W < 12,000$.

In £UK at current prices

$$I_{cc} = 4900\ (W)^{0.94} \qquad (4)$$

For a mild steel centrifugal compressor, I_{cc} (£ UK, current) is given by

$$I_{cc} = 1940 \ (W)^{0.94} \qquad (4a)$$

Reciprocating compressor costs follow a similar relationship but, according to (9) are about 15% greater in the above range.

Cost estimation is a risky process but the above result does enable approximate cost calculations to be performed.

2.4.2 Extraction towers. Cost depends on internal diameter, material of construction, internal pressure and method of closure. Some of the correlations in the literature (see Peters and Timmerhales (10) for example) present vessel costs as a function of total vessel weight, while others give costs directly in terms of vessel dimensions and material of construction. Over the restricted range of dimensions listed below, the graphical data given in (9) may be represented within ± 15% by

$$I'_{ET} = (2.9 \times 10^3 \ d^{0.9} h_T^{\ 0.76})(9.3 + 0.87 \ (P-5.0)^{0.78})$$

where d is the vessel diameter in m, $1 < d < 4$

h_T is the vessel length in m, $2 < h_T < 40$

P is the working pressure in bar (abs) $5 < P < 200$

I'_{ET} is installed cost for vertical stainless steel pressure vessels ($ US, Mid 1982).

In £ UK at current prices,

$$I_{ET} = 2.6 \times 10^3 d^{0.9} h_T^{\ 0.76}(9.3 + 0.87(P-5.0)^{0.78}) \qquad (5)$$

Costs for pressure up to 5 bar are taken to be identical to those at 1 bar.

2.4.3 The separator (condenser/vaporizer) unit. The liquid CO_2 stream conveying dissolved product from the extractor first passes through the expansion valve, where partial vaporisation occurs. The mixture of vapour and liquid then enters the separator where the remaining liquid CO_2 is boiled off using heat of condensation from the recompressed CO_2. The product thrown out of solution as the liquid CO_2 boils off is continuously collected. The heat energy transferred in this unit substantially exceeds the mechanical energy supplied to the system via the compressor and a large heat exchange area is required. This area (a) is given by

$$a = \frac{\Delta H_s (\Omega)(1-Y\sigma_\pi)}{YU \ \Delta T} \approx \frac{\Delta H_s (\Omega)}{YU \ \Delta T} \qquad (6)$$

where

ΔT = temperature driving force across the heat exchanger tubes, K

Ω = product flow rate, kg s^{-1}

U = overall heat transfer coefficient, kW $m^{-2}K^{-1}$

ΔH_s = excess of the specific enthalpy of the vapour carbon dioxide leaving shell side of the separator unit heat exchanger over that of the stream entering. ($H_C - H_B$ in Figure 3 and Table 1 for an evaporation pressure of 50 bar.)

The installed cost of the separator may be estimated from graphical data in reference 9. For stainless steel shell and tube heat exchangers over a restricted range of conditions these may be expressed as

$$I'_{CR} = 443a^{0.664}\ (1.90 + 3.676\ P^{0.09}) \qquad (7)$$

where

P = shell-side pressure in bar ($10 < P < 100$)

a = heat transfer area in m^2 ($100 < a < 1000$)

I'_{CR} = installed cost ($ US, mid 1982)

In £ UK at current prices,

$$I_{CR} = 385\ a^{0.664}(1.90 + 3.676\ P^{0.09}) \qquad (8)$$

3. EXTRACTION WITH SUPERCRITICAL SOLVENT

Published information for industrial scale supercritical extractions is scarce. The same general design considerations will apply as in the subcritical case, through mass transfer rates and equilibrium loadings can be higher. Capital costs are not so easily estimated since the pressures involved fall outside the cost correlations for some of the units. (Use of manufacturer's quotations is recommended, even in the early stages of costing.)

Energy requirements largely depend on the method of separation of solute from solvent. A possible process layout for extracting a bed of solids with supercritical CO_2 has been given in (3) and a generalised schematic of a supercritical extraction process in which solute/solvent separation is achieved by reduction in solvent power is shown in Figure 4. High density fluid passes through extraction column(s) which may contain packed beds of solids (as in hop extraction and decaffeination of green coffee beans) or may be continuously fed with a liquid stream which is to be separated. The solvent bearing extracted solute then undergoes a reduction in solvent power and the product which is deposited is separated off from the bulk of the solvent. The solvent power of the latter stream is then regenerated after which the stream is recycled to the extraction unit. The reduction in solvent power may be brought about by changes in temperature, pressure or a combination of both.

Solubility in near-critical solvents generally increases with solvent density at given temperature and with temperature at given density (Brogle (13)). A reduction in pressure is an obvious way of reducing solvent density and hence of achieving solute recovery. Isobaric temperature increase can also be used to decrease solvent power: the solubility of materials of low volatility in supercritical solvents at fixed pressures not greatly exceeding the critical pressure frequently decrease with increasing temperature (King et al. (5)). The expenditure of mechanical energy is substantially less than in the pressure reduction method, though there will be thermal energy requirements.

It may be noted that if the solvent is liquid under ambient conditions, the mechanical energy requirements of the pressure reduction method of solute recovery may be considerably reduced since repressurisation of the solvent in liquid form may be used rather than repressurisation of gaseous solvent. The pilot plant used by the UK National Coal Board for the supercritical extraction of coal was of this type (Fin et al. (14)). However the decrease in mechanical energy requirements is balanced by increased thermal requirements.

An alternative separation technique for the selective removal of components from supercritical solution is adsorption (Coenen and Kriegel (15), Shishikura et al. (19), Ikushima et al. (20) and Schaeffer et al. (21)). Although not discussed in detail here, this is potentially a very powerful method, particularly for the removal of trace components.

3.1 Separation by Pressure Reduction followed by Recompression

In this case (Figure 1a in (3)) the main energy cost will be recompressing the solvent (a trimming heat exchanger will be required to remove this energy from the system). Provided separation is virtually complete (which may not be so easy to achieve in the supercritical case), equations 1 and 2 of section 2.2 are applicable. By selection of suitable pressures and temperatures, substantially higher values of solubility (and hence of Y in Equation 2) than in the subcritical case should be available. This factor will tend to reduce energy consumption. However ΔH may also be greater.

3.2 Separation by Temperature Change

If separation of solute from solvent is brought about solely by near-isobaric changes in temperature, part of the desired temperature rise and subsequent temperature drop may be brought about by heat transfer between the solute-bearing stream entering the separator and the "solute-free" stream leaving it. However hot and cold trimming exchangers using process streams will be required to achieve the last few degrees of the desired temperature changes and energy consumption will be associated with these.

4. ENERGY AND OTHER COSTS IN SOME TYPICAL CASES

Energy and other costs for a hop extraction plant utilising subcritical carbon dioxide have been summarised in (3). This plant had a capacity of 500 kg hops/hour and was similar to that shown schematically in Figure 1. Three extractor vessels were used, two of which were on-line at a given time. The solute recovery unit comprised an inclined tubular heat exchanger. In this case the cost of the energy consumed by the compressor amounted to only 6% of the total process costs. These values were taken from data for an early plant.

In reference (3) also, preliminary costing calculations were presented for hypothetical plant in which a 10 weight % broth of ethanol in water was extracted to give an 80 weight % product using liquid CO_2 at 25°C as extractant. The plant layout was similar to that in Figure 2. Extraction took place at 65 bar with pressure reduction in the solute recovery unit to 50 bar.

The energy consumed by the compressor was very similar to one estimated by Moses and de Fillipi being about one half that for a similar enrichment carried out by "ordinary" distillation (i.e. distillation with no energy recovery from the condenser). The compressor energy consumption cost was 24% of the final process cost which was found to be 6.0p per kg, a figure comparable with ordinary distillation costs.

Provided equilibrium and mass transfer rate data are available, preliminary costing calculations such as the one above present no particular difficulty.

These data are increasingly available in the literature and, if not available, are not difficult to obtain with suitable bench-scale equipment. The extraction of material from a solid substrate using marginally subcritical liquid CO_2 as solvent is taken as a further example below.

5. EXAMPLE OF PRELIMINARY COSTING

For illustrative purposes we shall consider the hypothetical case of the extraction of oil from ground rape seed using liquid carbon dioxide at 65 bar and 25C as solvent. It is not claimed that this is a good way of isolating rape oil from the seed. The example has been chosen because equilibrium and mass transfer rate data on which to base the calculations are available and it is an extraction of a type which might well be carried out batchwise with respect to solids in equipment similar to that shown in Figure 1. As in the hop extraction plant, the separator unit is taken to consist of an inclined tubular heat exchanger. Evaporation of solvent takes place on the shell-side and condensation (at a more elevated pressure close to the extraction pressure) on the tube-side. Two evaporation pressures will be considered, 50 bar and 60 bar. Idealised solvent flow paths (ABCDF for 50 bar evaporation and Ab c d F for 60 bar evaporation) are shown schematically on the T/S diagram given in Figure 3. In this diagram the influence of solute on solvent stream properties is neglected, as is heat transfer between the system and the surroundings. The solvent enthalpies and other parameters at points A to F are given in Table 1. Rape seed and rape oil properties are given in Table 2. The extraction conditions (25C, 65 bar) are shown as point A. The liquid CO_2 at this point is slightly undersaturated (the boiling temperature of CO_2 at 65 bar is 25.5C). From point A the solvent containing dissolved solute passes through the expansion valve where

TABLE 1 - Solvent Enthalpies and other Parameters at Relevant Points on T/S Diagram (Fig. 3) for Evaporator Pressures of 50 bar and 60 bar Absolute

Point on T/S diagram	Condition	Pressure P (bar)	Temperature t (C)	Enthalpy H(kJ/kg)
A	L	65	25.0	577.3
B	L + V	50	14.3	577.3
L	L	50	14.3	540.0
C	V	50	14.3	721.0
D	V	65	34.3	729.7
E	V	65	31.7	721.0
F	L + V	65	25.5	-
b	L + V	60	22.0	577.3
c	V	60	22.0	705.5
d	V	65	27.8	707.9

(Thermodynamic data are from ref 11, Tables 2 and 5.)

isenthalpic expansion either to 50 bar, 14.4C or to 60 bar, 22.0C takes place. The solvent is then a mixture of vapour and liquid the liquid fraction being 0.79 in the first case (point B) or 0.91 in the second (point b). The mixed-phase carbon dioxide then passes to the shell-side of the inclined heat exchanger which constitutes the separator unit. Here the remaining liquid solvent is boiled off and the precipitated solute (in this case a liquid) is removed. The state of the vapour leaving the shell-side of the exchanger is given by point C (for a 50 bar evaporation) or by point c (for a 60 bar evaporation). This vapour is isentropically compressed to 65 bar. It is then cooled in the small heat exchanger H before passing into the tube-side of the large inclined heat exchanger where it condenses at 25.5C providing latent heat to boil the incoming stream from the expansion valve.

An approximate costing of the above process has been made for three plant throughputs (Table 5). These costings are based on an 8000 hour operating year, so the product flow rate (Ω kg/sec) is related to the annual product throughput (A tonnes/year) by

$$\Omega = 3.47 \times 10^{-5} A \qquad (9)$$

TABLE 2 - Ground Rape Seed Parameters

Average particle diameter	550×10^{-6} m
Bulk bed density (ρ_B)	440 kg m^{-3}
Bed voidage	0.45
Total oil content of seed	40%
Extractable oil content (F)	0.30 kg oil/kg seed
Loading (Y) of rape oil in CO_2 at 65 bar, 25C (assuming equilibrium)	8.4×10^{-4} kg kg^{-1}
Equilibrium concentration of rape oil in CO_2 at 65 bar and 25C (c_M)	0.6 kg m^{-3}
Solubility of CO_2 in the oil at 50 bar, 14C as weight ratio (σ_π)	0.25 kg kg^{-1}

5.1 Energy Consumption and Heat Exchanger Duties

The main energy user is the recompressor and the present estimate of energy consumption takes only this into account. Since Y and σ_π are both small fractions (Table 2), the energy consumption of this unit (for which an overall efficiency of 0.75 is here assumed) is given by Equation (2) to be

$$E_c = \Delta H/(0.75Y) \text{ kJ per kg product} \qquad (10)$$

For 50 bar evaporation $\Delta H = H_D - H_C$ while for 60 bar evaporation $\Delta H = H_d - H_c$ (see Table 1 and Figure 3). Based on an electricity cost of 3.3p/unit, the energy cost per tonne of product (C_E) is given by

$$C_E = 0.0092\ E_C \qquad (11)$$

Values of C_E for the two separator pressures considered are given in Table 5.

The ideal power consumption (WkW) is given by

$$W = \Delta H/Y \quad (12)$$

This is also the duty for the small heat exchanger H. The duty (QkW) for the separator unit exchanger is one order of magnitude greater. For evaporation pressures of 50 bar and 60 bar, Q is given by

$$Q = (H_C - H_B)S \quad (13a)$$

$$\text{and } Q = (H_c - H_b)S \quad (13b)$$

respectively where S ($\simeq \Omega/Y$) is the solvent flow rate entering the separator.

In constructing Table 5, values of H_B, H_b, H_C, H_c, H_D, H_d were taken from Table 1.

5.2 Installed Capital Costs of the Compressor, Extraction Towers and Product Separator

The installed capital costs for stainless steel and mild steel compressors were estimated from Equations 4 and 4a (with W given by Equation 12) and are given in Table 5. Installed costs of the extraction towers and separator unit were estimated from Equations 5 and 8 or, outside the range of these equations, from the original graphs in reference 9. Before tower cost can be estimated, tower dimensions are required.

5.2.1 Sizing the extraction towers

5.2.1.1 The cycle of operations. As in the existing hop extraction process, the three towers are operated in a cyclic mode with two towers on line at any time. During each cycle, the off-line tower is being emptied of exhausted rape seed, refilled with fresh ground seed and then with liquid CO_2 whilst the other two towers are being extracted in series. The downstream tower contains partly extracted (but not exhausted) seed while the upstream tower contains a mixture of exhausted and partly extracted seed.

At the end of a cycle, the upstream tower is exhausted and is taken off line. The previous downstream tower becomes the upstream tower, and the previous "off line" tower becomes the downstream tower. After a further cycle is completed, the upstream tower has again been exhausted, and the towers are switched again.

5.2.1.2 The mass transfer model. M. Barr (5), J. Chami (1) and M. Bernardo Gil (7) have carried out bench tests in which freshly ground rape-seed was extracted into liquid carbon dioxide and also into compressed supercritical carbon dioxide. They find that the rate of mass transfer for non exhausted seeds can, for empirical purposes, conveniently be represented by a simple plug flow model with a film resistance type of term giving the equations

$$dN = k_c a(c_M - c)\,\alpha\, dh \quad (14)$$

and

$$dN = S_v dc \quad (14a)$$

where

α	=	cross section area of extraction tower	m^2
dN	=	rate of extraction of solute in element dh of bed	$kg\ s^{-1}$
c	=	solute concentration at height h up tower	$kg\ m^{-3}$

c_M	=	equilibrium solute concentration	$kg\ m^{-3}$
a	=	interfacial area per unit volume of bed	m^{-1}
$(K_c a)$	=	empirical volumetric mass transfer coefficient	s^{-1}
S_v	=	volumetric flowrate of solvent	$m^3 s^{-1}$

When about 70% of the total oil in the seed has been extracted, c_M sinks towards a low value (taken here as zero) and the seed is effectively "exhausted". Prior to this c_M is roughly constant. Under conditions of constant c_M and $(k_c a)$, Equation 14 may be integrated from the inlet of an extraction tower packed with seed (where the concentration of oil in the solvent stream is taken to be zero) to the outlet giving

$$c_T = c_M(1-\exp(-k_c a\ \tau)) \qquad (15)$$

where c_T = concentration of oil in stream leaving tower

and τ = residence time in seed bed (on empty tower basis)

$$\tau = h_B/U_s \qquad (15a)$$

where h_B = bed height

U_s = superficial solvent velocity

The numerical data used in the present calculations are taken from the work of Chami who passed carbon dioxide through crushed rape seed beds of various heights at a set of superficial velocities and measured c_T. He estimated c_M to be about 0.6 $kg\,m^{-3}$ at 25C and 76 bar from his data for very long residence times. Other data for the seed are listed in Table 2. Both Chami's c_M values and his c_T values for the larger of the beds studied were smaller than obtain in earlier work by M. Barr (who used recirculation to obtain c_M and a different seed). Comparison is made in Figure 5. In common with the other workers Chami (1) found that the empirical mass transfer coefficient $(k_c a)$ calculated from Equation 15 increased with superficial velocity. At 25°C, 70 bar the increase was from $4.4 \times 10^{-3}s^{-1}$ at $1.5 \times 10^{-4}ms^{-1}$ to $2.8 \times 10^{-2}s^{-1}$ at $9.1 \times 10^{-4}m\ s^{-1}$. $(k_c a)$ normally increases with U_s and the value $2.8 \times 10^{-2}s^{-1}$ was taken to provide a lower limit to the values to be expected at the substantially higher superficial velocities used in the present calculations.

Equation 15 will only be obeyed for a period Δt_1. At the end of this period the seed at the foot of the tower (the position of maximum extraction rate for plug-flow conditions) will have become exhausted so that c_M at this level sinks abruptly to zero. Over the ensuing period Δt_2 the zone of exhaustion extends progressively further up the seed bed finally reaching the top when the entire bed is exhausted.

5.2.1.3 Design criteria and sizing equations. Two towers in series are used in the present design rather than the single tower discussed above. Their design (below) is such that the fluid leaving the downstream tower is very nearly saturated (better than 98%) at all times. The driving force $(c_M - c)$ for mass transfer being very small the seed towards the top of this tower will remain virtually unextracted at all times. A result of this is that, when this tower is placed in the upstream position, there is virtually no discontinuity between the state of the seed at the top of this tower and the seed at the bottom of the freshly charged tower which then occupies the downstream position. This continuity persists throughout the cycle. At the end of each cycle the zone of exhausted seed extends throughout the upstream tower and is just entering the downstream tower. The latter is then placed in upstream position whereupon the zone of exhaustion at once moves up this tower, reaching the top after a period Δt_2. The cycle time is Δt_2. The bed volume V should be such that the extractable oil contained in each freshly charged tower is equal to the annual production times the ratio of interchange period Δt_2 to the annual operating period, of 8000 hours or 2.88×10^7s. i.e.

$$V = (A \times 10^3(\Delta t_2)/(2.88 \times 10^7 \rho_B F) = \alpha h_B \quad (16)$$

where ρ_B = bed density (kg m^{-3})

F = weight fraction of extractable oil in seed

Other nomenclature as before.

The rate of removal of oil from the two towers in series, Ω, is given by

$$\Omega = S_v c_{TD}$$
$$\simeq S_v c_M$$
$$= (A \times 10^3)/(2.88 \times 10^7) \quad (17)$$

where $S_v = \alpha U_s$ (18)

c_{TD} is the solute concentration (kg m^{-3}) in the stream leaving the downstream extraction tower.

From Equations (17) and (18),

$$\alpha = (A \times 10^{-3})/(2.88 \times 10^7(U_s c_M)) \quad (19)$$

From Equations (16) and (19), $\Delta t_2 = h_B \rho_B F/(U_s C_M)$ (20)

The residence time τ of solvent in each bed is given by

$$\tau = (h_B/U_s) = C_M \Delta t_2/(\rho_B F) \quad (21)$$

The interchange period Δt_2 should (1) be convenient for plant operation (2) be sufficiently large to ensure near saturation of the solvent leaving the downstream tower. 8 hours (2.88×10^4s) was selected. Even in the worst situation (immediately before interchanging the towers) when the inlet to the downstream tower consists virtually of pure CO_2 this was found to give a better than 98% approach to saturation at the outlet. (This estimate was obtained by substituting τ given by Equation (21) in Equation (15) taking $2.8 \times 10^{-2} s^{-1}$ as a lower limit for $(k_c a)$ (see section 5.2.1.2).)

5.2.1.4 Selection of optimum tower dimensions. The following procedure was used to calculate the optimum tower dimensions and solvent superficial velocity. At each of a set of values of U_s

(i) h_B was calculated from Equation (20) with $\Delta t_2 = 2.88 \times 10^4$ s and c_M, ρ_B and F as in Table 2. To allow space at the top and bottom of the bed, the tower height h_T was arbitrarily taken to be 1m greater than h_B or

$$h_T = h_B + 1 \quad (22)$$

(ii) The tower diameter d was calculated from Equation (23)

$$d = 2(\alpha/\pi)^{1/2} \quad (23)$$

with α given by Equation (19).

(iii) The installed cost of the towers (I_E) was estimated as described in section 5.2.

The above procedure was repeated for annual production rates A of 5000, 5000/3 and 75 tonnes/annum, the values of U_s, h_T and d giving the lowest I_E being selected in each case. Results are summarised in Table 3. It may be seen from this table that at the two lower production rates, capital costs show shallow minima at superficial velocities close to 1.1 cm/s. This velocity with the corresponding tower dimensions, was selected as the optimum for these production rates.

At the highest production rate, the range of possible conditions is limited by availability of data on large diameter (> 4m) pressure vessels. According to the cost correlations used the lowest capital cost at this production rate is achieved when d approaches its highest practicable value of about 4m and occurs with a superficial velocity of 2.3 cm/sec. This was selected as the best operating velocity for this production rate.

It should be stressed that the above calculations are illustrative only and caution is required in interpreting them. For example, the superficial velocities given in Table 3 are substantially higher than those at which bench tests have yet been made and additional provision for containment of the bed may be required.

TABLE 3 - Tower Heights (h_T), Diameters (d) and Installed Capital Costs (I_E) as Functions of Superficial Velocity (U_s) for Three Production Rates

(1) 500 tonnes/year (0.173 kg s^{-1})

(2) (500/3) tonnes/year (0.058 kg s^{-1})

(3) 75 tonnes/year (0.0026 kg s^{-1})

(h_T = (h_B + 1)m where h_B = bed height)

The installed capital cost given is for 3 towers.

$U_s \times 10^3$ ($m\ s^{-1}$)	h_T (m)	d(1) (m)	d(2) (m)	d(3) (m)	$I_E(1)$ (£ x 10^3)	$I_E(2)$ (£ x 10^3)	$I_E(3)$ (£ x 10^3)
0.8	2.1	6.8	3.9	0.83	*	1404	355
0.9	2.2	6.4	3.7	0.78	*	1397	347
1.0	2.3	6.1	3.5	0.74	*	1389	347
1.1	2.4	5.8	3.3	0.71	*	1388	347
1.2	2.6	5.5	3.2	0.68	*	1389	385
1.5	3.0	5.0	2.9	0.61	*	1398	407
2.0	3.6	4.3	2.5	0.53	*	1433	444
2.3	4.0	4.0	2.3	0.49	2380	1453	481
2.5	4.3	3.8	2.2	0.47	2410	1470	495
3.0	4.9	3.5	2.0	0.43	2470	1504	518

5.2.2 Installed cost of product separator. The installed cost of the inclined heat exchanger used in the product recovery unit depends on the required heat exchange area a (see Equation (8)). a in turn depends on the heat exchanger duty Q (given by Equation (13)), the overall heat transfer coefficient U and the temperature driving force ΔT

$$a = \frac{Q}{U\Delta T} \qquad (24)$$

ΔT = (boiling temperature of CO_2 at 65 bar)
- (boiling temperature of CO_2 at separator pressure)
- (boiling point elevation due to presence of solute in the boiling CO_2).

The boiling point elevation is here neglected, in view of the very low solubility of the oil in the CO_2 (Table 2) giving

$\Delta T = (T_F - T_B) = 11.2C$ for a separator pressure of 50 bar

or $\Delta T = (T_F - T_b) = 3.5C$ for a separator pressure of 60 bar.

(T_F, T_B and T_{b1} are given in Table 1).

The design of the exchanger follows conventional principles, the condensing CO_2 (i.e. the higher pressure stream) flowing within the tubes. With symbols as in the nomenclature, U is given by

$$\frac{1}{U_o} = \frac{d_o}{d_1}\frac{1}{h_i} + \frac{1}{h_o} + \frac{d_o \ln(d_o/d_i)}{2k} \qquad (25)$$

(Fouling is considered to be negligible in this application)

The shell-side heat transfer coefficient (h_o) is calculated as the sum of the convective and nucleate boiling heat transfer coefficients (h_c and h_n respectively). Both h_i (the inside tube coefficient) and h_c are given in terms of tube diameters, thermal conductivities, vapour and liquid densities and Reynolds and Prandtl numbers in the appropriate streams by standard correlations (17). h_n is given as a function of heat flux, pressure and solvent critical pressure by the Mostinski equation (18).

In this example 45% cut baffles have been used, at a spacing equal to half the shell internal diameter. The tubes are arranged on a triangular pitch with a spacing of one and quarter times the outer tube diameter (d_o). The number of tubes (N_t) has been calculated from the empirical equation

$$N_t = 0.319\ (D/d_o)^{2.142}$$

Tube diameters were selected to maximise the overall heat transfer coefficient, which was largely controlled by the condensing heat transfer coefficient (tube-side pressure drop did not exceed 0.4 kPa). The shell diameter was reduced until the shell side pressure-drop (which was calculated assuming liquid flow only along the shell) become unacceptably large. The greatest allowable drop was taken as about one bar. Calculations were carried out for plant throughputs of 5000 (5000/3) and 75 tonnes/year and evaporation pressures of 50 and 60 bar. Typical final values for the heat transfer coefficients and exchanger dimensions for the two higher flow rates with evaporator pressure of 50 bar are given in Table 4. The corresponding installed capital costs (section 5.2) are shown in Table 5.

TABLE 4 - Optimum Inclined Heat Exchanger Dimensions for Evaporation Pressure of 50 bar

Annual Throughput (tonnes)	Exchanger Duty (MW)	Required Heat Transfer Area (m^2)	N_t	L (m)	Tube i.d. $\times 10^3$ (m)	Shell i.d. (m)	Tube Side ΔP (kPa)	Shell Side ΔP (kPa)
5000	29	2200	2650	10	2.18	1.5	0.4	111
(5000/3)	9.6	740	900	10	2.18	0.9	0.4	103

(U_o was calculated to be about 1.18 kW $m^{-2}K^{-1}$ in each case)

5.3 Estimation of Separation Cost

Electricity costs (given by Equation 11) and the installed capital costs of the three major items (the compressor, the separator and the extraction towers) are assembled in Table 5. The costs given for the extraction towers and separator are for stainless steel units.

Costs of both stainless steel and mild steel compressors are given.

TABLE 5 - Energy, Capital and Separation Costs (excluding labour costs) for two Evaporation Pressures and three Scales of Operation

(Values in brackets are for a mild steel compressor)

Annual Output A (tonnes)	Ideal Power Consn. W(kW)	Installed Capital Costs (£ x 10^3) Compressor	Separator	Extraction towers	Total	Charge on capital and other costs* (£/year x 10^3)	Separation cost (excluding labour) (£/tonne)
Evaporation Pressure 50 bar (ΔP = 15 bar) Electricity cost (C_E) = £127/tonne product							
5000	1790	5600 (2200)	460	2380	8446 (5040)	1689 (1008)	465 (329)
5000/3	598	2000 (800)	220	1390	3610 (2410)	722 (482)	560 (416)
75	27	108 (43)	35	347	490 (425)	98 (85)	1434 (1260)
Evaporation Pressure 60 bar (ΔP = 5 bar) Electricity Cost (C_E) = £35/tonne product							
5000	495	1670 (660)	940	2380	4990 (3980)	998 (796)	235 (194)
5000/3	165	600 (240)	450	1390	2440 (2080)	488 (416)	328 (285)
75	7.4	33 (13)	63	347	442 (423)	88 (85)	1208 (1168)

* maintenance and CO_2 replacement

The heat exchanger costs given by reference (9) appear to be somewhat conservative. Nevertheless the data in Table 5, taken overall, should provide a realistic guide to the total capital costs.

The combined annual charge for depreciation and return on capital was taken as 16% of the total capital cost and that for maintenance and carbon dioxide replacement as 4%.

Discussion

The worked example for the extraction of material from a solid matrix with marginally subcritical carbon dioxide complements a similar example for the extraction of an ethanol/water solution given in an earlier paper (3). The intention of the example was to illustrate the way in which a proposed near-critical extraction process can be costed from bench data and to highlight aspects in the design which are important in dictating the energy consumption and overall economics of such a process.

No attempt was made to select a type of extraction or operating conditions which would give an economically sound process. The selection of better operating conditions is discussed below.

Several conclusions may be drawn from an examination of Table 5.

(1) The energy consumption (and in many cases the separation cost also) can be significantly reduced by lowering the pressure drop ΔP used for evaporation. The effect is most marked for the high throughput plant with stainless steel compressor for which the separation cost is halved by reducing ΔP from 15 bar to 5 bar. However greater control problems would probably arise in the latter case.

(2) There is a marked overall economy of scale. For a pressure drop ΔP of 5 bar and a stainless steel compressor, the separation cost per tonne of product is about five times as great for the 75 tonne/annum plant as it is for the 5000 tonne/annum plant.

(3) The compressor cost, unlike the costs of the separator and extraction towers, does not show economy of scale to any marked degree, being approximately proportional to the capacity of the plant. The relative contribution made by this unit to the total capital cost increases with the size of the plant and for the largest throughput plant the use of a stainless steel, rather than the cheaper mild steel, compressor increases the separation cost significantly. Unfortunately stainless steel is required for this application since rape seed contains about 10 weight % moisture. This will be partially extracted and some will be present in the stream passing through the compressor, presenting a corrosion problem if mild steel is used (the equilibrium loading of water in saturated liquid CO_2 at 25C is similar to that of natural oil (16)).

(4) The lowest extraction cost listed in Table 5 is about £200/tonne which is very similar to the market price of bulk rape seed oil. As suspected at the outset therefore the extraction of rape-seed with saturated liquid CO_2 at 25C is unlikely to be economically attractive. If CO_2 extraction is to be contemplated for this extraction, different conditions are required. As stressed in Section 2.4, the solubility of solute in the near critical solvent plays a very important role in determining both energy and capital cost and factors which increase this will be beneficial. Increase in the extraction pressure at 25C is one possibility (Figure 5). Higher solubilities may also be achieved by operating with supercritical CO_2, though at the expense of using higher temperatures. (Complete miscibility between rape-oil and compressed CO_2 has been reported at 100C and 750 bar (22)).

Near-critical extraction is not a separation method of universal applicability but it does deserve careful consideration in some situations, for example as a step within an existing high pressure process or as a replacement for existing solvent extraction processes, particularly where non polar or slightly polar solvents are at present used. The near-critical process has the advantage that solvent removal from the product is virtually complete and not expensive in energy. This aspect may become particularly important in the food industry where permitted residual solvent levels are liable to re-examination.

The costing of alternative near-critical extraction process in cases such as these has been the subject of this paper.

Nomenclature

A	annual output of product	tonnes/annum
C_E	electrical cost per tonne of product	£(UK) $tonne^{-1}$
D	heat exchanger shell diameter	m
E_c	energy consumption per kg of product	$kJ\ kg^{-1}$
F	weight fraction of extractable oil in seed	
H	specific enthalpy (subscripts refer to points on T/S diagram in Figure 3)	$kJ\ kg^{-1}$
I	installed capital cost (subscripts CC, CR, ET and E refer to the compressor, separator unit, a single extraction tower and three extraction towers respectively)	£(UK)
L	length of heat exchanger tube	m
N_t	number of heat exchanger tubes	
P	absolute operating pressure	bar
Q	heat duty of separator heat exchanger	kW
S	entropy	$kJ\ kg^{-1}\ K^{-1}$
S	mass flow rate of solvent entering separator unit	$kg\ s^{-1}$
S_v	volumetric flow rate of solvent	$m^3 s^{-1}$
T	absolute temperature (subscripts refer to points on T/S diagram (Figure 3))	K
U_o	overall heat transfer coefficient based on outside area	$kW\ m^{-2}K^{-1}$
U_s	superficial solvent velocity	$m\ s^{-1}$
V^G	specific volume of saturated solvent vapour	m^3kg^{-1}
V^L	specific volume of saturated liquid	m^3kg^{-1}
W	ideal power consumption	kW
Y	weight ratio of solute to solvent at entry to separator	$kg\ kg^{-1}$
a	(in product $k_c a$) interfacial area per unit volume if bed	m^{-1}
a	heat exchanger area	m^2
c	solute concentration (subscript M denotes equilibrium value)	$kg\ m^{-3}$

c_T	solute concentration in stream leaving specified tower	$kg\ m^{-3}$
c_{TD}	solute concentration in stream leaving downstream tower	$kg\ m^{-3}$
d	vessel diameter	m
dN	rate of extraction of solute in element dh of bed	$kg\ s^{-1}$
d_o	outside diameter of heat exchanger tube	m
d_i	inside tube diameter	m
h	distance from inlet to extraction vessel	m
h_B	total bed height	m
h_T	height of extraction vessel	m
$(k_c a)$	volumetric mass transfer coefficient	s^{-1}
k	thermal conductivity of tube walls of heat exchanger	$kW\ m^{-1}K^{-1}$
t	temperature	C
ΔH	Specific enthalpy difference across compressor	$kJ\ kg^{-1}$
ΔH_s	increase in specific enthalpy of solvent on passage through shell side of separator heat exchanger, - (given by (H_C-H_B) for 50 bar or (H_c-H_b) for 60 bar evaporation)	$kJ\ kg^{-1}$
ΔH_v	enthalpy of vaporization of solvent	$J\ kg^{-1}$
ΔP	pressure difference between condensing and evaporating streams in product recovery unit	bar
ΔP_I	increase in pure solvent saturation pressure required to produce increase ΔT_I in boiling temperature	bar
ΔT	temperature difference between condensing and evaporating streams in product recovery unit	K
ΔT_I	elevation in solvent boiling point due to presence of solute	K
Δt_2	interchange period	s
Ω	product flow rate	$kg\ s^{-1}$
α	cross section area of extraction tower	m^2
ε	compressor energy consumption per kg solvent flow	$kJ\ kg^{-1}$
η	compressor efficiency	
ρ	bed density	$kg\ m^{-3}$
σ_π	weight ratio of solvent to solute in product stream leaving separator	$kg\ kg^{-1}$
τ	"empty tower" residence time	s

References

1. King, M.B., Bott, T.R. and Chami, J.H., 1987, Chapter 22 in 'Separations for Biotechnology' (Verral, M.S. and Hudson, M.J. Eds.) Ellis Horwood.

2. Nelson, W.L., 1958, Petroleum Refinery Engineering, p. 362, McGraw Hill, 4th Edn.

3. King, M.B., Boyes, A.P., Bott, T.R. and Mubarak, A.D., 1987, I.Chem.E. Symposium Series No. 103, 361.

4. Gardner, D.S., 1982, Chem. Ind. (London), 402.

5. King, M.B., Bott, T.R., Barr, M.J. and Mahmud, R.S., 1987, Separation Science and Technology, 1103.

6. Gahrs, H.J., 1984, Bunsenges Phys. Chem., 88, 894.

7. Bernardo Gil, M.G., King, M.B. and Bott, T.R., 1988, Proceedings of the International Symposium on Supercritical Fluids, Nice, France, October, Vol. 2, p. 651 (Published by Institut Polytechnique de Lorraine, ISBN2 - 905267-13-5).

8. Happel, J. and Jordan, D.G., 1975, "Chemical Process Economics", 2nd Ed. Marcel Dekker, p. 231.

9. Ulrich, G.D., 1984, "A Guide to Chemical Engineering Process Design and Economics", Wiley, pp. 295 and 300.

10. Peters, M.S. and Timmerhales, K.D., 1981, "Plant Design and Economics for Chemical Engineers", 3rd Edn., McGraw Hill, p. 574.

11. Angus, S., Armstrong, B. and de Reuck, K.M. (Eds.), 1976, "Carbon Dioxide. International Thermodynamic Tables of the Fluid State No. 3", International Union of Pure and Applied Chemistry.

12. Ergun, S., 1952, Chem. Eng. Prog., 48, No. 2, 89.

13. Brogle, H., 1982, Chem. Ind. (London), 385.

14. Fin, M.J., Bower, C.J. and Hughes, R.D., 1983, Fluid Phase Equilibria, 10, 327.

15. Coenen, H. and Kriegel, E., 1984, Ger. Chem. Eng., 7, 335.

16. Francis, A.W., 1954, 58, 1099.

17. Coulson, J.M., Richardson, J.F. and Sinnott, R.K., 1983, "Chemical Engineering", Volume 6, Chapter 12, Pergamon.

18. Mostinski, I.L., 1963, Brit. Chem. Eng., 8, 850.

19. Shishikura, A., Fujimoto, K.Kaneda, T. Arai, K. and Saito, S., 1986, Agric. Biol. Chem., 50, 1209.

20. Ikushima, Y., Hatakeda, K. Saito, N., Ito, S., Asano, T. and Cioto, T., 1988, Proceedings of the International Symposium on Supercritical Fluids, Nice, France, October, Vol. 2, 735.

21. Schaeffer, S.T., Zalkow, L.M., Teja, S.J., 1989, Ind. Eng. Chem. Res., 28, 1017.

22. Klein, T. and Schulz, S., 1989, Ind. Eng. Chem. Res., 28, 1973.

23. Quirin, K.W., 1982, Fette Seifen Anst., 84, 460.

24. de Fillipi, R.P. and Vivian, J.E., 1981, UK patent 2,059,787.

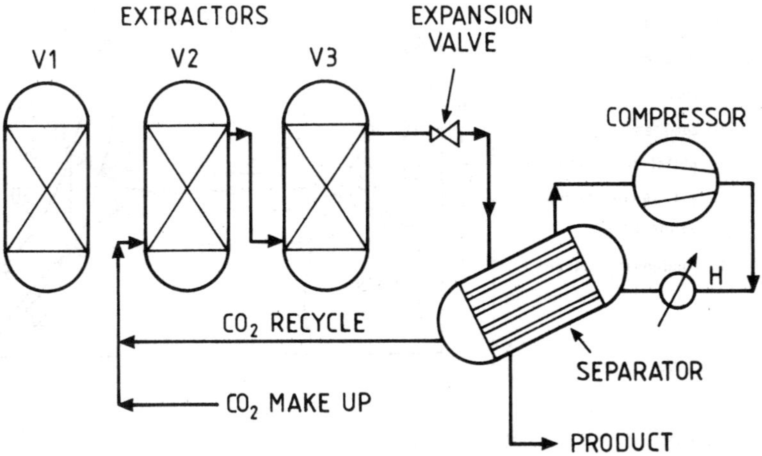

Figure 1. Layout of process for semi-continuous extraction of material of low volatility from bed of solids using liquid CO_2.

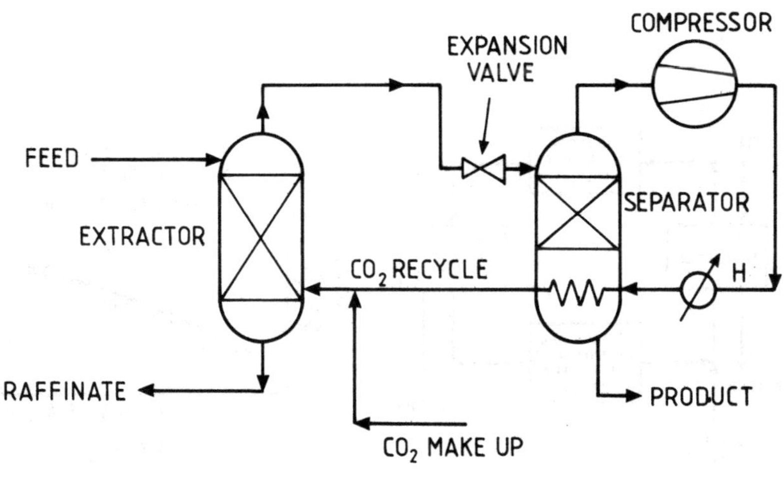

Figure 2. Simplified layout of process for extracting a liquid stream with liquid CO_2.

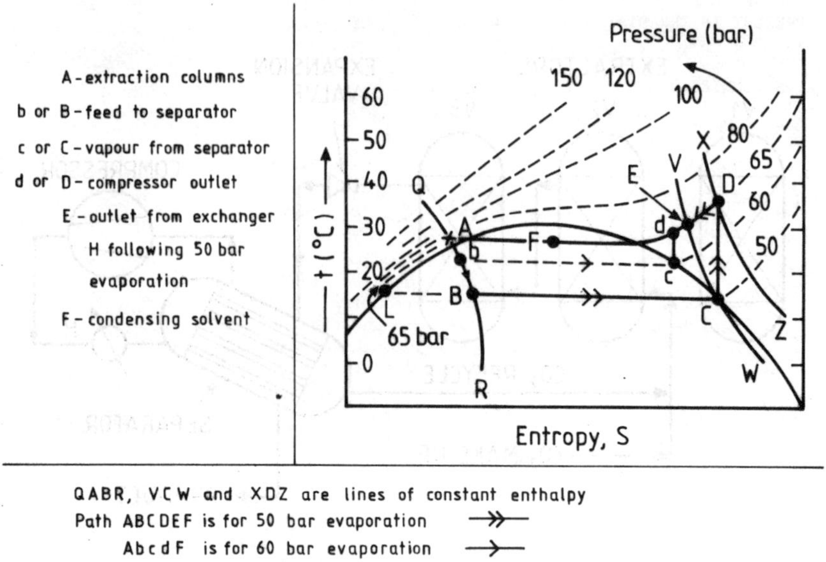

Figure 3. State of solvent in processes shown in Figs. 1 and 2. (Extraction is with marginally subcritical CO_2 at 25C, 65 bar)

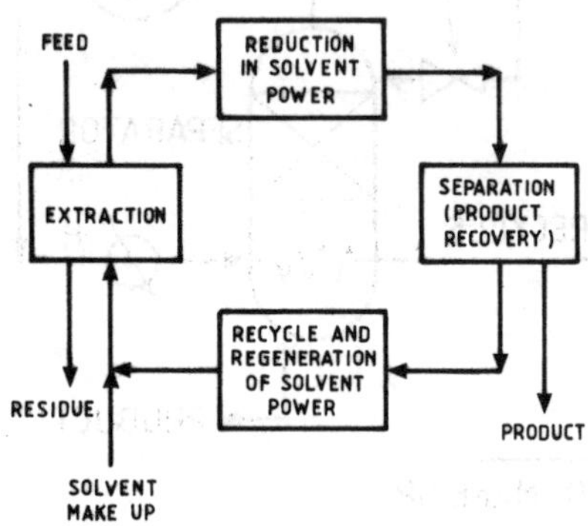

Figure 4. Generalised schematic of supercritical extraction process.

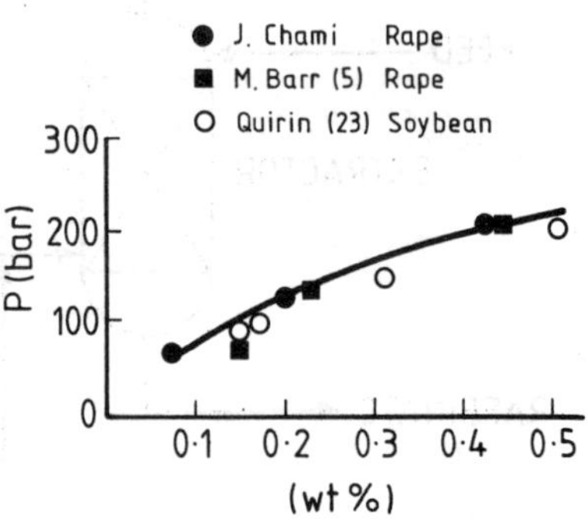

Figure 5. Equilibrium solubilities of rape and soybean oil in compressed CO_2 at 25C.

THE DEVELOPMENT OF A FLOWSHEET TO SEPARATE URANIUM AND PLUTONIUM PRESENT IN IRRADIATED OXIDE FUEL

I S DENNISS and C PHILLIPS. R&DD, BNFL SELLAFIELD

The solvent extraction and redox behaviour of technetium was shown to interfere with the original U/Pu separation flowsheet by causing excessive consumption of the hydrazine used to stabilise the process. Various methods of reducing the consumption of hydrazine were then examined and the adoption of a solvent continuous uranium/plutonium separation column was shown to be the best option. Extensive work was then carried out to develop and prove the new process.

INTRODUCTION

Magnox fuel has been reprocessed at the Sellafield site of British Nuclear Fuels for twenty-five years using the Purex process. Recently a new plant (THORP) to reprocess oxide fuel from Advanced Gas-cooled and Light Water Reactors has been developed using a range of pilot plant facilities.

The major innovations in this new plant are the adoption of pulsed columns to process the plutonium-bearing stream (to aid criticality control) and the use, wherever possible, of chemicals which leave no residues in the waste streams (salt-free reagents).

The flowsheet for the first solvent extraction cycle, where plutonium and uranium are separated first from fission products and then from each other, is shown in Figure 1. The active feed containing U, Pu and fission products is fed to a combined extract and scrub column (HA) where U, Pu and some fission products are extracted into solvent (30% tributyl phosphate in odourless kerosene). The loaded solvent is then passed to a heated scrub column (HS) where further fission products are removed from it.

Plutonium is then converted to the inextractable three valent form in the 1BX column where it transfers into the aqueous phase which is passed into the 1BS column. In the 1BS, residual uranium is extracted from the plutonium stream and recycled to 1BX.

Uranium leaves 1BX in the solvent phase which is sent to the 1BXX "plutonium-polish" mixer settler, any plutonium removed in this contactor is recycled to 1BX.

The most significant use of a salt-free reagent is in the uranium/plutonium separation contactor (1BX) in the first solvent extraction cycle (Figure 1). Here uranous nitrate (U(IV)) is added to reduce Pu(IV) to inextractable Pu(III) together with a stabiliser, hydrazine, to react with any nitrous acid generated and hence prevent oxidation of U(IV) and Pu(III).

Early trials of the flowsheet in miniature scale (1/6000) alpha and fully active pilot plants showed that the fission product technetium behaved in an unexpected manner. In the active feed, technetium is present as Tc(VII) which forms extractable complexes with uranium and the fission product zirconium. The latter causes the complete extraction of Tc in the HA(E) column and it is carried forward to the IBX column where it is reduced to inextractable Tc(IV) which follows the plutonium product. It also takes part in a complex series of redox reactions in the IB system in which it catalyses the destruction of the hydrazine stabiliser by nitrate ions. This behaviour was reported by Garraway at Extraction '84 (1).

The initial design of the process included aqueous continuous (solvent dispersed) 1BX and 1BS columns operating at 35°C. In the case of 1BX this was to maximise the aqueous residence time since the rate of reduction of Pu(IV) to Pu(III) in this phase was believed to control the degree of separation of Pu from U (DF Pu) achievable across the contactor. The combination of the aqueous continuous column with a long aqueous residence time and the technetium hydrazine reaction therefore led to the requirement for a relatively large concentration of hydrazine to stabilise the process which was largely consumed by a parasitic reaction.

The amount of hydrazine required could be safely added to the process but it was judged to be undesirable that such a large proportion of the essential stabiliser should be consumed in the process. It was also thought that the large concentrations of hydrazine which might be present in the process when technetium was absent (eg at start up and shut down) could pose operational problems.

Methods to limit the extent of the technetium hydrazine reaction in the 1BX and 1BS columns were therefore examined. This paper describes some of the work carried out to investigate the alternatives and develop and demonstrate the chosen method.

METHODS TO LIMIT THE TECHNETIUM-HYDRAZINE REACTION

The alternatives available are to reduce the amount of technetium getting through to the 1BX column or to reduce the rate of reaction or time for it to occur within the 1BX/1BS system.

Reagents such as thiocyanate, thiourea and ascorbic acid which reduce technetium and render it inextractable were rejected because they can complex or reduce uranium and plutonium and will also form undesirable breakdown products. Zirconium is complexed by fluoride, sulphate and oxalate (1,2) which would prevent the co-extraction of technetium and zirconium. However potential plant corrosion problems and the complexing of plutonium with oxalate led to the rejection of these methods.

Introduction of further solvent extraction steps to backwash technetium prior to the 1BX column was also considered but rejected because of the increased mechanical complexity introduced and the production of an extra waste stream.

The next option examined was to reduce the rate of reaction in 1BX and 1BS by chilling the system to 20°C. This gave about a three-fold reduction in the rate of hydrazine destruction in miniature alpha and fully active pulsed columns. At the same time however, DFs for Pu and Tc across 1BX were also reduced by factors of 20-50. The resulting increased amounts of Pu and Tc in the solvent product from the IBX column (1BXSP) were backwashed in the "plutonium polish" mixer settler (1BXX see Figure 1) and so did not adversely affect the rest of the process, but the increase in the amounts of these species reaching the mixer settler and being recycled to the IBX column was judged to be undesirable.

Parallel trials using a uranium-only test rig which simulates the HA cycle at full size and which has been described elsewhere (3,4) also indicated difficulties with the operational control and mass transfer efficiency of the aqueous continuous 1BX column at both standard and lower temperatures.

In summary, these trials investigated the mass transfer efficiency and throughput capacity of the HA cycle pulsed columns. Attention was paid to the operational stability of this series of linked pulsed columns, particularly the behaviour of the dispersed phase hold-ups (DPHUs) (3). The most relevant findings are summarised below.

It had long been realised that axial mixing in the 1BX aqueous phase was likely to be significant because of the unfavourable phase ratio and hence low aqueous superficial velocity and that limitation of this axial mixing would be necessary to obtain satisfactory plutonium DFs. Trials of the 1BX column at 35°C on the full size rig showed that control of the dispersed phase hold-up (DPHU) and interface position was also difficult because of the high solvent to aqueous and hence dispersed to continuous phase ratio.

It was found that the 1BX column would operate stably in either high (more than 50%) or low (less than 30%) DPHU regimes, depending on throughput and pulse energy (Figure 2). Operation in, or passage through the transition region between the two regimes, however, led to wide and erratic variations in DPHU which made column control very difficult and also increased the axial mixing. A similar phenomenon has been reported by other workers (5). If the transition region was to be avoided, it would clearly be necessary to operate the plant in the low DPHU regime. This, however, limited the pulse energy that could be applied, and hence the liquor throughput achievable to values that gave a rather limited margin above the plant requirement. Reducing the 1BX temperature to 20°C exacerbated these difficulties by lowering the pulse energy and throughput boundaries at which the transition region was entered. These difficulties could be resolved by increasing the diameter of the column, but this would of course increase the aqueous residence time, thus defeating the object of reducing the temperature.

Reducing the temperature of the 1BX column was therefore rejected as unsatisfactory, but the measure was adopted for the 1BS column where there were no deleterious consequences. This left reducing the time for reaction in 1BX as the only remaining option available.

Reducing aqueous phase residence time

The most satisfactory means of reducing aqueous residence time is to operate the 1BX with solvent as the continuous phase. Initial scoping trials on the full size test rig showed that the aqueous DPHU would be approximately 10% giving a low aqueous residence time. There are other advantages too, the superficial velocities of both phases in the column are high, leading to a minimisation of axial mixing and it was found that the column operated very stably indeed with none of the DPHU transitions and interface control difficulties experienced with aqueous continuous operation.

To counterbalance these advantages there was initially uncertainty as to whether acceptable decontamination factors could be achieved for plutonium and technetium across 1BX and whether hydrazine, which is inextractable, could adequately scavenge nitrous acid from the bulk solvent phase and hence stabilise the system.

The results of the initial trials in the 1/6000 scale alpha active test rig were however encouraging; the flowsheet was chemically stable and good decontamination factors were obtained. A full development programme was therefore carried out.This involved experiments in miniature alpha and fully active pulsed columns, intermediate alpha active pulsed columns and a purpose built, full scale uranium active 1BX operated in the solvent continuous mode. The major findings of this programme are summarised below.

DEVELOPMENT OF THE SOLVENT CONTINUOUS 1BX PULSED COLUMN

The consumption of hydrazine and stability of the system was examined in a series of alpha active batch and miniature counter-current trials. These showed that the time-dependent reaction between technetium and hydrazine occurs in both aqueous and solvent continuous columns but there is an additional consumption independent of technetium and time in the latter case. The same reaction occurs in mixer settler U(IV) flowsheets (6) and is caused by the oxidation of U(IV) in solvent catalysed by Pu(IV). The chemical stability of the 1BX was subsequently confirmed in the miniature fully active pilot plant and in the intermediate (1/200) scale alpha active pulsed column facility at Dounreay. In the latter case it was shown that varying the dispersed (ie aqueous) phase hold-up had no effect on the chemical stability of 1BX. This demonstrated that hydrazine can adequately stabilise the solvent continuous process. The flowsheet has now been operated for over 1000 hours and has been shown to be chemically stable with a remarkable tolerance to minor feed variations.

The most detailed information on the mass transfer efficiency of the 1BX column was obtained from the 1/200 scale rig at Dounreay where column profiles and the dispersed phase hold-up (DPHU) and drop sizes could be measured. The pulsed columns in this rig were 5.3m high (packed height) and 25.4mm internal diameter. Each was constructed from 7x0.75m glass sections with full sampling facilities installed between each section. Each section was packed with stainless steel nozzle plates (3.2mm diameter holes, 23% free area) suspended on a tie rod at 5 cm intervals. The rig was operated over a range of flowrates and pulse amplitudes to investigate the effect of changes in throughput and pulse energy on the mass transfer efficiency of 1BX.

Measurement of average DPHU showed a characteristic dependence on the pulse amplitude (Figure 3) over a range of hydrodynamic conditions. Drop size measurements over the same range of conditions showed no significant variation of specific surface area (Table 1) and so a constant value was used in the analysis of mass transfer data.

Under active conditions the rig was operated with two sets of flow rates chosen to duplicate the specific throughputs to be used in the THORP 1BX column. A linear relationship was found between log DF Pu and DPHU for each set of flow rates and if an allowance is made for residence time the relationship is independent of flow rates. This is illustrated in Figure 4 which assumes that mass transfer is the rate controlling factor in a pseudo first order process; the logarithm of the DF is therefore plotted against the product of the inter-facial area and solvent residence time in the column. The results have been analysed in terms of simplified mass transfer or chemical rate controlled processes (see Appendix 1) and found to give better agreement with the former assumption (Table 2). This is supported by the further finding from the Dounreay rig of a relationship between the DF Pu and acid profile across 1BX (see Figure 5). This figures shows that the acid present in the solvent feed to 1BX is backwashed more rapidly, giving a steeper acid profile, as the DF for Pu across the column increases.

A purpose-built solvent continuous 1BX column which replicates the THORP design has been installed and operated in the full scale uranium active test rig at Sellafield. This column is taller than that used at Dounreay (ca 10 and 5.3 m packed heights respectively) and so, since both are operated at the same specific throughputs, the residence times are longer in the full scale column. Trials using this column have confirmed the hydrodynamic stability and ease of control shown previously.

Preliminary investigations of the DPHU values achievable in the full scale 1BX column have shown similar dependence on pulse amplitude and throughput to that observed in the Dounreay column. Similar average DPHUs can also be obtained (Figure 6) and this, together with the fact that there is a longer residence time in the full scale column indicates that satisfactory DFs will be obtained for Pu and Tc in the THORP column.

This prediction is based on a comparison of the hydrodynamic characteristics of the Dounreay and full scale columns and this is confirmed by the acid profiles measured in the full scale column which are "steeper" than those measured at Dounreay (Figure 7). Acid is backwashed more efficiently in the full scale column and so it should also backwash plutonium and technetium at least as efficiently as the Dounreay 1BX column.

CONCLUSIONS

This work has demonstrated the value of carrying out flowsheet development work in parallel in small and full scale facilities to test the chemical and engineering design of the process. It has also shown the importance of fully active trials to demonstrate unexpected effects caused by the mix of components in irradiated fuel.

The mechanisms and rates of reactions involving the hydrazine stabiliser in the uranium-plutonium separation contactor have been fully investigated as described in reference (1) and this has allowed the hydrazine feed requirement for the reprocessing plant to be established.

. Study of the behaviour of plutonium in relation to the hydrodynamic condition of the uranium-plutonium separation column has indicated strongly that separation performance is governed by plutonium mass transfer. Comparison of the results from the 1/200 scale alpha active and full scale uranium active pulsed columns has demonstrated that the THORP 1BX column will achieve the degree of separation of uranium and plutonium specified in the flowsheet.

ACKNOWLEDGEMENTS

The contributions made to this work by Dr J Garraway, Mr B N Prest and the members of the Solvent Extraction Group, R&DD, BNFL Sellafield and Dr P J Thompson, Mrs K Cartwright, Mr S Castling and the experimental and analytical staff of UKAEA Dounreay are gratefully acknowledged.

REFERENCES

1. J GARRAWAY, The behaviour of technetium in a nuclear fuel reprocessing plant. Extraction '84. ICE Symposium Series Number 88.

2. E VIALARD and M. GERMAIN, Technetium behaviour control in the Purex process. ISEC, 1986. International Solvent Extraction Conference, Munich. Volume 1, pp 137 - 142.

3. C PHILLIPS, Chemistry and Industry, 2 September 1985, pp 577 - 582.

4. C PHILLIPS, Proceedings International Solvent Extraction Conference, Toulouse, France, 1985.

5. H SCHMIDT, Sep Sci and Tech, 18 (14 and 15), 1595. 1983.

6. B BOULLIS and P BARON, Modelling of uranium/plutonium splitting in Purex process. Extraction '87. ICE Symposium Series Number 103.

TABLE 1 Measurement of representative 1BX drop-size photographs

DPHU (%)	DF Pu	Total drop area (mm^2)			Total drop volume (mm^3)			Specific surface area (mm^{-1})		
		Top	Middle	Bottom	Top	Middle	Bottom	Top	Middle	Bottom
6.9	54	675	1569	1902	217	563	662	3.11	2.79	2.87
9.0	290	1397	1387	1433	483	372	416	2.89	3.73	3.45
11	1062	1089	1852	1911	318	632	580	3.42	2.93	3.29

Average of 11 sets 3.21 3.27 3.39

TABLE 2 Apparent first order rate constants calculated assuming mass transfer and chemical control

	Mass transfer	Chemical
Dounreay S cont	5×10^{-3} cm s^{-1}	2.6×10^{-2} s^{-1} (35°C)
Sellafield S cont	1.9×10^{-3} cm s^{-1}	4×10^{-2} s^{-1} (35°C)
Sellafield Aq cont	1.2×10^{-3} cm s^{-1}	2×10^{-3} s^{-1} (35°C)
Literature	5×10^{-4} -5×10^{-3} cm s^{-1}	0.12×10^{-2} s^{-1} (25°C)

APPENDIX 1

Kinetic Analysis of Mass Transfer Assuming Simple Pseudo First Order Equation

$$\text{Ln DF} = kT$$

1 Chemical Reduction Control

k = reaction rate constant

T = aqueous residence time = DPHU/aqueous flow

$$\therefore \text{Ln DF} = \frac{k.\text{DPHU}}{\text{aqueous flow}}$$

2 Mass Transfer Control

$$\text{Ln DF} = k^1 \frac{A\,T_s}{V_s}$$

k^1 = mass transfer rate constant

A = interfacial area = DPHUxS (specific surface area)

V_s = volume of solvent

T_s = solvent residence time = V_s/solvent flow

$$\therefore \text{Ln DF} = \frac{k^1.\text{DPHUxS}}{V_s} \times \frac{V_s}{\text{solvent flow}}$$

$$= k^1 \frac{\text{DPHUxS}}{\text{solvent flow}}$$

FIG 1 THORP HA CYCLE AND PULSED COLUMN TEST RIG

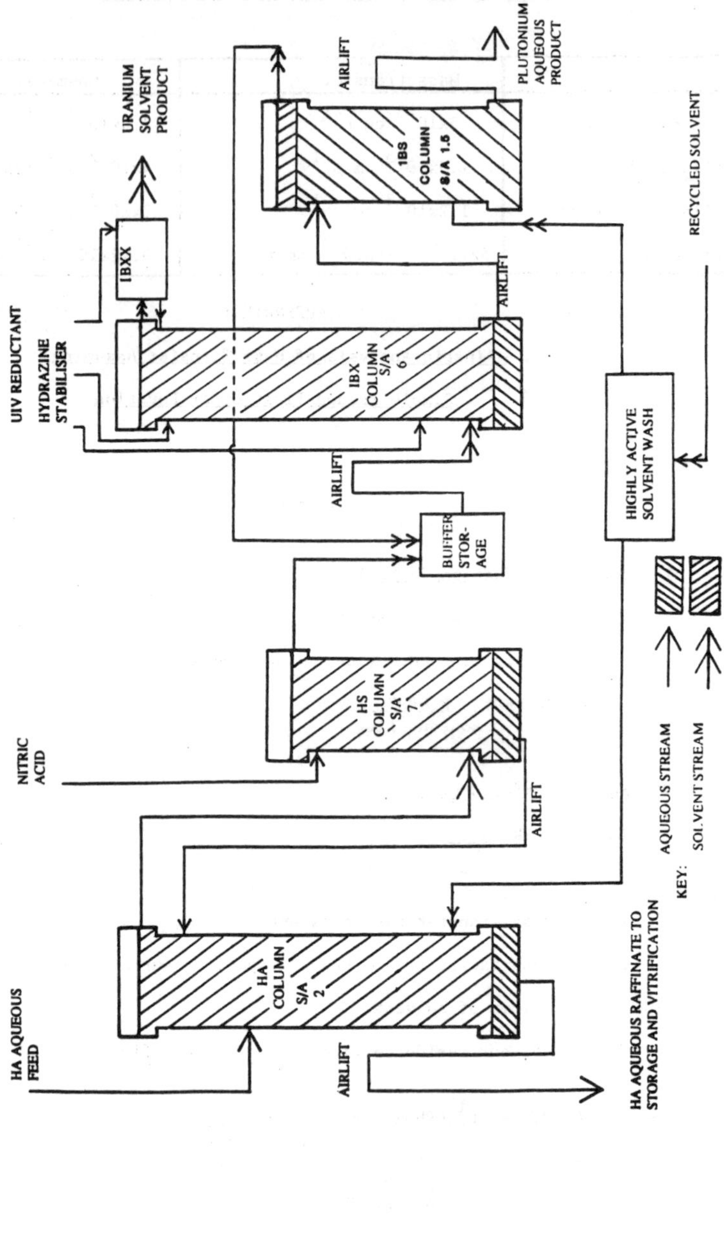

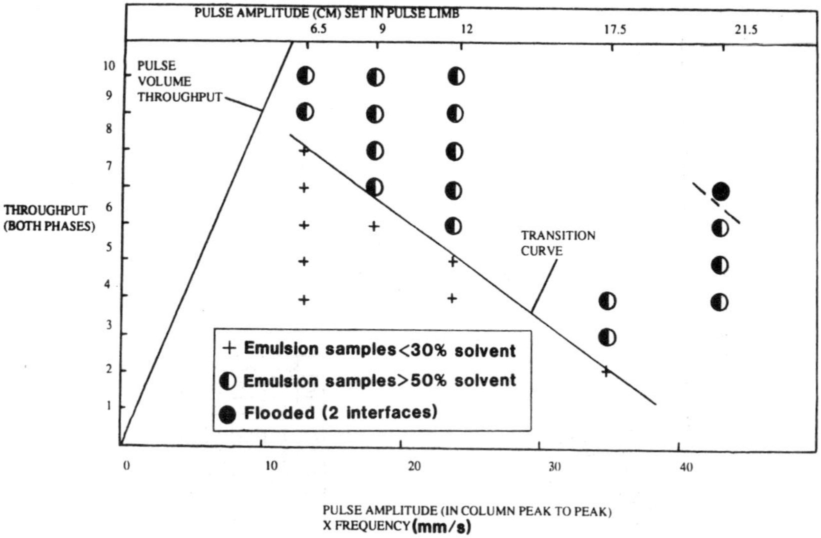

FIG 2 TYPICAL IBX COLUMN AQUEOUS CONTINUOUS FLOODING DIAGRAM

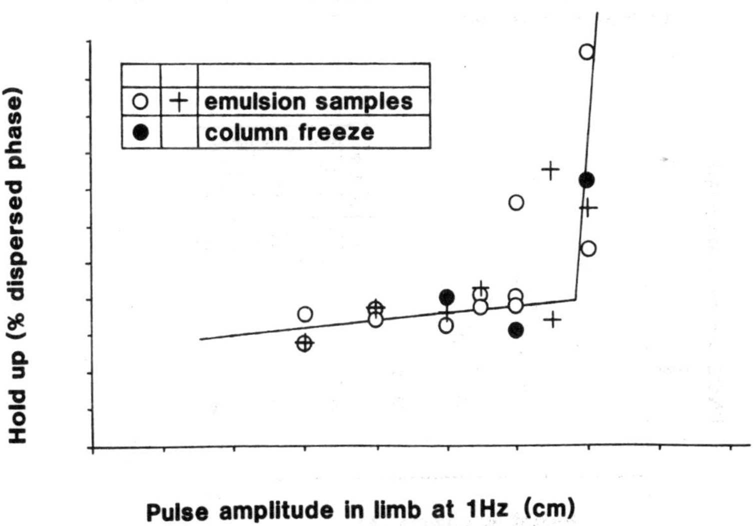

FIG 3 Dounreay solvent continuous 1BX column hydrodynamics in uranic operation 35°c,1 Hz operation, specific throughput 34 m/hr

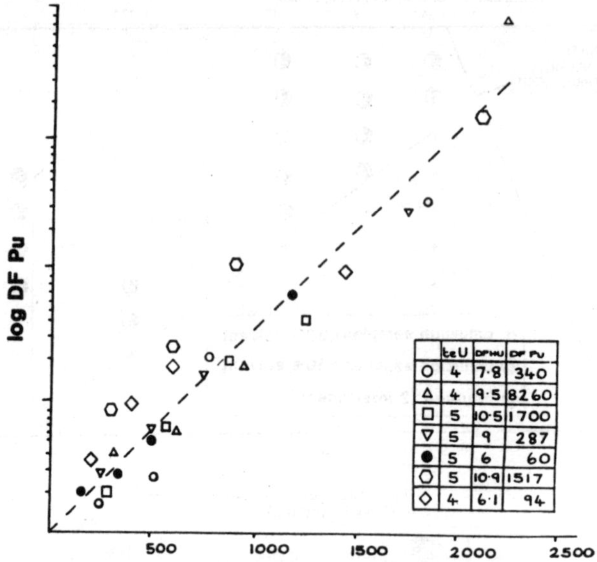

FIG 4 Dounreay solvent continuous 1BX column mass transfer performance

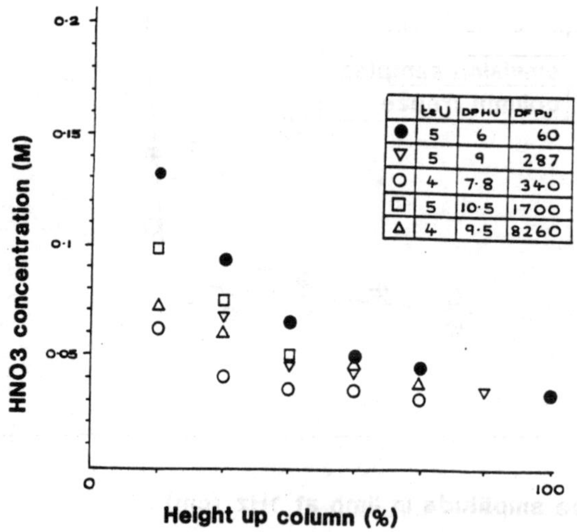

FIG 5 Solvent acid profiles in the solvent continuous Dounreay 1BX column

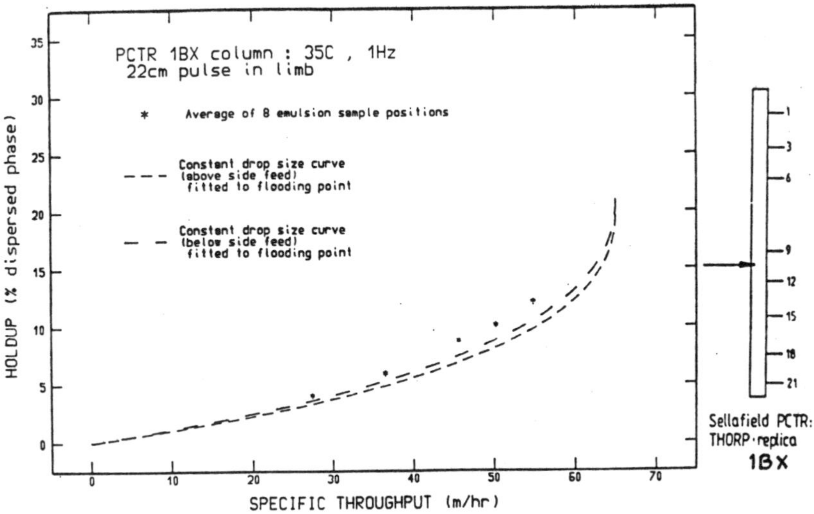

FIG 6 THORP replica solvent-continuous 1BX Hydrodynamics in uranic operation

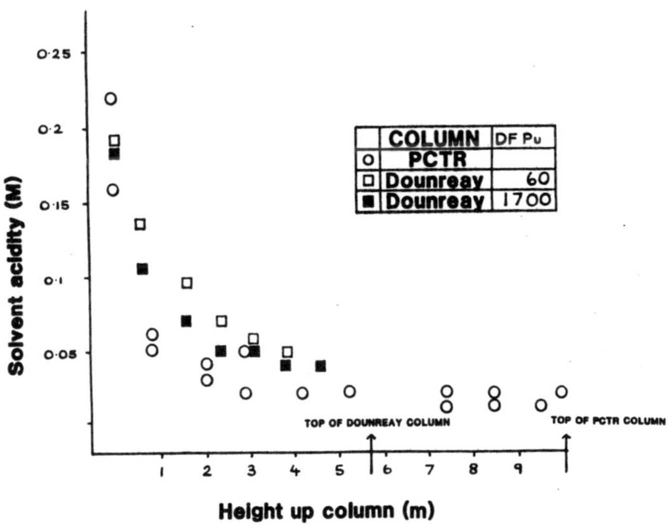

FIG 7 Solvent acidity profiles in the Dounreay and Full Scale 1BX columns

U/Pu SEPARATION IN AN ORGANIC CONTINUOUSLY OPERATED PULSED COLUMN

J. Schön*, H. Schmieder*, B. Kanellakopulos*, H.-J. Bleyl*, D. Ertel*, P. Dressler*

In the MINKA pulsed column system of the Hot Chemistry Institute U-Pu separation was studied in an organic continuously operated BX column. It is reported here about experience gathered in operation of this separation system, about the influence of technetium on the U-Pu separation efficiency, and about the extration behavior of the elements Pu, U and Tc under these conditions of separation.

INTRODUCTION

Within the framework of flowsheet studies on the PUREX process U-Pu separation in a continuously operated organic BX partitioning column was investigated in the MINKA miniature pulsed column facility, installed at the Hot Chemistry Institute of the Karlsruhe Nuclear Research Center, as a potential alternative (Philipps (1) to electrochemical U-Pu separation (Petrich (2), (3), Schmieder et al (4), Baumgärtner and Schmieder (5).

It is reported here about experience gathered with the organic continuous partitioning system, with external U IV feed and about the influence exerted by technetium on the separation efficiency. Moreover, hydrazine destruction with technetium as catalyzer is discussed. The influence of technetium on Pu reduction was studied and by the example of the concentration profiles plotted of the organic and aqueous phases of the BX separation column the extraction behavior of U, Pu and Tc is indicated.

---

*Institut für Heiße Chemie, Kernforschungszentrum Karlsruhe GmbH D-7500 Karlsruhe, Postfach 3640

EXPERIMENTAL

Interconnected Columns.

U-Pu separation was studied under the conditions prevailing in the first extraction cycle of the PUREX process. In MINKA four pulsed columns were interconnected for this purpose to form a closed extraction cycle. These interconnected columns are represented in Fig. 1 and the details of the columns are listed in Table 1.

In Fig. 1 P11, P15, etc. denote the metering pumps for the aqueous and organic process solutions, respectively, and B16, B36, etc. denote the transfer vessels. Mass flowmeters (MM) or volume flowmeters (VM) and balances, respectively, are used for flow measurements.

The MINKA pulsed columns have an extraction length of 3.5 m. At 0.5 m spacing each they are equipped with sample collectors for the two phases and with thermocouples.

Moreover, additional process solutions can be fed into the columns at these positions. The feed points are marked in Fig. 1 by the numbers in the columns, e.g. for K_1 feed into the column at levels 1, 7, 8, etc.

The A and BX columns are operated as organic columns in the continuous mode whereas the BS and C columns are operated as aqueous columns in a continuous mode. A detailed description of the extraction cycle with the related service units is given in (Schön et al (6), (7), (17).

Flowsheet.

The composition of the organic feed solution of the BX column (Fig. 1: BXF) and the operating parameters of the BX separation column have been adapted to the usual design of the PUREX process for LWR fuel; see Table 2.

In order to be able to study the influence of technetium on U-Pu separation under unfavorable process conditions preseparation of technetium in the extraction and scrubbing parts of the A column was dispensed with. Therefore, technetium was continuously metered into the organic product of the A column during the experiment (Fig. 1: HAP) as an aqueous Tc VII stock solution. This resulted in a Tc concentration of 60 - 70 mg Tc/l in BXF. On the basis of earlier results on Pu accumulation (6), (7) and on U-Pu separation (Schmieder et al. (8) the feed point of the BXX-U IV solution was fixed at a level 1 m above the BXF inlet (Fig. 1: BX column, level 3).

The experiments were performed in four steps. The individual steps differ by the routing of the organic scrubbing stream of the BS column (Fig. 1, BSU, either return into the BXF stream or into the C column) and by the interconnection of the BX and BS columns. The respective variants of the flowsheet and the related experimental periods have been compiled in Table 3.

RESULTS AND DISCUSSION

BX Separation Efficiency.

The separation efficiency of the organic continuously operated BX column is evaluated by the decontamination factor (DF) (9).

In Fig. 2 the development versus time of DF has been plotted together with the plutonium content in the organic discharge of the BX column (BXU).

A surprising finding is the clear drop in plutonium concentration in the BXU stream to values of about 0.2 - 0.5 mg Pu/l in the course of transition from single column operation (part 1 of the experiment) to BX-BS interconntected column operation (part 2 of the experiment beginning in the 34th hour).

In the third part of the experiment technetium was metered in (beginning in the 64th hour). The presence of technetium in the partitoning column was visible from considerable gas evolution as early as two hours after the onset of dosage which continued until the end of the experiment. However, the U-Pu separation efficiency was adversely influenced only at the end of 12 hours after the onset of technetium feed by a marked rise in Pu concentration in the BXU stream. As the experiment proceeded, the plutonium content in BXU showed the tendency to fall and disconnection of the BS column (part 4 of the experiment, 104th - 113th hour) had no influence on this tendency. Approximately starting in the 96th hour plutonium concentrations established in BXU which varied between 0.7 and 1.4 mg/l corresponding to a DF of 900 to 400. Hence, these values are clearly above the Pu contents of BXU which were analyzed in the absence of technetium (parts 1 + 2 of the experiment, up to the 64th hour).

Degradation of Hydrazine.

With the onset of technetium feed the nitric acid concentration in the aqueous phases of the BX and BS columns decreases due to oxidation of hydrazine with nitric acid as the catalyzer (Garraway and Wilson (10), Akopov et al (11), Boukis (12), Wilson and Garraway (13), Garraway (14), Zelverte (15). The development versus time of the hydrazine concentration in the BXP and BSP solutions has been represented in Fig. 3 for this MINKA experiment.

In the absence of technetium both solutions have almost the same hydrazine concentrations (Fig. 3, 0-64th hour).

By contrast, the values differ clearly in the presence of technetium (beginning in the 64th hour). After only 8 min. residence time of the aqueous phase in the BX column approximately 50% of the hydrazine feed has degraded. (Fig. 3: symbol: square). The subsequent residence time of about one hour duration in the bottom decanter of the BX column causes the hydrazine content in the aqueous phase to decrease to 0.06 to 0.08 M/l N_2H_4 (Fig. 3: symbol: circle). In the BS column connected in series the residence time of the aqueous phase is again about 5 hours which leads to a further degradation of hydrazine down to values of $\leq 10^{-2}$ M/l (Fig. 3: symbol: cross). The low hydrazine concentration in the aqueous product solution of the BS column is still sufficient to stabilize the reduced Pu (17).

After the BS column has been disconnected in the 4th part of the experiment (104th - 113th hour) the hydrazine concentration in the aqueous phase of this column is further reduced. Approximately 8 hours after disconnection the hydrazine content in the aqeuous phase of the BS column (level 3 at 1 m column height) has decreased to $\leq 5 \times 10^{-5}$ M/l N_2H_4 and in the solution which, in addition to ≤ 1 g U/l and 0.41 g Tc/l, contains 3.6 g Pu/l there is a clear reoxidation of plutonium.

During catalytic destruction of hydrazine by nitric acid ammonium (NH_4^+) and hydrazoic acid (HN_3) are formed as the products (10 - 14). The development versus time of these reaction products in the aqueous phases of the BX and BS columns of MINKA was shown in (17). Under the conditions prevailing in this MINKA experiment an NH_4^+ concentration of $8 - 12 \times 10^{-2}$ M/l established in the presence of technetium in BXP and BSP, respectively. The corresponding values for HN_3 were measured to be 2×10^{-2} M/l in BXP and $7 - 10 \times 10^{-3}$ M/l in BSP. The organic BSU solution even contained $3 - 4 \times 10^{-2}$ M/l HN_3.

Under the conditions prevailing in the MINKA experiment specific rates of formation for the hydrazine daughter products of 0.21 -0.31 mole HN_3 per mole degraded N_2H_4 and for NH_4^+ values of 0.29 to 0.45, respectively, can be calculated from the amount of hydrazine consumed and from the measured NH_4^+ and HN_3 values, respectively.

In the absence of technetium about 2 mole N_2H_4 were consumed per mole of plutonium. In the presence of technetium the amount of N_2H_4 consumed in the BS column increased to 7.6 mole N_2H_4 per Mole Pu and in the BX column to 8.2 mole N_2H_4 per mole Pu.

Supplementing the analyses of the process samples in a simulated aqueous solution the extraction behavior of the reduced Tc and Pu species was investigated with 30% TBP/kerosene, and the degradation of hydrazine was measured in addition (Dressler et al. (16). The results have been plotted in Fig. 4. The concentrations of hydrazine in the aqueous phases and of the Pu and Tc species, respectively, in the organic phase have been plotted versus the contact time of the solution until extraction and hydrazine analysis, respectively. The concentrations have been plotted as relative values which are related to the composition of the initial solution = 100%. The simulated solution contained at the beginning of the investigation 1.4 M HNO_3 and 0.116 M N_2H_4 as well as 0.7 g Pu/l, 4.7 g U IV/l, 1.4 g U VI/l, and 0.33 g Tc/l.

It is clearly evident from the figure that the hydrazine concentration diminishes with the length of holdup of the sample. The initial degradation of hydrazine down to concentrations of ~ 10^{-2} M/l can be described formally by a half-life $t_{1/2}$ = 120 min. This value supplements the measurements made on real samples taken from MINKA (17) in which for comparable concentrations of U, Pu, HNO_3 and N_2H_4 half-lives of $t_{1/2}$ = 80 min were determined for 0.43 g Tc/l and $t_{1/2}$ = 160 min for 0.24 g Tc/l, respectively.

Another phenomenon worth noticing is the reduction behavior of plutonium and technetium in the simulated solution.

After e.g. 1 min. holdup about 95% of plutonium and about 30% of technetium have undergone reduction. This means that the presence of U IV speeds up the reduction of technetium (13), (15), Koltunov et al. (25), Ramazano et al (26) compared with the course of the reaction in the HNO_3-N_2H_4 system (10 - 15). After about 30 min. the reduction of technetium has attained its final level. Any intermediates which might be formed during the process of technetium reduction through U IV are indicated in (15).

On the other hand, the reduction of plutonium in the presence of technetium seems to be incomplete which is evident from a comparison with literature values on plutonium reduction through U IV in the HNO_3-N_2H_4 system Biddle et al. (18), Menchero (19), Schlea (20), McKay (21). We suppose that this change in the reduction behavior of plutonium in the presence of technetium is the principal cause of the poorer U-Pu separation efficiency in the BX column.

It is further evident from Fig. 4 that with diminishing N_2H_4 content in the solution reoxidation starts of the reduced plutonium and technetium species, respectively. In the simulated solution fast reoxidation started at hydrazine contents below 2 x 10^{-2} M/l N_2H_4.

Concentration Profiles.

In Fig. 5 the concentrations have been plotted of HNO_3, N_2H_4, U, Pu and Tc in the aqueous and organic phases, respectively, of the BX column. A BX system free of technetium (left side of Fig. 5) is compared with a system containing technetium (right side of Fig. 5).

On both sides of Fig. 5 the concentrations in mole/l and g/l, respectively, have been plotted versus the column height. "Da" stands for the bottom aqueous decanter and "Do" for the top organic decanter of the BX column. The positions 1 -8 correspond to the column levels: level 1 = feed at 0 m column height, level 3 = BXX feed at 1 m column height, and level 8 = BXS feed at 3.5 m height.

The profile on the left side of Fig. 5 was taken in the absence of technetium after 32 hours duration of the experiment. The profile on the right side of the figure was measured 25 hours after the onset of Tc feed (=89th hour). This profile is reproduced here as an example and it shows the characteristic changes in the concentration plots in the presence of technetium.

In the top part of Fig. 5 the N_2H_4 and HNO_3 concentrations have been plotted. In the presence of technetium N_2H_4 undergoes strong degradation already in the top part of the column. In the bottom part of the column further reduction of N_2H_4 concentration is inhibited by the BXX feed.

The concentration profiles of uranium have been plotted in the figures in the middle. For the system free of technetium the U IV concentration profile of the aqueous phase has been entered in addition. It was determined by spectrophotometry (Schmieder et al (22). In the presence of technetium this method of analysis cannot be applied in routine operation because the reduced technetium, in the range of 500 - 700 nm, gives rise to an enhanced background not constant with the time (12) so that the measurement of U IV is disturbed at 648 nm. Spectrophotometric measurements of technetium in the presence of U VI, U IV and Pu III are given (15), Borovinskij et al (24).

In the bottom part of Fig. 5 the concentration profiles have been plotted for plutonium and technetium. The comparison of the profiles shows that in the presence of technetium the plutonium concentration of the organic phase decreases at a much less steeper rate than in the profile free of technetium. In the presence of technetium about 1 m more column length is needed to attain the same plutonium concentrations as compared with the system free of technetium.

All the organic and aqueous Tc concentration profiles, respectively, have a distinctly flatter course than the respective plutonium profiles. Moreover, all organic technetium profiles exhibit in the zone of BXX feed a more or less pronounced jump in concentration. Moreover, the Tc profiles measured in the aqueous phase take a very flat course in the bottom part of the BX column (0 - 1.5 m height) so that nearly no changes in concentration occur from one sampling point to the next.

A possible cause of this unusual appearance of the profile, compared to the plutonium profile, could be the slow reduction of Tc VII by U IV (15), (25) (see Tc plot in Fig. 4) and/or internal reoxidations (see jump in Tc concentration in the organic phase; Fig. 5).

As already shown earlier (8) the separation efficiency of the organic continuously operated BX column is greatly influenced by the U IV concentration in both phases of the column. Of the U IV present an equivalent fraction is consumed for the reduction of Pu IV. In the presence of technetium another fraction is obviously consumed to reduce Tc VII so that the overall conditions of reduction in the BX column are adversely affected and contribute to a deterioration of the U-Pu separation efficiency. Therefore, an improved preseparation of technetium in the extraction and scrubbing columns, respectively, of the first cycle of the PUREX process should clearly improve the partitioning conditions in the BX column. Multiple feed of the BXX solution at different levels of the BX column should produce the same effect.

It is planned to test the improved separation of technetium in the next campaign to be performed in MINKA within the framework of flowsheet studies on the IMPUREX process (Schmieder and Petrich (23).

Acknowledgement

The authors wish to thank the operators of MINKA and the PUTE operation team for excellent experimental work and Mr. U. Galla for support in the interconnected operation of the PUTE-MINKA-MILLI facilities. We would like to extend our thanks to the staff of the analytical group who performed quickly and precisely the analyses of the numerous process samples.

LITERATURE

1. Ch. Philipps, 1985, Summer School on Extraction, Toulouse.
2. G. Petrich et al, 1984, Extraction 84, JCE Symp. Series No. 88.
3. G. Petrich, U. Galla, H. Goldacker, H. Schmieder, 1986, Chem. Engineering Science Vol. 41, 981.
4. H. Schmieder et al, 1987, RECOD 87 Vol. 1, 393.
5. F. Baumgärtner, H. Schmieder, 1978, Radiochimica Acta 25, 191.
6. J. Schön et al, 1987, Extraction 87, J.Chem.E. Symp. Series No. 103, 103.
7. J. Schön et al, 1989, Nuclear Fuel Cycle Vol. 4, 45.
8. H. Schmieder, J. Schön, K. Haberland, K. Eiben, Jahrestagung Kerntechnik 1988, 341.
9. Reactor Handbook Vol. II (1961), 109.
10. J. Garraway, P.D. Wilson, 1984, J. of Less Common Metals 97, 191.
11. G.A. Akopov et al, 1982, Sov. Radiochemistry 24, 442.
12. N. Boukis, KfK-4362 (1988).
13. P.D. Wilson, J. Garraway, 1984, ANS Meeting on Fuel Reprocessing and Waste Management, Jackson Hole, USA, Vol. 1, 467.
14. J. Garraway, 1984, Extraction 84, J.Chem.E. Symp. Series No. 88, 47.
15. A. Zelverte, 1988, CEA-R-5443.
16. P. Dressler, B. Kanellakopulos, D. Ertel, personal communication.
17. J. Schön, H. Schmieder, B. Kanellakopulos, 1989, 6. Symp. on Separation Science and Technology, Knoxville, USA, in the press.
18. P. Biddle et al, 1966, J. Inorg. Nucl. Chem. 28, 1736.
19. Lopez Menchero, 1966, ETR 184.
20. C.S. Schlea, 1963, DP 808.
21. H.A.C. Mc Kay, 1963, AERE-R-4381.

22.H. Schmieder, E. Kuhn, W. Ochsenfeld, 1970, KfK-1306.

23.H. Schmieder, G. Petrich, Radiochimica Acta, in the press.

24.V.A. Borovinskij et al, 1981, Sov. Radiochem. 23, 127.

25.V.S. Koltunov et al, 1980, Sov. Radiochem. 22, 510.

26.L.N. Ramazano et al, 1981, Sov. Radiochem. 23, 181.

TABLE 1 - Design of MINKA columns

Column	A	BX	BS	1C
Function	extraction + scrub	separation	Pu-scrub	U-strip
Mode	organic	organic	aqueous	aqueous
Heating [°C]	25-60	25-60	40	25-60
Diameter [cm]	2.6	2.6	3.3	3.5
Height [cm]	300+50	350	350	350
Sieve plates:				
Spacing [cm]	3	3	5	5
Holes [cm]	0.15	0.15	nozzle	0.3
Pulsation:				
Amplitude [cm]	1.5	1.5	1.5	2.5
Frequency [Hz]	1.2	1.5	1.5	1.0

TABLE 2 - Flowsheet conditions of U-Pu-separation

Process stream	BXF	BXS	BXX	BSX
Relative flow rate	1.0	0.14	0.03	0.17
M HNO_3	0.07	0.1	1.0	0.01
M N_2H_4	-	0.2	0.2	-
g U/l	80	-	140*	$< 10^{-3}$
g Pu/l	0.8	-	-	$< 10^{-4}$
g Tc/l	0.08	-	-	$< 10^{-4}$
% TBP	30	-	* > 70% U IV	30

TABLE 3 - Parts of the MINKA experiment

Part	Time [h]	Mode of Operation	
1	33	BX, BS	as single columns, BXP transfer into the BS column without BSU recycling into BXF
2	31	BX + BS	columns interconnected, BXP transfer into the BS column and BSU recycling into BXF
3	39	Part 2	plus Tc
4	7	BX, BS	as single columns, no BXP transfer into the BS column and no BSU recycling into BXF

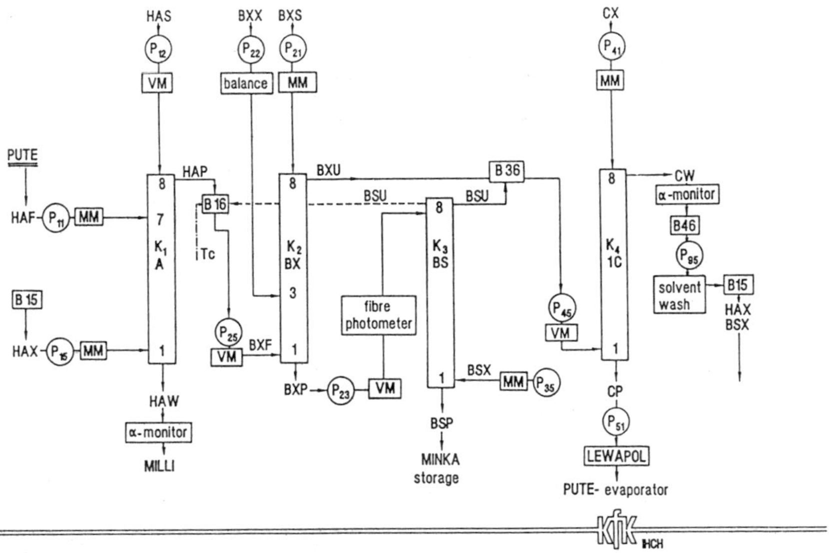

Fig. 1 MINKA

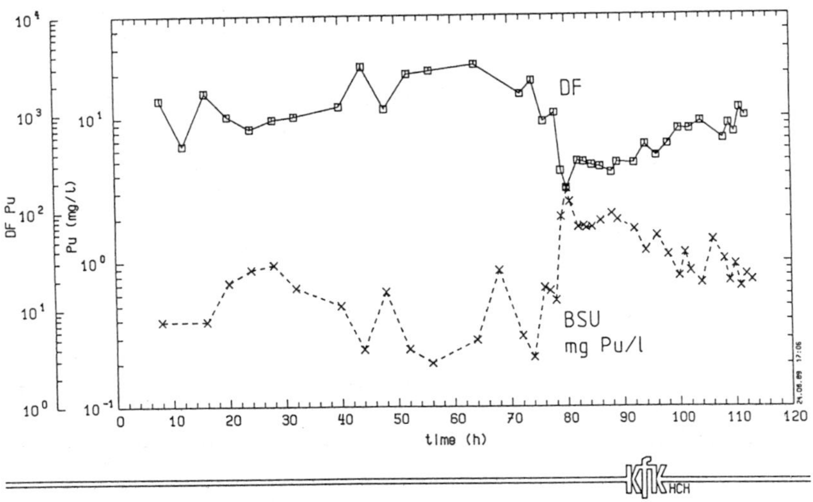

FIG.2 DFpu AND ORGANIC CONCENTRATION OF PU IN THE BXU

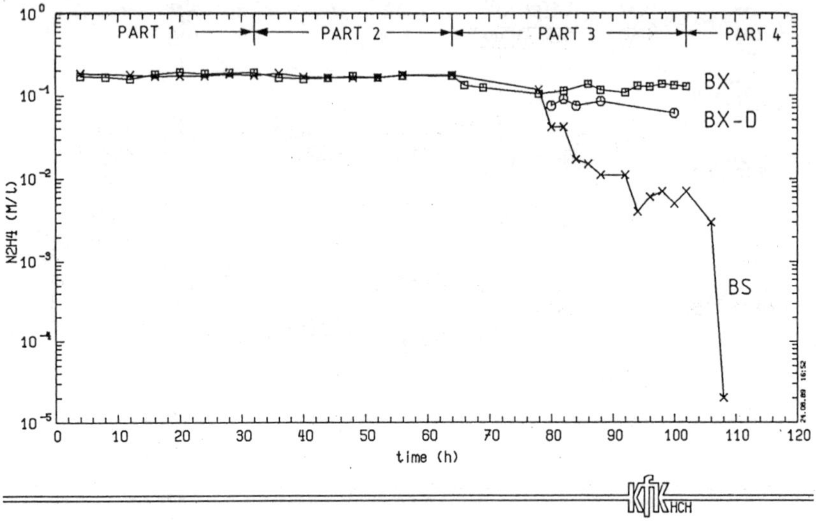

FIG.3 AQUEOUS CONCENTRATIONS OF N2H4 IN THE BX-BS COLUMNS

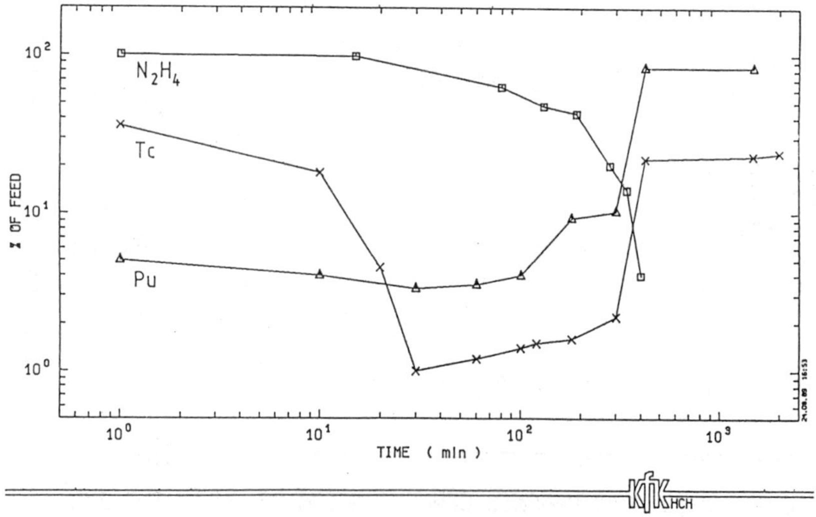

FIG.4 BEHAVIOUR OF N2H4 , Pu AND Tc

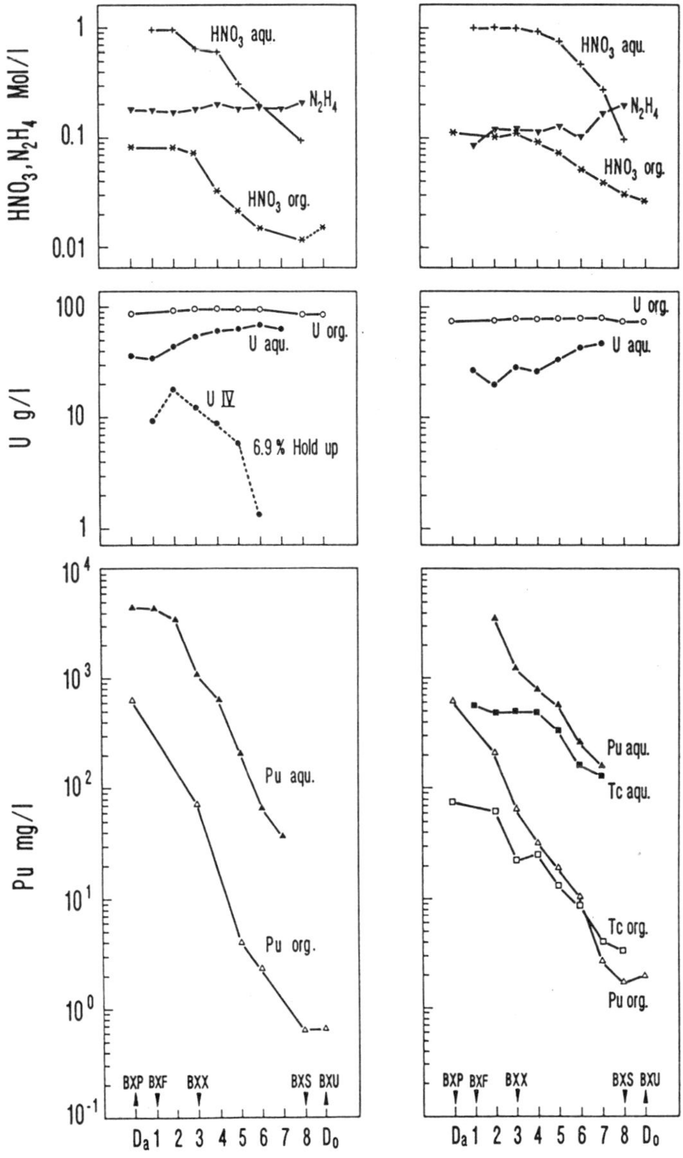

BX- CONCENTRATION PROFILES

KfK IHCH

A TECHNETIUM REJECTION FLOWSHEET

R. Baker*, J.H. Miles*, P.T. Roberts*

A single contactor unit has been designed which enables Tc to be removed from a TBP/diluent stream bearing U and Pu, by means of a 5M HNO_3 wash. A Tc waste stream is produced which is virtually free from U and Pu. The flowsheet has been tested firstly with U and Tc, then with U, Pu and Tc, and finally in a highly active facility with real PWR fuel solution. About 97% of the Tc was removed from the organic phase and U and Pu levels in the Tc waste have usually been below 0.04% of those in the product stream.

INTRODUCTION

The element technetium has many rare properties and one that is unique. It shares with promethium the distinction of being a fission product element not found in nature, and it shares with selenium, tellurium, antimony, molybdenum and perhaps germanium the distinction of being a fission product element which forms stable anions in nitric acid solution. It appears to be unique however in that this anion forms a complex with uranium which is extractable into TBP/diluent solutions.

It can be present in nitric acid solutions in up to 4 different valency states (IV, VI, VII and perhaps III); this number being rivalled only by that of plutonium (III, IV, V and VI). The 99 isotope has a long half life (2.12 x 10^5 years) and the pertechnetate anion, which is the stable form in natural oxidising media, is not absorbed on natural siliceous materials. This makes it a very important isotope in determining the dose to man from fission product material entering the biosphere in the far future (Hill (1)).

At the Extraction '84 Meeting Garraway (2) described the behaviour of technetium in a Nuclear Fuel Reprocessing Plant. Extraction of Tc in the extract and scrub stages is almost complete due to the involvement of extractable complexes with zirconium. In reductive U/Pu separation contactors on the other hand Tc is reduced to a mixture of the IV, and VI valent species, which are inextractable, but which react with hydrazine in a cyclic manner giving rise to excess consumption of this reagent.

Flowsheet Design

High concentrations of uranium in a TBP/diluent phase usually result in

*Process Services Divisiion, AEA Fuel Services, B.220, Harwell, Oxon OX11 ORA

reduced extraction of fission products (FP) because there is less "free TBP" remaining to extract them. In the case of Tc however this will not be so since

$$TcO_4^- + UO_2(NO_3)_2 2TBP \rightleftharpoons UO_2\ (TcO_4)(NO_3).2TBP + NO_3^- \quad (1)$$

uranium is required to form the extractable complex, its extraction in the absence of uranium being very small indeed. It will be clear from equation (1) however that high concentrations of nitrate ion will push this equilibrium back to the left and prevent the extraction of Tc. It is very convenient that the high concentrations of nitrate ion which prevent the extraction of Tc are ideal for the extraction of U and Pu which have maxima of extraction in the 4-8 M HNO_3 range.

If we therefore backwash an organic phase with strong HNO_3 we expect the Tc to revert to the aqueous phase, but the U and Pu to remain in the organic phase. This is carried out in stages 1 to 3 of the contactor in Figure 1. In order to prevent any trace of U and Pu leaving in this aqueous waste stream however, it is necessary to "strip" it with a low volume TBP/diluent wash, which is added to the main organic stream. (Stages 4-12 of this figure.)

High concentrations of HNO_3 in the organic phase leaving the contactor will be unwelcome in any U/Pu split contactor. All the known reducing and complexing reagents used for this purpose give poorer separation in the presence of high concentrations of HNO_3. We have therefore experimented with the use of a single stage of low volume, low concentration, acid wash to reduce the amount of acid going forward. This is stage 0 in Figure 3.

Figure 1 lists the distribution coefficients (D) and extraction factors (E) for U(VI), Pu(IV) and TcO_4^- in all sections of the rejection contactor. We note that U as U(VI) is well extracted at all points in this flowsheet and that Tc(VII), (TcO_4^-) is equally well rejected ($E < 0.2$). For the strip section we have included two values for each element; the first is for the top stage of the strip section with high U concentration in the organic phase, the second is for the bottom stage where U concentrations will be very low indeed. We see that for Pu the extraction factor for Pu(IV) at the phase volume ratio chosen, falls just below unity in the high U concentration stage. This probably occurs over only one stage and does not affect the operation of the flowsheet. In all other stages the extraction factor will be above 2, and the Pu well extracted.

We have developed this flowsheet in three stages; starting with a simple uranyl nitrate-pertechnetate-nitric acid system, and then repeating this process with plutonium present. These stages were carried out with mixer-settler equipment in a glove box. Finally, the flowsheet was tested with real PWR fuel in a fully shielded α, β, γ cell.

EXPERIMENTAL

Counter-current contacting equipment

All the flowsheet trials were carried out in miniature mixer-settlers. For the "U + Tc active" and "U + Pu + Tc active" trials these were purchased from the Ceries organisation in Lyon, France, and had been prepared by drilling in a block of Altuglas plastic (Ceries type "Pollux A"). For the fully active trials similar mixer-settlers were used which were made at Harwell Laboratory from radiation resistant materials. The block was of stainless steel and incorporated windows of CeO_2-doped radiation resistant glass*. These windows were sealed onto a gasket of peroxide cured EPDM compound** (BNFL 30H) which is also resistant to radiation and to nitric acid and to TBP/OK.

For the "U + Tc active" trial an organic feed solution was prepared by extracting uranyl nitrate crystals and then ammonium pertechnetate solution‡ into purified TBP/OK. Only the Tc rejection contactor was operated.

For the "U + Pu + Tc active" trial an extract contactor was operated (Figure 2) and the feed solution was an aqueous solution of U and Pu nitrates in HNO_3 with NH_4TcO_4 added. Zr was not added to the feed solution and it was realised that this would greatly reduce the amount of Tc extracted in a normal extract and scrub contactor. The scrub section was therefore omitted and the organic phase passed direct to the Tc rejection contactor apart from passage through 3 "dead" stages of mixer-settler to allow aqueous haze to separate out.

A low acid backwash contactor was used after the Tc rejection contactor, in order to recover Pu and U in an aqueous phase.

The Fully Active Trial

In the fully active trial a sample of 510 g of long cooled (approximately 12 years) high burn-up (30,000 MWd/te) PWR fuel was dissolved in HNO_3 with careful scrubbing of the gases evolved. The method has been described previously by Brown (3,4). The solution was filtered through 0.45μ millipore type HV Durapore filters fabricated from polyvinylidene fluoride. In this case an extract-and-scrub contactor and the Tc rejection contactor only were operated, and the aqueous wastes (highly active and Tc product) were collected in sealed cans and later disposed of in concrete. The organic product was absorbed in "Mor-Dri" which was later mixed into cement for disposal. Accurate solution flowrate control did not rely on metering pumps. The metering pump outlet was fed to a constant head cell and from there via a pre-set needle valve and electronic flow measuring equipment (Phase Separations Ltd. catalogue 900503), to the mixer-settlers.

The waste and product solutions were pumped out of the ends of the mixer settlers through a 3-way valve to the waste drum. Samples were taken from the valve at regular intervals. The working area was divided into "red" and "white" highly active and low active areas and different manipulator jaws used in each area. Control of the spread of contamination was excellent. Equipment removed from the cell after use averaged only 10 $\mu Svhr^{-1}$ whereas the levels in the cell were about 2 $Svhr^{-1}$.

The B.220.29 facility at Harwell allows short periods of very highly active work to be done at comparatively reasonable cost. Hire charges for the facility in this work are cheap when compared with the very high charges for the construction of fixed pilot plants, which use heavy shielding and occupy a whole laboratory suite.

Analysis. Analytical samples were taken with an evacuated tube from hypodermic needles mounted in the top of the settlers. Analyses for Tc were

---

* Manufactured by Pilkington Ophthalmic and Special Glasses, St. Asaph, Clwyd.

**Manufactured by Rapra Technology Ltd., Shawbury, Shrewsbury, Salop.

‡Amersham International catalogue no. TCS 1.

carried out by liquid scintillation in the first run but by the colorimetric technique of Howard and Weber (5) based on thiocyanate complexes for the other two runs. Uranium analyses were carried out by a fluorimetric procedure while acid determinations on organic phases used a potentiometric "gran plot" method, described by Rossotti and Rossotti (6).

Pu analyses were carried out by liquid scintillation techniques using Packard Insta-gel scintillant and a Beckmann LS 3800 instrument. This does not distinguish between α particles arising from Pu and those from Am but it was expected that the latter would be absent in samples from the last two contactors. In the case of the "U + Pu + Tc active" run this proved to be incorrect and it was shown by α-spectrometry that 93% of the α-count in the Tc waste was due to Am. The Pu concentration figures were corrected accordingly.

RESULTS AND DISCUSSION

The "U + Tc active" trial

Only one contactor was operated: it being fed with a synthetic TBP/OK phase. Only 2% of the Tc remained in the emergent organic phase and only 0.0006% of U came out in the aqueous "Tc waste" stream. The only unexpected feature was the hold up of some of the ^{238}U daughter isotopes (^{234}Th and ^{234}Pa) in the small quantities of organic phase left in the Tc rejection contactor in the stages below the feed point, after shut-down. This process is, if anything, an advantage because the associated γ activity is removed, for a time, from the uranium product. The isotopes will decay away in the organic phases from the contactor with a half-life of 24.1 days, and of course re-appear in the U product on the same time-scale.

The "U + Tc + Pu active" trial

This trial was carried out with 3 contactors according to the flowsheet of Figure 2. The first contactor was an extract contactor using an aqueous phase feed, and the third a backwash contactor to recover U and Pu in the aqueous phase. In order to encourage extraction of Tc the aqueous feed solution was made 1M in HNO_3 instead of the usual 3M so as to reduce interference by NO_3^- ion in the formation of the U-Tc complex (see Equation 1). The scrub section was dispensed with since in the absence of zirconium this would tend to backwash the Tc, but a short dead leg was incorporated in the hope that this would allow any haze of the aqueous feed solution, to drop out of suspension in the organic phase.

In this trial the proportions of Tc found in the emerging organic phase and in the final aqueous phase from the third contactor were below the limit of detection (estimated at 0.26% of the Tc going into the rejection contactor). The α-count profile across the Tc rejection contactor (Figure 4) fell rapidly across the stages 4-7 where, as explained above, Pu might not be expected to extract well, because of the high U loading. In stages 7-12 the α-count reaches a steady value and this has been shown by α-spectrometry to be 93% Am. Clearly the "dead stages" at the top of the extract column have not been sufficient to prevent some of the aqueous phase feed solution, with Am from the Pu stock, reaching the Tc rejection contactor. The use of the extract contactor without a scrub section was however vindicated by achieving 54% extraction of the Tc. The Pu remaining in the Tc waste is only 0.036% of that in the final aqueous product which is an adequate removal.

Acid concentration. The performance of the low acid scrub was disappointing. Analysis of the organic phase entering the Tc rejection contactor showed a level of only 0.16M HNO_3 which is not unexpected in view of the low (1M) aqueous acid concentration with which it has just been in contact in the extract contactor. The emerging organic phase however averaged 0.36M.

A McCabe-Thiele diagram suggests that the organic phase leaving the 5M acid in stage 1 should be 0.4M in HNO_3 but that this should fall back to 0.23M by operation of the low acid scrub. It is probable that the top stage of this mixer-settler was giving poor disentrainment of phases at the high phase volume ratio used.

The Fully Active Trial

The fully active trial was carried out in two sections. The first used 5M HNO_3 for the Tc rejection contactor but did not employ the low acid scrub (stage 0). In the second section the low acid scrub was included but the strong acid scrub was increased to 5.5M so as to keep the acid concentration in stages 1-11 at 5M.

The concentration of Tc at the inlet to the Tc rejection contactor was 49.5 mg ℓ^{-1} and this fell to an average of 2.23 mg ℓ^{-1} (4.5%) in the first section of the run and 1.53 mg ℓ^{-1} (3.1%) in the second. The better performance in the second half reflects the additional stage in use although it is not optimised for Tc rejection.

The U concentration profile across the Tc rejection contactor fell gently from 7 g ℓ^{-1} at stage 0 to about 4 g ℓ^{-1} at stage 5 and then very rapidly to 0.3 mg ℓ^{-1} at stage 10 (see Table 1).

Pu concentrations. The Pu concentration showed very little change in stages 0-8 and then fell by a modest factor of 3 (Table 1). (All these concentrations were determined on the aqueous phases.)

The failure to decontaminate from Pu is probably due to the absence of a conditioning step to convert Pu to the IV state before the run. The conditioning step was omitted because of the likelihood that it would lead to a spread of contamination. The boiling of the solution in the dissolver will have produced a high proportion of Pu(VI) and the concern expressed above about the extractability of Pu(IV) in the stage below the feed plate, will apply much more to Pu(VI), which is less extractable.

It may be noted however that all these aqueous phase concentrations are considerably below the organic phase product (840 mg ℓ^{-1}) so that a modest decontamination factor of about 20 does apply.

Acid concentration. The acid concentrations in the organic stream entering the rejection contactor averaged 0.33M. This rose to 0.50M in the emerging stream when no low acid scrub was applied but fell back to 0.39M in the second section of the run when the low acid scrub was in use. It is not expected that the increase from 0.33 to 0.39M in this stream entering a U/Pu partition contactor, will have a serious effect on its performance.

CONCLUSIONS

Further attention needs to be paid to the conditioning of the feed solution and behaviour of the low acid scrub. Nevertheless the work provides a very satisfactory demonstration of a simple process for removing Tc from highly active organic streams containing U and Pu.

ACKNOWLEDGEMENTS

This work was supported financially by the UK Departments of Energy and of the Environment, British Nuclear Fuels plc and the Central Electricity Generating Board. It received essential support from many different groups at Harwell Laboratory. The radiation resistant mixer-settlers were made in the workshops of the Engineering Division with help from the external organisations noted in the text. The highly active operations relied heavily on the B.220.29 Operations Team in the Chemistry Division and α-spectrum and uranium analyses, were carried out by members of Actinide Analysis Group in the same Division. Mr. C. Mason of our own Group carried out acidity determinations on organic phases.

REFERENCES

1. Hill, M.D., 1979, NRPB Report R86.

2. Garraway, J., 1984, Inst. Chem. Eng. Symp. Ser. 88.

3. Brown, P.E., 1965, Atomic Energy Document AERE-R 4933.

4. Brown, P.E., 1971, Atomic Energy Document AERE-R 5357.

5. Howard, O.H., and Weber, C.W., 1962, Anal. Chem. 34, 530.

6. Rossotti, F.J.C., and Rossotti, H., 1965, J. Chem. Educ. 42, 375.

TABLE 1 - U and Pu Concentration Profiles across the Tc rejection Contactor Aqueous phase Samples in the fully active Trial

Stage No.	0	1	2	3	4	5	6	7	8	9	10	11
U conc. (g ℓ^{-1})	6.62	-	4.51	4.00	3.72	4.39	1.07	0.89	0.012*	7×10^{-4}*	3×10^{-4}*	-
Approx** Pu conc. (mg ℓ^{-1})	156	-	95	109	201	163	136	115	122	-	68.0	42.1

* Results obtained by fluorimetry. Remaining U figures by coulometry.

**Liquid scintillation counting results. Converted on the assumption that 7.4×10^6 Bq ≡ 1 mg Pu

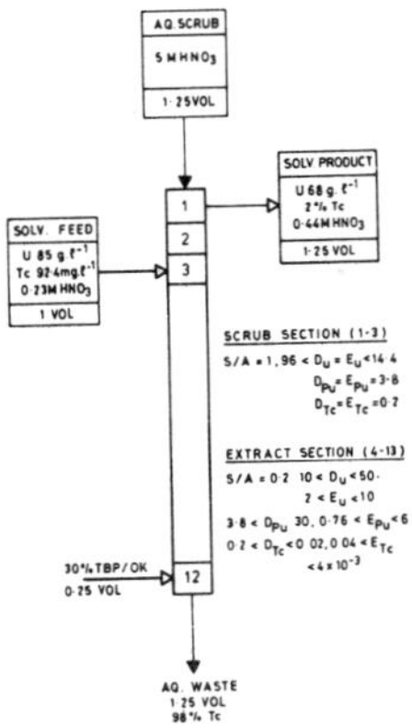

Figure 1. U + Tc active trial

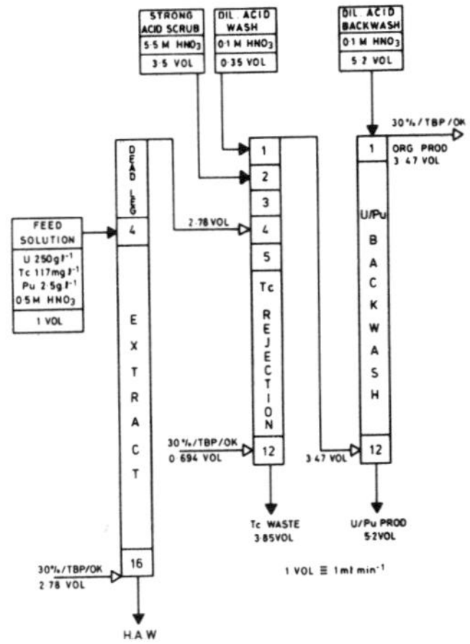

Figure 2. U + Tc + Pu active trial

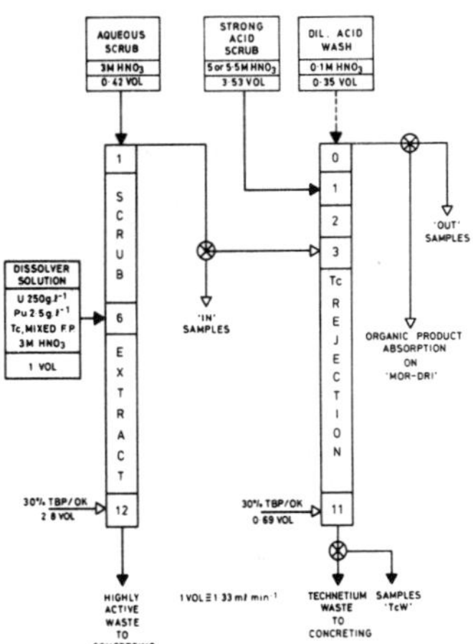

Figure 3. Fully active trial

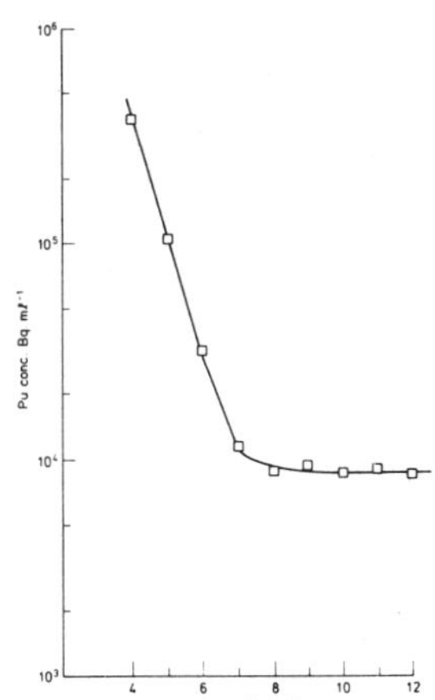

Figure 4. Pu concentration profile
Rejection contactor.
U + Tc + Pu active trial

SOME PLUTONIUM IV POLYMERS PROPERTIES IN PUREX PROCESS

H. SCOAZEC ; J.Y. PASQUIOU ; M. GERMAIN

Metabolism of plutonium polymers in fuel reprocessing using Purex process with TBP as solvent, and its practical consequence in real operation conditions are examined. Precipitation with dibutylphosphoric acid, a solvent degradation product, occurs both in extraction and stripping units when polymers are present.

INTRODUCTION

The increase of irradiation of nuclear fuels, or of their ^{238}Pu content, causes severe damage to the tributylphosphate solvent TBP used in the Purex process. The main degradation products, alkylphosphoric and phosphoric acids form more or less soluble complexes with plutonium, and the risk of solid formation in the process is enhanced. In the context of the general study of these compounds, we have examined the peculiar case of Plutonium (IV) polymers. A special attention must be paid to this plutonium species because their low extractability by TBP and their precipitation with different anions may result in loss of Plutonium, plugging of process equipment and can cause critical problems as it has been often reported (1, 2). Some studies have been developped, and results sometimes patented to avoid the presence of Pu(IV) polymers (1, 2, 3).

We have, ourselves, observed polymerised pluotnium precipitates in pilot process experiments when the feed solution of simulated fuel contained a few percent of Plutonium in the polymerised state.

Polymers of Pu(IV) have been extensively studied since the forties, but to our knowledge, metabolism of these compounds in a Purex process using TBP as solvent was never published. It is then our purpose to characterise the extractive properties of Pu(IV) polymers, written as Pu_n, and specially in presence of dibutylphosphoric acid HDBP, and their behaviour in multi-stage counter current extraction cycle (extraction, stripping). A third Plutonium purification cycle was chosen to perform this study, because of the high Plutonium concentration in this cycle.

EXPERIMENTAL

Pu(IV) polymers are prepared by adding a mother nitric solution of Pu(IV) (50 $g.l^{-1}$, HNO_3 2.5 M) to hot water (75° C), sot that the final concentrations are $Pu_n \simeq 1$ $g.l^{-1}$, HNO_3 = 0.05 N. In these conditions an emerald green solution is obtained immediately. After an hour, a spectrophotometric determination shows that more than 95 % of Pluto-nium is polymerised. Further adjustements of the nitric acid concentration can be made without depolymerisation.

Reagents HNO_3, HDBP, TBP are commercially available. TBP is purified by alcaline and water washing to remove acidic impurities (degradation products).

Experiments are made in test tubes at 23 ± 2° C or in a thermostated cell.

The acid concentration is determined by potentiometry and Pu by alpha radiometry.

The multi-stage counter current experiment is made in a glove boxe, specially designed for pilot experiment studies. It includes five mixer settler banks whose main functions are : U.Pu extraction, Pu stripping, Uranium stripping, alcaline wash of the solvent, and diluent wash of aqueous Pu(III) solution. The organic phase is continuously recycled. The feed solution of Pu is prepared in a sufficient quantity for the time of the run as described.

POLYMERS FORMATION AND PROPERTIES IN AQUEOUS PHASE

Formation reactions and chemistry of Plutonium polymers have already been largely reviewed in the literature (4, 5), and we shall only mention some aspects related directly to this study.

Formation

The Plutonium at valence four gives in low acid medium, polymeric species Pu_n. In the general scheme, polymerisation reaction is initiated by hydrolysis of the solvated cation M^{n+} according to :

$$[M(H_20)_m]^{n+} \rightarrow [M(OH)(H_20)_{m-1}]^{(n-1)+} + H^+$$

the solvatation water is converted in hydroxyl groups.

The first complex compound obtained is the dimer :

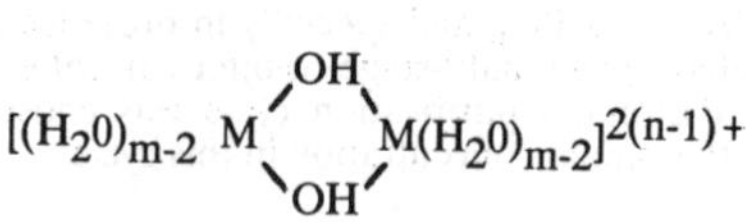

A long chain polymer is formed by continuation of this reaction. It is a polynuclear aggregate, bridged by hydroxyl groups. This is the well known Thomas model for the Plutonium polymerisation (6, 7).

For Plutonium there exists a difference between a fresh polymer and an aged one. It is generally admitted that the "ol" form (olation reaction) corresponds to the fresh polymer, while "oxo" form (oxolation reaction) corresponds to the aged polymer (5) obtained by ageing (or heating) (2). The aged polymer is obtained by dehydration of the "ol" form. We can summarize the reactions as follow :

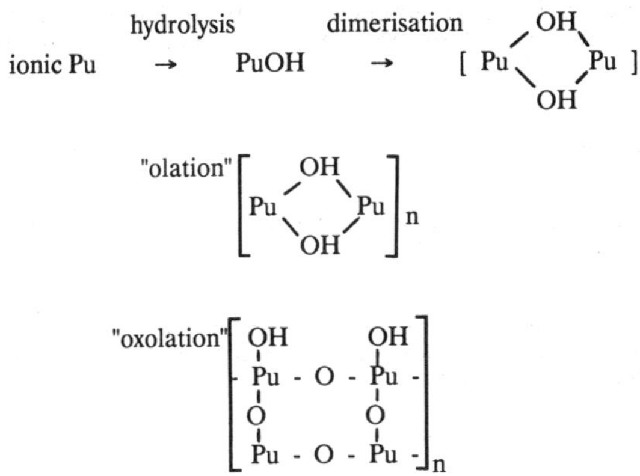

The system evolves towards PuO_2 formation, which is the thermodynamically stable form (8).

This is the crystalline form obtained after heating fresh polymer, and identified by XRay diffraction analysis.

The polymers can deposit by natural sedimentation or by centrifugation. Their molecular weight vary from 4000 to 10^{10}, according to the preparation method (9).

In a nuclear fuel reprocessing solution, if a Pu polymer is formed, it is a mixture of many species of different weights and formulas.

Dissolution in nitric acid aqueous solutions

This dissolution is more easy for a fresh polymer, because an aged one has stronger bondings. Dissolution occurs in the reverse manner to that of formation "oxo" form → "ol" form → soluble ionic form.

This depolymerisation occurs in concentrated hot nitric acid and the rate is increased in presence of fluoride ions (a more efficient depolymerisation agent) (2).

As it is very difficult to depolymerise and dissolve Pu polymers in real reprocessing conditions, polymer formation must be avoided.

Electrical charge

These species have a low positive charge. Generally 0.15 to 0.5 elementary charge is found by atom of Plutonium (10, 11).

Formation of insoluble compounds occurs for low anion quantities. It was shown for iodate, sulfate, oxalate, phosphate ions. Precipitation is partial with nitrate ions between 0.5 and 4 M (2).

Extractive properties

Plutonium polymers are described as inextractable by TBP, in a 20 % mixture of TBP in kerosene, but are extracted by 5 % dibutylphosphoric acid-kerosene solutions (12).

Plutonium polymers are extracted by TBP-diluent solutions only if traces of dibutylphosphoric acid are present. One considers generally that polymers cause plutonium loss in extraction units.

Conclusion

Plutonium polymers are rather well known in the area of their structure, formation or destruction. Studies give empirical data to avoid polymers formation, as relations between Pu, HNO_3 concentrations and temperature. But once formed, what is the behaviour of polymers in the process, especially in presence of HDBP which seems an important agent ?

RESULTS

Analysis of Pu_n

We have measured the ionic charge of the Pu_n solution previously described, by a precipitation technique. When an insoluble compound with a slightly charged Pu_n and a convenient anion A^- is formed, the ration A^-/Pu in the precipitate indicates the charge of Pu. If the solid is highly insoluble, we may only measure the ratio $[A^-]$ added/ΔPu_n in the solution. Pu-DBP complexes can permit such determinations.

By adding various amounts of HDBP in the Pu_n solution, the slope of the curve [HDBP] added = fΔ[Pu] gives the mean charge of Pu in Pu_n. The figure n° 1 indicates that the solid Pu_n $DBP_{0.27n}$ precipitates therefore the mean positive charge of the polymer is 0.27 by plutonium atom.

TABLE 1 - Extraction of Pu_n in presence of HDBP

Initial conditions = Pu_n 5.04 10^{-3}M ; HNO_3 0.05M

Φ_o/Φ_a = 1 ; temperature 23 ± 2° C ; Solvent 30 % TBP in TPH

Contact time 5 minutes

HDBP / Pu_n	Pu_n 10^{-3}M aqueous	Pu_n 10^{-3}M organic	KD Pu_n	% Pu_n extracted	% Pu_n preci-pitated
0.085	2.51	1.69	0.67	33	17
0.17	1.37	3.30	2.40	65	7.5
0.26	1.28	3.79	2.96	75	0
0.34	1.21	3.80	3.14	75	0

TABLE 2 - Extraction of Pu_n in presence of HDBP

Initial conditions = Pu_n 3.76 10^{-3}M ; HNO_3 0.5M

Φ_o/Φ_a = 1 ; temperature 23 ± 2° C ; Solvent 30 % TBP in TPH

Contact time 5 minutes

HDBP / Pu_n	Pu_n 10^{-3}M aqueous	Pu_n 10^{-3}M organic	KD Pu_n	% Pu_n extracted	% Pu_n preci-pitated
0	3.58	0.08	0.022	2.2	0
0.11	1.65	1.61	0.98	43	13
0.28	0.24	3.25	13.5	87	7
0.56	0.11	3.66	33	98	0
1.099	0.0007	3.75	> 1000	100	0

This value is characteristic of the Pu_n solution for our experiment. Change in the conditions of preparation causes a different charge (we have obtained other charges, up to 0.5 in other conditions).

Distribution of Pu_n between phases

The distribution of Pu_n is measured by contacting nitric aqueous solutions 0.05 M and 0.5 M with organic phase 30 % TBP in TPH (hydrogenated tetrapropylene). The mixing time is five minutes, about the same as used in the mixers of the pilot plant. Dibutylphosphoric acid is added separately for each experiment. Results are given on

Tables 1 and 2. We confirm that Pu_n is poorly extracted when HDBP is absent, the coefficient of distribution of Pu is $2.6\ 10^{-2}$. The exact coefficient is probably lower because we cannot exclude the presence of others Plutonium species like Pu(III), Pu(IV) ..., and HDBP traces in the purified organic phase.

If HDBP is introduced into the organic phase, the extraction of Pu_n becomes important, and for equimolar quantities the coefficient of distribution of Plutonium polymers is probably over one thousand.

The figure 2 shows the amount of Pu_n extracted as a function of the added HDBP. We can subsequently conclude that :

* a precipitate appears at the interface for low HDBP values. The default of mass balance decreases when HDBP increase. The mass balance is correct for HDBP/Pu = 0 or HDBP/Pu > 0.2 to 0.5 according to the acidity. This also means that such Pu_nDBP precipitates may occur in extraction unit of pilot experiment.

* the slope at the beginning of the curve and the mass balance indicate that the complex extracted has the same formula than the precipitate obtained in aqueous phase : a neutral complex $Pu_n(DBP)_{n/4}$.

Back extraction of Plutonium polymers

The solution used for back extraction contains HNO_3 0.1 M, hydroxylamine nitrate 0.5 M and hydrazine nitrate 0.16 M. The temperature is 45° C and contact time five minutes, the volumetric ratio organic phase/aqueous phase is two. Results are reported Table 3.

TABLE 3 - Back extraction of Pu_n in presence of HDBP

Initial conditions = Pu_n $3.76\ 10^{-3}$M ; HNO_3 0.05M

$\Phi_o/\Phi_a = 1$; temperature 23 ± 2° C ; Solvent 30 % TBP in TPH

Contact time 5 minutes

$Pu_n 10^{-3}$M organic	$DBP^- 10^{-3}$M organic	DBP^- / Pu_n	$Pu_n 10^{-3}$M aqueous	$Pu_n 10^{-3}$M organic	KD Pu	$\%Pu_n$ back extracted	$\%Pu_n$ precipi-tated
1.7	0.33	0.20	0.33	1.17	3.54	9.7	21
3.3	0.75	0.22	0.43	2.42	5.6	6.5	20
3.8	1.2	0.32	0.25	3.33	13.3	3	9
3.8	1.6	0.42	0.38	3.64	9.6	5	0

As for extraction experiments, a precipitate appears at the interface for low ratios HDBP/Pu. Plutonium is poorly back-extracted, the apparent coefficients of distribution are high. Aqueous phase is blue, which indicates that Pu(III) is present, and reduction kinetics of $Pu(IV)_n$ appears to be the limiting phenomena in back extraction.

Precipitates obtained during extraction or back extraction are lipophilic, they stand at the interface and move preferentially in the organic phase. In a counter-current process, precipitates should be carried by the organic phase.

Pilot plant experiment

The behaviour of Plutonium (IV) polymers has been studied in the conditions of a Plutonium purification cycle.

Plutonium (IV) is extracted from a nitric acid solution and back extracted by a reducing solution of hydroxylamine nitrate, HAN, using mixer settler banks according to the chemical flowsheet of figure 3.

Isotopic composition of Plutonium of the initial feed solution is 3.4 % ^{238}Pu, while polymeric Plutonium added is 0.35 % ^{238}Pu and represents about five percent of total Plutonium of the second feed solution.

Run was started without polymers and when steady state was reached, polymers were added to prepare the second solution. Two concentration profiles were determined.

In presence of polymers, we observed no change in Plutonium loss in the aqueous raffinate and the formation of a white precipitate at the interface level of the settlers of the extraction bank.

Their importance increase from stage 1 to 6 and with running time. After twenty hours the accumulation is so important that hydraulic operation becomes difficult because of virtual blocage due to precipitate formation.

This precipitation is attributed to the reaction of Plutonium polymers with dibutylphosphoric acid present in the organic phase.

In the scrub stages, some slight deposits are observed at the interface in the settlers.

In the stripping unit, no important change was observed in the Plutonium loss of the discharged solvent but examination of the Plutonium profiles shows that profile in presence of polymers is shifted towards high values (see figure 4).

At the end of the run we have observed, in the stainless steel mixer-settler bank, an accumulation of precipitate at the interface in decreasing quantity from stage 1 to stage 12. The estimated mass of these compounds is about twenty percent of total Plutonium polymers.

This compound has been identified as Plutonium (IV) polymer dibutylphosphate $Pu_n(DBP)_{n/4}$ (Pu content = 70 %). Its Plutonium isotopic composition is that of the initial Plutonium polymers.

Two conclusions may be drawn from this result :

* no isotopic exchange is observed between polymeric and ionic Plutonium ;

* no significant polymers formation occurs in the process.

This run has shown that polymeric Plutonium species cause mainly precipitation in the extraction units whose main consequences may be hydrodynamic troubles in the extractors. Within the 20 hours duration run, we did not observe any Plutonium loss in extraction raffinate, probably due to the presence of traces of HDBP present in the solvent.

In the Plutonium strip we observed no loss of Plutonium polymers in the solvent although test tube experiments showed that the efficiency of Plutonium polymers stripping was bad. We believe that the two following reactions occur simultaneously.

1 - Polymeric Pu(IV) reduction by HAN

If we assume that limitating step in the back extraction is the reduction of Pu_n to trivalent state Plutonium, then total residence time of the aqueous phase (mixer + settler) has to be considered. The consequence should be that reduction time will be more important than expected from test tube experiments where only mixing time was considered (5 minutes compare to 17 minutes).

2 - Partial precipitation of plutonium polymers.

Its relative importance seems too limited to explain the good efficiency of plutonium stripping.

CONCLUSION

We have pointed out the behaviour of Plutonium polymers in the Purex process using TBP as solvent.

Compounds HDBP-Pu_n are very insoluble in the aqueous phase. In the presence of solvent, the behaviour is very complex. With low HDBP/Pu_n ratios they are slightly extracted and precipitate at the interface. When the HDBP content of the solvent increases the solubility in the organic phase increases, and no precipitate occurs.

A major problem issuing from the presence of Plutonium polymers in nuclear fuel reprocessing, is the formation of a precipitate with dibutylphosphoric acid. It may be responsible for hydraulic troubles in extraction or back extraction units and clogg the apparatus. In our experiment of a plutonium purification cycle, no significative plutonium losses were measured, in the aqueous raffinate or in the discharged solvent.

The only way to avoid these problems is to avoid Plutonium polymer formation. The conditions to do it are currently known, they are good control of three main parameters : plutonium and acid concentrations, and temperature. When concentration of plutonium solutions by evaporation is necessary, conditions must be carefully controlled because the risks of polymers formation are very important. In actual or future plants these problems are generally taken into consideration and should not occur.

BIBLIOGRAPHY

1. Tallent et al., 1988
"Method of removing Pu(IV) polymer from nuclear fuel reclaiming liquid"
U.S Patent 4, 316, 776

2. Costanzo, B.A., Biggers, R.E., 1963
"A study of the polymerisation, depolymerisation, and precipitation of tetravalent plutonium as function of temperature and acidity by spectrophotometric methods : preliminary report" ORNL TM 585

3. Savage, D.J., Kyffin T.W., 1968
"Oxidative breakdown of polymeric Pu(IV) by Ce(IV)"
Polyhedron, 5, 3, 743

4. Cleveland, J.M., 1970
The chemistry of plutonium
Gordon and Breach science publishers, New York

5. Johnson, G.L. and Toth L.M., 1978
"Plutonium (IV) and thorium (IV) hydrous polymer chemistry"
ORNL/TM-6365

6. Gimblett, F.G.R., 1963
Inorganic Polymer chemistry, Chapter 3, Butterworths ed, London

7. Pokras, L., 1956
J. Chem. Education 33, 152

8. Ryand and Bray, 1979
"Dissolution of PuO_2"
ACS Symposium series 117, Hawaï

9. Ichikawa, F., Sato, T., 1984
"On the particle size distribution of hydrolyzed plutonium (IV) polymer"
J. of Radionuclear Chem., 84, 2, 269

10. Katz, J.J. and Seaborg, G.T., 1957
The chemistry of the actinides elements
Methnen and Co, London

11. Bell, J.T., Coleman, C.F., Costanzo, D.A., Biggers, R.E., 1973
J. Inorg. Chem., 35, 629

12. Ockenden, D.W., Welch, G.A., 1956
"The preparation and properties of some plutonium compounds"
Part. V. Colloïdal quadrivalent plutonium
J. of the Chem. Soc. 3358

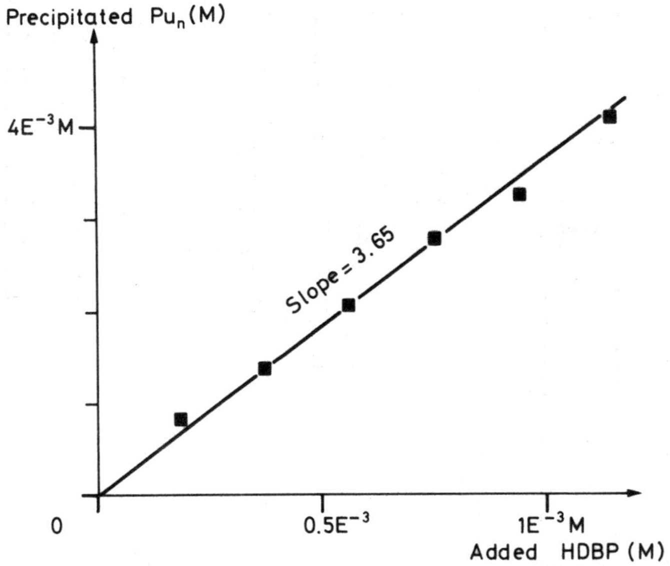

Figure n°1 - PRECIPITATED Pu_n VERSUS ADDED HDBP
HNO_3 = 0.05N, Pu_0 = 4.56 E^{-3}M

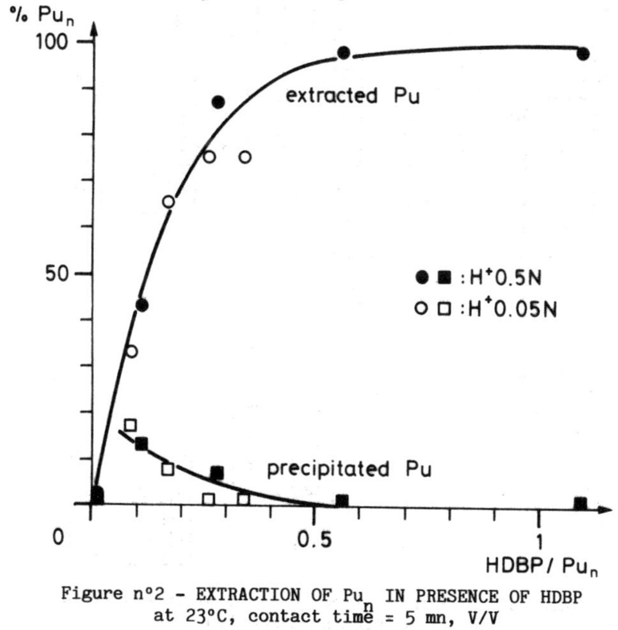

Figure n°2 - EXTRACTION OF Pu_n IN PRESENCE OF HDBP
at 23°C, contact time = 5 mn, V/V

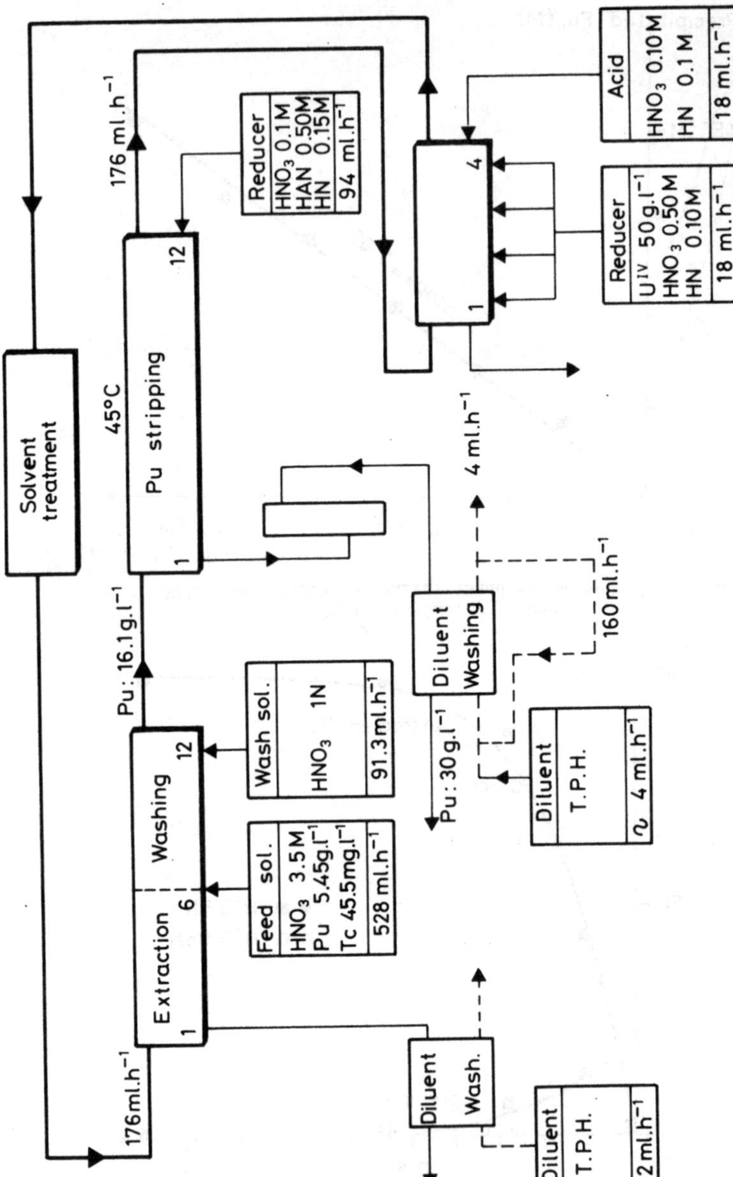

Figure n°3 - Chemical flow sheet of pilot experiment
Plutonium purification

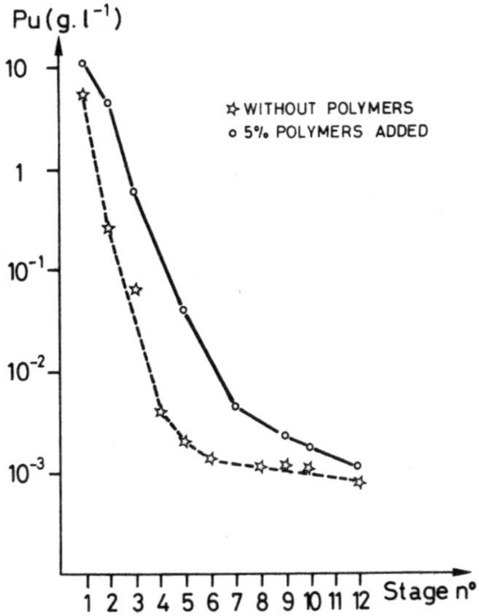

Figure n°4 - PLUTONIUM PROFIL IN ORGANIC PHASE, BACK-EXTRACTION BANK

COUMPOUNDS OF PLUTONIUM (IV) WITH ALKYLPHOSPHATES AND PHOSPHATE IN NUCLEAR REPROCESSING

H. SCOAZEC ; J. LIVET ; J.Y. PASQUIOU ; M. GERMAIN

The nature of Pu(IV)-alkylphosphates and phosphate formed, especially in the PUREX process chemical conditions, has been studied.
(i) With HDBP, two solid compounds : $Pu(DBP)_3NO_3$, xH_2O and $Pu(DBP)_4$ have been identified in aqueous phase, and only one $Pu(DBP)_3NO_3$ in organic phase,
(ii) With H_2MBP one compound was found in aqueous phase : $Pu(MBP)_2$ yH_2O, while three compounds are formed with H_3PO_4, proposed formulas are $Pu(HPO_4)(NO_3)_2$, $Pu(HPO_4)_2$ and $Pu_2H(PO_4)_3$.
Solubility of some of these compounds were examined in the fuel reprocessing conditions, in aqueous or organic phase (30 % TBP-diluent) excluding mixed biphasic systems.

INTRODUCTION

Nuclear fuels reprocessing using PUREX process with tributylphosphate (TBP) as solvent is universally used. It is the most efficient industrial process to recover and separate Uranium and Plutonium from spent fuels. However reprocessing of highly irradiated fuels causes higher radiolytic decomposition of the solvent, that means that degradation products concentrations should increase. These compounds : alkylphosphoric and phosphoric acids, form, especially with tetravalent elements (Pu(IV), Zr(IV), ...), soluble complexes and insoluble precipitates according to the chemical conditions (1, 9). In the organic phase soluble complexes are responsible for increased metal retention in back extraction units.

Insoluble compounds in the organic and aqueous phases, may either deposit on surface equipments or lay as third phase at interface levels of the extractors. These compounds decrease extractors efficiency by changing the surface properties and hindering the separation of the phases. Examples of solubilities are given in Table 1 for nitric acid solutions.

These compounds are not currently well known and we have undertaken an extensive study of their compositions and solubilities to have a better knowledge and control of the risks of solids formation in industrial plants.

EXPERIMENTAL

A nitric acid solution of plutonium (IV) was used.

TABLE 1 - Solubilities of Pu(IV) in nitric acid solutions in presence of TBP degradation products - Initial conditions $Pu_0 = 4.18\ 10^{-4}$ M

Time	HDBP 2 10^{-3} M		MBP 2 10^{-3} M		H_3PO_4 2 10^{-3} M	
Days	HNO_3 1M	HNO_3 4M	HNO_3 1M	HNO_3 4M	HNO_3 1 M	HNO_3 4M
0.2	4.6 10^{-6}M	2.8 10^{-5}M	1.09 10^{-5}M	≥4.18 10^{-4}M	4.0 10^{-4}M	≥4.18 10^{-4}M
2	4.6 10^{-6}M	1.8 10^{-5}M	0.95 10^{-5}M	4.18 10^{-4}M	4.0 10^{-4}M	" "
6	5.0 10^{-6}M	1.8 10^{-5}M	0.96 10^{-5}M	≥4.18 10^{-4}M	4.6 10^{-4}M	" "
8	--	--			3.3 10^{-4}M	" "
14	--	--			3.0 10^{-4}M	" "
21	--	--			2.5 10^{-4}M	" "

Plutonium dioxide (PuO_2) is dissolved in nitric acid and then purified, especially for americium content, by solvent extraction using 30 % TBP in TPH (Hydrogenated Tetra Propylene diluent) with hydroxylamine nitrate back-extraction.

Traces of TBP are removed by contact with pure diluent.

Pu solution is then concentrated by evaporation and adjusted to Pu(IV) and nitric acid concentration.

Plutonium valency is determined by spectrophotometry.

Pu content of mother solution is about 50 $g.l^{-1}$ and acidity 2.5 M HNO_3. Concentrations are adjusted for all experiments with appropriate acid medium. The ^{238}Pu content is low (0.3 %) to minimise α radiolysis effects.

The organic phase is a 30 % mixture of TBP in TPH.

Impurities such as dibutylphosphoric acid (HDBP), monobutylphosphoric acid (H_2MBP) and other acids are removed by alkaline and pure water washes until neutrality of aqueous phase is reached. Definite acidity of organic phase is obtained by HNO_3 extraction from an appropriate aqueous phase and checked by pH measurements. Organic plutonium solutions of various concentrations are prepared by mixing the organic phase with concentrated organic solution of plutonium of same acidity.

HDBP, H_3PO_4, HNO_3 are commercial analytical grade reagents.

H_2MBP is prepared by extraction from a 50-50 % HDBP-H_2MBP mixture (MERCK).

PROCEDURE

All experiments were achieved at room temperature 23 ± 2° C. Phosphoric complexants are dissolved in the same phase as the plutonium solutions to avoid biphasic systems and wrong interpretations of precipitation phenomena. Once chemical equilibrium is reached (soluble plutonium remains constant with time), the mixture is filtered on a 1.2 μm pore size filter to separate the solid.

Solids and solutions are then analysed. We point out two important considerations : the equilibrium time can be as long as one month, especially in organic phase ; the precipitate must be washed with the same medium that the precipitation (except Pu and complexant) to avoid any change in solid compound formula, as was observed.

Precipitation curves are obtained by adding increasing quantities of complexants in Pu solutions, in separate experiments (r = initial molar ratio of complexant to plutonium IV). Solutions were measured for plutonium and HDBP (only for aqueous phase study).

Solids were identified by X-rays diffraction for HDBP compounds, some compounds were analysed after dissolution (mainly for plutonium content). We usualy operate with a plutonium concentration in large excess compare to the complexant to be representative of the process conditions and permit easier interpretation of the results.

The slope of the curves (Δ complexant/ Δ Pu) indicates the ratio R = complexant/Pu in the precipitate ; this can be verified by solid analysis (X-rays diffraction, thermogravimetry, ...).

A special investigation of a new compound $Pu(DBP)_3NO_3$, by XRays diffraction and comparison with the structure of the isomorphic compound $Th(DBP)_3NO_3$ was made.

Solubilities of precipitates are achieved by redissolution and plutonium is measured by α radiometry. HDBP is measured by gas chromatography.

The presence of nitrate ions in the solid was assumed by studying the effect of nitrate concentration in the aqueous phase on solubility.

Proposed formulas are finally, based on these results and literature data.

EXPERIMENTAL RESULTS

Compounds formation has been studied as a function of plutonium IV, nitric acid, nitrate and phosphate concentrations.

TABLE 2 - R values for Pu(IV) dibutylphosphate in aqueous phase

HNO_3 M		0.5	1	2	3	4	8	Mean Value
$[DBP]_0$	$.10^2$M	0.913	1.01	1.22	1.89	1.52	2.02	
$[Pu]_0$	$.10^2$M	0.451	0.451	0.451	0.783	0.451	0.451	
$[DBP]_{eq}$	$.10^2$M	$\leq 2\ 10^{-3}$	$\leq 2\ 10^{-3}$	0.005	0.003	0.0182	0.620	
$[Pu]_{eq}$	$.10^2$M	0.116	0.054	0.0472	0.095	$1.26\ 10^{-4}$	$2.7\ 10^{-4}$	
$R = \frac{DBP}{Pu}$ (mass balance)		2.73	2.54	3.02	2.75	2.97	3.10	2.86
$R = \frac{DBP}{Pu}$ (analysis of the precipitate after washing)		2.00	1.99	1.78	2.08	1.77	1.76	1.9

HDBP-Pu compounds in aqueous phase

In aqueous phase, solubility of plutonium dibutylphosphates is very low, and precipitation occurs rapidly. A pink precipitate is observed at the bottom of the test tube. The figure 1 shows an example of the ratio ligand/metal for which the precipitation occurs.

Solubility determination is not possible by this technique because of its low value. Solubility of $Pu(DBP)_3NO_3$ in 8 M HNO_3 has been measured by redissolution and found to be $2.6\ 10^{-4}$ M.

Table 1 gives an example of the solid composition as a function of nitric acid concentration. Three solids compounds were observed (XRays diffraction and chemical analysis).

$$Pu(DBP)_2\ (NO_3)_2\ H_2O\ ;\ Pu(DBP)_3\ (NO_3)\ xH_2O : Pu(DBP)_4$$

according to the concentrations of the reagents. As an example, analysis of initial concentrations and concentrations after precipitation for different nitric acidities is given Table 2. The mass balance indicates that the mean value of the ratio R is 2.85.

TABLE 3 - Pu content of Pu-HDBP precipitates obtained in experiments of Table 2

Remark : in $Pu(DBP)_2(NO_3)_2$ Pu = 30.6 %
in $Pu(DBP)_3(NO_3)$ Pu = 25.7 %

HNO_3 M Value	1	2	4	8	Mean
Weight of precipitate mg	39.0	41.2	34.8	44	
Pu content mg	10.48	10.99	9.45	11.62	
Pu per cent	26.8	26.7	27.2	26.4	26.8

If precipitates are washed by water (even with low acid content) the R value is 1.9 which indicates a change in the solid composition. Table 3 reports analyses of plutonium content in the solids obtained, they are close to the composition corresponding to $Pu(DBP)_3NO_3$, and different from $Pu(DBP)_2 (NO_3)_2$. It appears clearly that washes and evolution with time of such compounds could lead to erroneous interpretations.

HDBP-Pu compounds in organic phase

Precipitation as a pink violet solid occurs slowly and sometimes after several days. Chemical equilibrium may take more than one month, especially with low reagent concentration. It means that measurements made after several hours, as previously published (3) should be considered carefully.

By raising the complexant to plutonium IV ratio (r) successive soluble complexes are formed and observed by spectrophotometry (fig. 2) with r = 1 to 4 before precipitation (for $r \geq 3$). The solid formation occurs slowly and only when $r \geq 3$ in nitric acid concentration range from 0.08 M to 0.8 M and Pu 4.10^{-4} M to 0.17 M.

Solid composition has always been found to be the same : $Pu(DBP)_3NO_3$ (anhydrous compound).

XRays diffraction indicates that the network is hexagonal type (6 helicoïdal axis) with following parameters : a = 15.20 Å ; c = 30.20 Å and Z = 6 (number of patterns in the cell).

After separation by filtration this compound appears to be unstable under radiolysis, and is transformed into several products including $Pu(DBP)_4$ according to XRays diffraction analysis.

Solubility of $Pu(DBP)_3NO_3$ is much higher in organic phase than in aqueous phase and increases with nitric acid concentration.

Approximately for 0.08 M HNO_3 solubility S = 4 10^{-4}M, in 0.4M HNO_3 S = 2 10^{-3}M while in 0.8 M HNO_3 S = 4 10^{-3}M.

Other compounds have been proposed in the literature ; $Pu(DBP)_2(NO_3)_2$, $Pu(DBP)_4$ they could be formed in other chemical conditions that we did not examine (biphasic system for example).

As a matter of fact, for Th(IV) we have prepared and identified by XRays diffraction and chemical analysis three different compounds : I = $Th(DBP)_4$; II = $Th(DBP)_3NO_3$, xH_2O ; III = $Th(DBP)_2(NO_3)_2$, H_2O.

The appearance of these compounds is determined by the ionic product $[Th^{4+}]\ [NO_3^-]^4$ as previously reported (5). For example in 0.5 M HNO_3 when thorium concentration increases, we have successively the compounds I → II → III. The thorium concentrations for these transitions are :

$$I \rightarrow II \quad [Th] = 0.915\ M$$

$$II \rightarrow III \quad [Th] = 1.9\ M$$

In 2.0 M HNO_3 these values become :

$$I \rightarrow II \quad [Th] = 0.5\ M$$

$$II \rightarrow III \quad [Th] = 1.595\ M$$

The corresponding values for Pu have not been determined yet.

H_2 MBP-Pu compounds in aqueous phase

In the aqueous phase solubilities of these compounds are low and equilibrium is reached rapidly (< 1 hour). Figure 3 shows that a solid compound with R = 2 is obtained in the range of acidity from 0.5 to 8 M HNO_3. It is a pinky white solid.

Moreover, we observe that when nitric acid concentration increases, precipitation occurs for higher H_2MBP values, probably due to the effect of H^+ on H_2MBP dissociation.

Solubility increases from 10^{-5} M to 4 x 10^{-4} M when nitric acid concentration varies from 1 M to 8 M.

We conclude, as reported by Moore (7) that there is formation of a single solid complex whose formula could be $Pu(MBP)_2\ yH_2O$.

Filtration and washing of this compound caused a change in color : greenish or dark pink compounds corresponding to other formulas like $Pu(HMBP)_2(NO_3)_2$ or $Pu(HMBP)_4$ are obtained.

H_2MBP-Pu compounds in organic phase

When adding H_2MBP to the organic plutonium solution, precipitation occurs very slowly (equilibrium time ≥ 1 week). Solubility appears to be higher than in aqueous medium. We suppose a mechanism similar to the precipitation with HDBP, namely formation of an insoluble complex $Pu(MBP)_2$ when H_2MBP to Pu(IV) ratio is higher than 2. This study is still in progress.

H_3PO_4-Pu(IV) compounds in aqueous phase

In aqueous solutions, addition of H_3PO_4 to a Pu(IV) solution may cause either the appearance of a white precipitate or gel formation according to chemical conditions. Equilibrium is reached slowly (several hours to several days) and kinetics are mainly dependent upon Pu(IV) and H_3PO_4 concentrations (figure 4).

Table IV shows that :

* Insoluble plutonium phosphates appear at low nitric and high plutonium concentrations, while gel formation occurs at high nitric acid concentrations (even a solid gelatin).

* Several plutonium phosphates are suggested and characterised by R ratio from 1 to 2, some of them containing nitrate ions.

* Range of formation of these phosphates are very narrow and may correspond to very precise nitric acid and plutonium concentrations.

Presence of nitrate in the solid is checked by variation of the nitrate content of the solution and by measuring its effect on solubility.

Proposed formulas are : $Pu(HPO_4)(NO_3)_2$, xH_2O ; $Pu(HPO_4)_2$, zH_2O ; Pu $H(PO_4)_3$, yH_2O.

The second compound is gelatinous and pink.

TABLE 4 - R values for Pu(IV) phosphate precipitates in aqueous phase

HNO_3 M→ Pu $g.l^{-1}$ ↓	0.15	0.25	0.5	1	4	8
0.2	nd	R = 1 NO_3^- not analysed	nd	Sol	Sol	Sol
1	R = 1 $\frac{NO_3^-}{Pu(IV)} = 2$	R = 1.5 $NO_3^- = 0$	R = 1.5 $NO_3^- = 0$	R = 2 $NO_3^- = O$	Sol	Sol
5	nd	R = 1.5 S = 8 10^{-5}M	R = 1.5	R = 2	Sol	Sol

nd : not determined ; Sol : Soluble ; S : Solubility

H_3PO_4-Pu(IV) compounds in organic phase

In organic phase we have observed precipitation with initial plutonium concentrations of about 4 10^{-3} M and 4 10^{-2} M (0.8 M nitric acid) for very low H_3PO_4 concentrations. Equilibrium needs a long time to be reached (more than one week) and a compound with R = 4 has been found, which suggests the formula $Pu(H_2PO_4)_4$.

This is a quite different solid from those obtained or reported for aqueous phase or biphasic system (8, 9). This study has to be completed by variation of nitric acid content of the solution, because other solids containing nitrate ions may be formed.

The compound obtained is slightly soluble in the organic phase S 10^{-4} M at equilibrium.

CONCLUSION

Solid compounds of Pu(IV) with phosphate and alkylphosphates are numerous and formed very easily from aqueous and organic (30 % TBP-TPH) solutions.

In the aqueous phase, systems are very complex because many compounds are formed. Schematically plutonium (IV) solubility increases from dibutylphosphates to phosphates and when nitric acid concentration increases. Precipitation occurs rapidly, except for phosphoric acid, at low phosphate concentrations. Nitrate ions may be present in the precipitate with HDBP and H_3PO_4, but were not found in these experiments with H_2MBP.

In organic solutions, the main differences with aqueous solutions can be summed up as follows :

* Plutonium (IV) solubility is much higher than in aqueous phase except for H_3PO_4 for which it remains comparable.

* Precipitation occurs slowly and only when the ratio r is higher to a definite value, which is the same as the ratio R, in the precipitate, for HDBP and H_2MBP complexes.

Successive soluble complexes are formed for lower values of r, for which no precipitation was observed.

More generally the behaviour of HDBP and H_2MBP complexes with Pu(IV), are comparable and may be due to the presence of butyl chains whose lipophilic nature characterises their properties (kinetics, solubility ...), while H_3PO_4 complexes behave quite differently (Table 5).
In fuel reprocessing conditions, it appears that in the extraction cycles plutonium (IV) precipitation should occur preferently in aqueous phase with dibutyl and monobutylphosphoric acid.

In fact, their complexes with Pu(IV) are strongly extracted into the organic phase where solubilities are much higher, so that usually no precipitation is observed.

Phosphoric acid is poorly extracted into the organic phase and could be responsible for precipitations in aqueous or biphasic system. Some of the compounds suggested in this study have been reported earlier (8, 9). However its concentration is always low in normal process conditions and no precipitation should occur.

When aqueous phases are stored for a long time or evaporated for concentration, risks appear more important and it is necessary to eliminate very efficiently the main phosphoric acid precursor : TBP.

The authors are indebtful to B. GUILLAUME for his contribution to the studies and fruitful discussions.

TABLE 5 - Characteristics of Pu(IV) precipitation with HDBP, H_2MBP and H_3PO_4

Complexant		HDBP	H_2MBP	H_3PO_4
Aqueous phase	Solubility (S)	Very low $S=10^{-6}M$	Very low $S=10^{-5}M$	Low $S=10^{-4}M$
	Effect of HNO_3	- Increases S - Nitrated salts	Increases S	Increases S until gel formation
	Beginning of precipitation	For low HDBP/Pu	For low H_2MBP/Pu depends upon HNO_3	For low H_3PO_4/Pu
	Kinetics of precipitation	Quick	Quick	Slow
	Composition of precipitate	Probably hydrated R = 3 ou 4 NO_3^- = 1 or 0	R = 2	R = 1 to 2 NO_3^- = 2 or 0
Organic phase	Solubility	10^{-4} to $10^{-3}M$ according to HNO_3	High $S=10^{-3}$ to $10^{-2}M$	Low S - $10^{-4}M$
	Effect of HNO_3	Increases S	Not examined yet	Increases S - Nitrated salts
	Beginning of precipitation	For HDBP/Pu(IV) ≥ 3	Probably for $H_2MBP/Pu \geq 2$	For low H_3PO_4/Pu(IV)
	Kinetics of precipitation	Slow according to concentration	Slow according to concentration	Slow according to concentration
	Composition of precipitate	$R=3, \frac{NO_3^-}{Pu(IV)}=1$	Probably R = 2	R = 4

BIBLIOGRAPHY

1. Sokhina, L.P., Rovnyi, S.I., and Goncharuk, L.V., 1988
"Neptunium and Plutonium coprecipitation with iron and zirconium dibutyl phosphates" Translated from Radiokhimiya, 30, 4, 457

2. Sokhina, L.P., Solovkin, A.S., Teterin, E.G., Bogdanov, F.A., and Shesterikov, N.N., 1978
"Dibutylphosphates of tetravalent plutonium, zirconium and thorium, formed in solutions of tri-n-butyl phosphate (TBP)"
Translated from Radiokhimiya, 20, 1, 28

3. Solovkin, A.S., 1978
"Mono- and di-n-butylphosphates of certain metals, important in regeneration
processs of irradiated nuclear fuels"
Translated from Radiokhimiya, 24, 1, 56

4. Sokhina, L.P., Rovnyi, S.I., Goncharuk, L.V., and Bogdanov, F.A., 1988
"Solubility of the di-n-butylphosphate of tetravalent plutonium"
Translated from Radiokhimiya, 30, 3, 418

5. Pasquiou, J.Y., Livet, J., Germain, M., Musikas, C., 1987
"Pu(IV)-dibutylphosphate complexes in the purex process"
Extraction'87 : the recovery of high value materials
EFCE Publication Series N° 347, 207

6. Rovnyi, S.I., Boikova, I.A., Goncharuck, L.V., Milovanova, A.S., Fedoseev, D.A. and Sokhina, L.P., 1985
"Interaction of plutonium (IV) with mono-n-butylphosphoric acid in nitric acid solutions and in the system 30 % TBP + n-dodecane"
Translated from Radiokhimiya, 27, 4, 429

7. Moore, R.L., 1950
"The solubility of plutonium (IV) monobutyl phosphate"
USAEA Report HW 18188

8. Fedoseev, D.A., Boikova, I.A., Milovanova, A.S., Fedotov, S.N., Naumov, M.A., and Vladimirova, M.V., 1986
"Radiation chemical behavior of actinides in extraction systems IV. Interaction of Pu(IV) with H_3PO_4 in the system 30 % TBP + n-dodecan"
Translated from Radiokhimiya, 28, 5, 580

9. King, E.L., 1949
"The solubility of plutonium (IV) phosphates and the phosphate complexes of plutonium (IV)"
The transuranium elements, National Nuclear Energy Series, IV, 14-B, 638

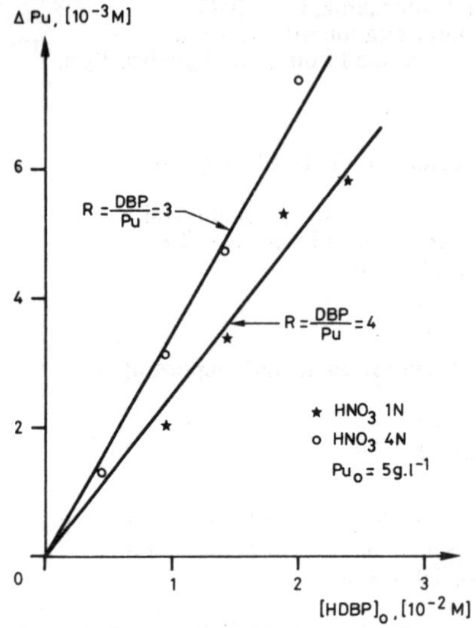

Figure 1 _ Example of precipitation curves for Pu - HDBP complexes in aqueous phase

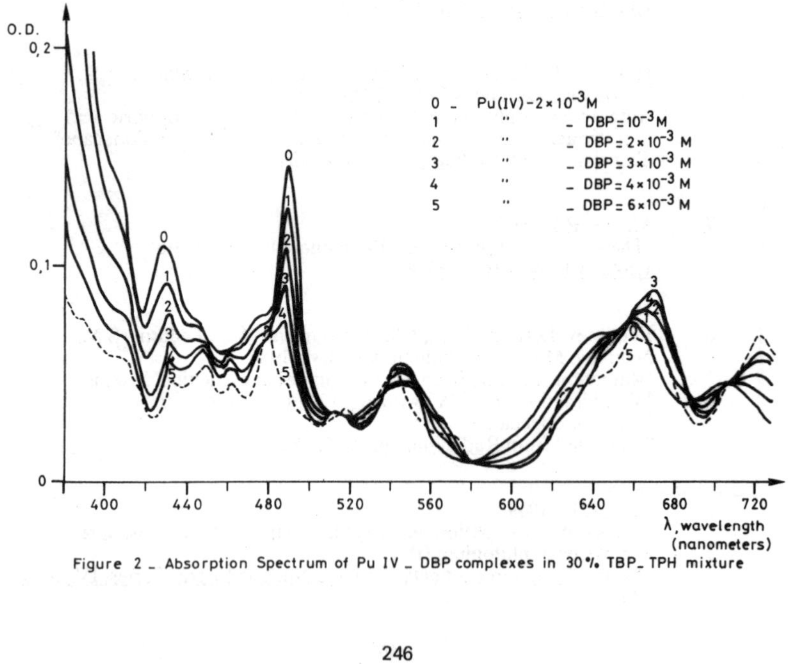

Figure 2 _ Absorption Spectrum of Pu IV _ DBP complexes in 30 % TBP_TPH mixture

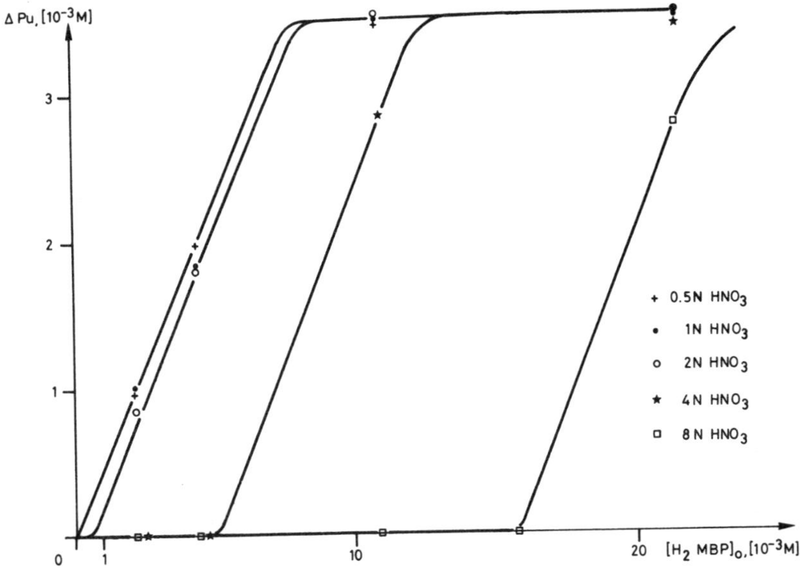

Figure 3 _ Kinetics of precipitation in aqueous-phase, example for HNO_3 1M

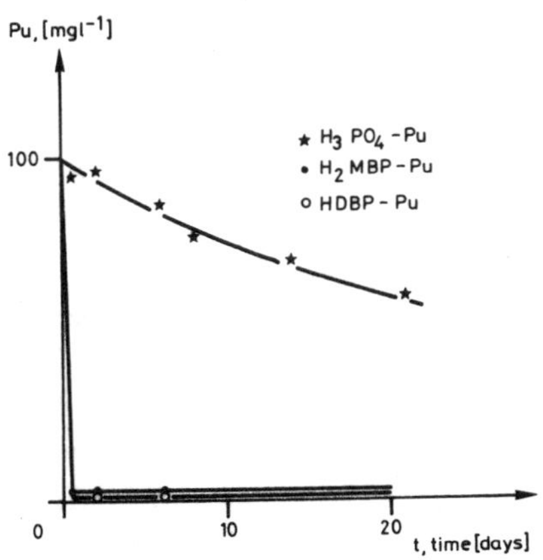

Figure 4 _ Kinetics of precipitation in aqueous_phase example for 1M HNO_3

CALCULATION OF THE DYNAMIC PRESSURE DROP IN PULSED SIEVE-PLATE COLUMNS

M. Lorenz*, H. Haverland**, A. Vogelpohl*

An analytical model which allows to calculate the dynamic pressure drop in a liquid-pulsed sieve-plate extraction column is presented. The model is based on equations known from the literature and does not require any experimentally determined parameters. The applicability of the model has been verified by pressure drop measurements for a broad range of operating parameters. The model permits the computation of the dispersed phase hold-up from dynamic differential pressure curves and allows to calculate the pressure peaks.

INTRODUCTION

Pulsed sieve-plate columns are frequently employed for the extractive separation of liquid mixtures. Because of the input of mechanical energy, this type of extractor offers greater operational flexibility in comparison with unpulsed extraction columns. It allows a relatively high throughput and gives a very high efficiency of separation at the same time.

Information on the periodical behaviour of the differential pressure as a result of the pulsation is of vital importance for the design of the apparatus (in particular the pulsator) on the one hand, and for monitoring during operation (on-line measurement of the disperse phase hold-up by the differential pressure method) on the other hand. Within the scope of a research project on fluid dynamics in pulsed sieve-plate columns, pressure drop measurements were performed during one- and two-phase operation with plates of different geometry under varying pulsation conditions and liquid throughput in the column.

On the basis of these measurements, a mathematical model developed by Jenkins and Myers (1) has been modified in such a way that the dynamic pressure drop in pulsed sieve-plate extractors can be calculated in advance without the need of supplementary experimental determination of parameters.

---

* Institut für Thermische Verfahrenstechnik, Technical University of Clausthal, Clausthal-Zellerfeld, FRG

** Schott Glaswerke, Mainz, FRG

During our measurements, the analytical model was applied predominantly for eliminating the dynamic and frictional pressure drop contributions from the differential pressure signal recorded during the two-phase operation in the active section of the column, and for determining the disperse phase hold-up from the remaining hydrostatic differential pressure component.

TEST APPARATUS AND MEASURING TECHNIQUE

The test apparatus employed is illustrated schematically in figure 1. The central part of the system is a pulsed sieve-plate extraction column with a nominal diameter of 80 mm and an active column height of 4.3 m. The active section of the column consists of four calibrated glass sections with a length of 1 m each. Stainless steel rings with threaded connections have been installed between the glass sections for connecting the measuring probes. The geometrical data of the column are given in table 1. The toluene-water system proposed by the EFCE (2) was employed as the test system. The physical properties of this system are presented in table 2. During the measurements, the test liquids were circulated through the system (storage vessels, flow meters, column, storage vessels). The lower-density organic phase (toluene) was dispersed and supplied to the column below the sieve-plates by means of a distributor. The continuous phase (water) flowed countercurrently with respect to the drop swarm ascending in the column. A sinusoidally pulsating motion was generated by means of a piston-type metering pump attached to the bottom of the column.

The measurements were performed with sieve-plates whose geometrical data are detailed in table 3. The sieve-plates were punched from stainless steel plates and suspended on support rods in the column. Because of the construction method, the edges of the holes in the sieve-plates are sharp on one side and somewhat rounded on the other. The sieve-plates were installed with the sharp-edged side facing upward, that is, in the direction of flow of the dispersed phase. Experience has shown that installation in this direction ensures proper shearing of the nonwetting drops at the edges of the holes in the stainless steel sieve-plates.

Two different types of liquid-liquid differential pressure transducers were employed for measuring the differential pressure. The technical specifications of the transducers are presented in table 4. The two types of differential pressure transducers differ with respect to their response time, which has an average value of 0.5 s for type 1 (Foxboro 823 DP) and 0.04 s for type 2 (Setra C 228-1). The signals were recorded and processed by a personal computer which was equipped with additional signal converters (analogue-digital, digital-analogue, voltage-frequency). The response time of the converters was decidedly shorter than that of the differential pressure transducers.

THEORETICAL CONSIDERATIONS

The basis for the theoretical considerations is provided by a model developed by Jenkins and Myers (1) for calculating the dispersed phase hold-up in a pulsed sieve-plate extraction column from the differential pressure signal measured during operation. The differential pressure signals are recorded by means of the measuring system shown schematically in figure 2. It consists of a differential pressure transducer and two pipelines filled with the continuous phase. These lines connect the respective chambers of the differential pressure transmitter with the upper and lower ends of the test section.

Dynamic pressure drop in a pulsed sieve-plate extraction column

During two-phase operation, a total pressure difference is recorded with the differential pressure transducer. This overall value consists of a static component Δp_S, and a dynamic component. The dynamic component is due to the frictional pressure drop at the sieve-plates Δp_F and the acceleration of the liquid content Δp_I.

Thus, the total differential pressure recorded by the transmitter Δp_T is given by

$$\Delta p_T(t) = \Delta p_S + \Delta p_I(t) + \Delta p_F(t) \tag{1}$$

where

$$\Delta p_S = (\varrho_c - \varrho_m) \cdot g \cdot H , \tag{2}$$

$$\Delta p_I(t) = \varrho_m \cdot y''(t) \cdot H , \tag{3}$$

$$\Delta p_F(t) = N \cdot \xi \cdot \frac{\varrho_m}{2} \cdot y'(t) \cdot |y'(t)| , \tag{4}$$

$$\varrho_m = \varrho_c - x \cdot \Delta\varrho = \varrho_c - x \cdot (\varrho_c - \varrho_d) . \tag{5}$$

With the instantaneous velocities of the dispersed and the continuous phase given by

$$y'_c(t) = \pi \cdot f \cdot a \cdot \cos(2 \cdot \pi \cdot f \cdot t) + \frac{v_c}{1-x} \tag{6}$$

and

$$y'_d(t) = \pi \cdot f \cdot a \cdot \cos(2 \cdot \pi \cdot f \cdot t) - \frac{v_d}{x} \tag{7}$$

the instantaneous velocity of the fluid is given by

$$y'(t) = \pi \cdot f \cdot a \cdot \cos(2 \cdot \pi \cdot f \cdot t) + v_c - v_d . \quad (8)$$

For the total instantaneous pressure difference $\Delta p(t)$ between the measuring points of the active column section follows

$$\Delta p(t) = \Delta p_T(t) + \varrho_c \cdot g \cdot H. \quad (9)$$

The following assumptions have been made in the equations (1) to (9):

- In comparison to the frictional pressure drop the internals of the column, all other frictional pressure drops (i.e. between the two phases or between the fluid and the interior wall of the column) are negligible.
- The pressure drop associated with the drop formation is neglected.
- The liquid contained in the column follows the sinusoidal oscillation of the pulsator.
- The resistance coefficient depends only on the geometrical dimensions of the sieve-plates and is equal in both directions of the flow.

Drag coefficient of the sieve-plates

In order to apply equation (9) for calculating the dynamic pressure drop, the drag coefficient of the sieve-plates during the nonsteady-state flow has to be known in addition to the temperature-dependent physical properties of the system, the geometrical dimensions of the measuring equipment, and the hold-up. Since no information on these are available from the literature, it is assumed in the following calculations that the drag coefficient of the sieve-plates at any time of the pulsating motion corresponds to that of the instantaneous velocity for steady-state flow. Semiempirical equations are available from the literature for the determination of this drag coefficient.

For deriving these equations, the sieve-plate is regarded either as a single-hole plate or as a venturi nozzle. Application of the momentum theorem and the Bernoulli equation results in the following relationship for the drag coefficient of the sieve plate in the case of the single-hole plate

$$\xi = \frac{1}{\varphi^2} \cdot \left[\frac{1}{\alpha} - 1 \right]^2 + \xi_F(Re) + \left[\frac{1}{\alpha \cdot \varphi} - 1 \right]^2 . \quad (10)$$

The first term of equation (10) describes the drag coefficient for a sudden constriction of the cross-sectional area. The second term accounts for the frictional pressure drop at the hole wall, and the third term represents the drag coefficient for a discontinuous enlargement of the cross-section.

On the basis of the theoretically derived equation (10), Zelfel (3) determined the matching parameters of the model equation for a total of 34 different sieve-plates. The investigations were conducted with a column of a semiindustrial scale with air-water system. In the course of these investigations, Zelfel found that the frictional and constrictional pressure drop can be neglected for thin sieve-plates (ratio of plate thickness to hole diameter far less than unity, $b/d_h \ll 1$). The drag coefficient is thus calculated from the pressure drop for a discontinuous enlargement in the cross-section with due consideration of the jet constriction by the following equation

$$\xi = \left[\frac{1}{\alpha \cdot \varphi} - 1 \right]^2 . \tag{11}$$

For the calculation of the contraction ratio α, the relationship

$$\alpha = 0.88 \cdot (t_h/d_h)^{-0.2} , \tag{12}$$

determined empirically by Zelfel, provides a good approximation.

Other equations for the calculation of the drag coefficient which are based on the theoretically derived equation (10) are presented in the publications of Bohl (4), Mersmann (5) as well as Hunt, Hanson, Wilke (6).

If the sieve-plate is regarded as a venturi nozzle, the Bernoulli equation can be employed for demonstrating that the drag coefficient is given by the following relationship

$$\xi = \left[\frac{1 - \varphi^2}{C^2 \cdot \varphi^2} \right] . \tag{13}$$

The quantity C is the orifice coefficient and takes into account in an integral manner the pressure drop contributions caused by the cross-sectional constriction, the friction and the subsequent cross-sectional enlargement. The investigations conducted by Hughmark, O'Connell (7), Kolodzie et al. (8) and Smith, van Winkle (9) had as an objective the determination of this coefficient as a function of the plate geometry, the properties of the drilled or punched sieve-plate orifices as well as the flow velocity of the medium under investigation. A typical example of these equations is the relationship between the orifice coefficient and the ratio of the hole diameter to the plate thickness, as determined by Hughmark and O'Connell (7) and presented in figure 3.

Determination of the hold-up from the dynamic differential pressure curves

The average hold-up in the measuring zone under consideration can be determined from the dynamic pressure curves recorded by the differential pressure transducer. Insertion of equations (2) to (8) into equation (1) and subsequent integration over a pulsation period yields the following relationship for the calculation of the hold-up

$$x = \frac{\dfrac{\overline{\Delta p_T}}{H} - N \cdot \xi \cdot \dfrac{\varrho_c}{2} \cdot I_m}{g \cdot \Delta\varrho - N \cdot \xi \cdot \dfrac{\Delta\varrho}{2} \cdot I_m} \qquad (14)$$

where

$$I_m = f \int_0^{1/f} y'(t) \cdot |y'(t)| \cdot dt \qquad (15)$$

and

$$\overline{\Delta p_T} = f \int_0^{1/f} \Delta p_T(t) \cdot dt \ . \qquad (16)$$

For a phase ratio of unity, equation (14) reduces to

$$x = \frac{\overline{\Delta p_T}}{g \cdot \Delta\varrho \cdot H} \ . \qquad (17)$$

VERIFICATION OF THE MODEL FROM MEASUREMENTS

The verification of the model thus established for calculating the dynamic pressure curves in a pulsed sieve-plate extraction column was performed in three steps. The drag coefficients of the sieve-plates employed were first measured for steady-state flow and compared with the equations from the literature. With the use of the drag coefficient thus obtained, the model concept was examined on the basis of one- and two-phase measurements of the dynamic pressure. Finally, the hold-up calculated from the dynamic measurements was compared with the hold-up determined by the displacement method. These individual steps are explained in the following.

Drag coefficients of the sieve-plates

The drag coefficients of the sieve-plates used were measured by means of single-phase, steady-state experiments without pulsation at a superficial velocity between 0.0 and 6.8 cm/s for the continuous phase. This range of the superficial velocity corresponds to the maximum instantaneous velocities of the continuous phase during two-phase operation of the column up to a pulsation intensity of about 2.0 cm/s. The differential pressure was measured with transducer 1 over 21 sieve-plates (see table 4). The signals from the differential pressure transducer were detected with a voltage-frequency converter and evaluated by means of equation (4).

In figure 4, the measured pressure difference is plotted as a function of the square of the established superficial velocity. For all sieve-plate versions, a linear relationship holds for the two variables over the range under investigation. The drag coefficient of an individual sieve-plate, which is independent of the superficial velocity, can be determined from the slope of the curve ($N \cdot \zeta \cdot \varrho_m / 2$). This drag coefficient is independent of the direction in which the sieve-plates are installed. However, attention must be paid to the fact that the drag coefficient thus obtained describes not only the pressure drop at the sieve-plate, but also the pressure drop caused by wall friction. An estimate indicates that the wall friction is appreciable only in the case of sieve-plates with a hole diameter of 6 mm at low superficial velocity (less than 1.7 cm/s). In all other cases, the wall friction can be neglected.

In table 5, the drag coefficients determined for the sieve-plates employed are compared with the drag coefficients calculated with equations from the literature. The experimental results correlate with the equations given by Zelfel (equation (11) and (12)). In contrast, equation (13) which is usually applied in the Anglo-Saxon literature in combination with the orifice coefficient determined as by Hughmark (7) yields considerably higher drag coefficients. For this reason, the drag coefficient is calculated in accordance with Zelfel for the remaining verification of the model.

Dynamic pressure drop measurements for one- and two-phase operation

After evaluation of the model with respect to the drag coefficient of the sieve-plates, numerous dynamic pressure drop measurements have been performed for one- and two-phase operation of the pulsed sieve-plate column.

During the single-phase experiments, the pulsation amplitude was varied between 6 and 12 mm in the range of pulsation intensity between 0.75 and 2.0 cm/s as usually encountered in practice. The superficial velocity of the continuous phase was also varied. The investigations were conducted on sieve-plates with a hole diameter of 4 mm. The differential pressure was measured with transducer 2. The electrical signals from the differential pressure transducer were converted with a 12 bit analogue-digital converter and recorded by the personal computer. The recording of the differential pressure signals was synchronized with the pulsator and was started at the instant of zero-point passage for the pulsating motion during the downward stroke. This corresponds to t = 0 in equation (6). The signals were recorded in cycles of 10 ms for the duration of a pulsation cycle.

In figures 5 and 6, differential pressures measured as functions of time for two selected operating points are compared with the differential pressures calculated from equation (1). The measured curves are typical for all single-phase experiments. The reproduction of the measured curves by the model equation is excellent. Similar good agreement between measured and calculated curves has been obtained for all combinations of amplitude and frequency investigated for different superficial velocities of the continuous phase.

A high-frequency oscillation is superimposed on all measured pressure curves. This oscillation results from the high sensitivity of the differential pressure transducer and is probably due to the vibration transmitted from the pulsator to the column framework, on which the differential pressure transducer was also mounted.

The two-phase experiments were conducted at various liquid loads and different phase-ratios. The acquisition and processing of the signals proceeded in a manner analogous to that for the single-phase experiments. For calculating the pressure difference from equation (1), the hold-up must be known. For this reason, the pressure difference was first measured for a specified operating point and the hold-up was subsequently determined by the displacement method. With this method of hold-up measurement, the dispersed phase component in the column is determined from the difference in the height of the interface levels during operation and after shutting off all inlets and outlets and subsequent collection of the dispersed phase volume at the top of the column. With the hold-up thus measured, the pressure differences for this operating point are calculated from equation (1) and compared with the measured pressure differences.

Figure 6 shows a comparison of the differential pressure measured during a two-phase experiment at a phase ratio v_d/v_c of 3 with the pressure curve calculated from the model equation. The model calculation reproduces the measured pressure differences with sufficient accuracy. However, the recorded maximum of the differential pressure during the upward stroke is slightly higher than the value computed from the model. A possible cause of this deviation is the neglect of the additional pressure drop resulting from drop formation during the upward stroke in the analytical model.

In conclusion, it can be stated that the model describes the measured differential pressure curves with sufficient accuracy for one- and two-phase experiments.

Comparison between measured and calculated hold-up

After ensuring that the differential pressure signals measured during two-phase operation can be described by the analytical model presented, the technique for measuring differential pressure was applied to the hold-up measurement during hydrodynamic investigations. The test programme encompassed measurements with three different systems (toluene-water, n-butylacetate-water, n-butanol-water) and the sieve-plates specified in table 3 with a variation of the pulsation intensity (between 0.0 and 2.0 cm/s) and throughput (60 and 80 per cent of the maximum throughput) at a constant phase-ratio of unity.

During the experiments, the pulsation frequency was first determined with a light barrier and the temperature of the liquid was measured with a thermocouple. The densities of the liquids were then calculated for the measured temperature. The hold-up was finally calculated from equation (17) by subsequent integration of the differential pressure curves measured by the computer by means of transducer 1 with a voltage-frequency converter. The calculated hold-up values were compared with the dispersed phase contributions determined by the displacement method.

In figure 8 the hold-up curves measured with the two methods for various combinations of the liquid/liquid system and the plate geometry are compared. It is evident that the differential pressure method is equivalent to the displacement method over the entire range of pulsation intensity. In the mixer-settler range too, the differential pressure method always yields good results. A hold-up range from 5 to 25 per cent was checked by variation of the throughputs. The relative deviation between the hold-up values measured by the two methods was between 0 and a maximum of 15 per cent.

In applying the differential pressure method for the hold-up measurement, it must be ensured that the tubing is always completely filled with the continuous phase. If this is not the case, the tubing must be flushed before operation with the continuous phase until they are free of solvent residues and bubbles. This can be checked by measuring with the column filled with continuous phase where the integrated differential pressure must yield a value which differs from zero only slightly when the pulsator is switched on and the inlets and outlets of the column are closed.

CONCLUSIONS

On the basis of a model proposed by Jenkins and Myers a reliable method has been developed for calculating the dynamic pressure drop in pulsed sieve-plate extraction columns over a wide range of operating parameters. The model can be employed to determine hold-up with the differential pressure method as well as in the design of pulsators and in the strength analysis of sieve-plates.

ACKNOWLEDGEMENT

The authors wish to thank the Arbeitsgemeinschaft Industrieller Forschung e. V. for financial support of the test programme and Dipl.-Ing. K. Eckert and Dipl.-Ing. V. Nitsche for performing the experiments.

NOMENCLATURE

a pulse amplitude (stroke) (m)
b sieve-plate thickness (m)
C orifice coefficient
D column diameter (m)
d_h hole diameter (m)
f pulse frequency (s^{-1})
g acceleration due to gravity (m/s^2)
H length of the test section (m)
h_p plate distance (m)
N number of plates in the test section
p pressure drop (Pa)
Re Reynolds number
t time (s)
t_h pitch of the sieve-plate holes (m)
v superficial velocity (m/s)
y' velocity of the liquid in the column (m/s)
y'' acceleration of the liquid in the column (m/s^2)
x dispersed phase hold-up
ϱ density (kg/m^3)
φ fractional free area
ζ drag coefficient
α contraction ratio

Subscripts

c continuous phase
d dispersed phase
F friction
I inertia
m mean
T transmitter
Δ difference

REFERENCES

1. Jenkins, J. A., Myers, P. E., 1984, EFCE Publication Series No. 43, Extraction '84, 217-230.

2. Misek, T. (Editor), 1985, "Standard Test Systems for Liquid Extraction", EFCE Publication Series No. 46.

3. Zelfel, E., 1966, "Experimentelle Untersuchungen des Druckverlustes an Siebböden", Ph.D. Thesis, TU München, FRG.

4. Bohl, W., Wagner, W., 1978, "Technische Strömungslehre Aufgaben und Lösungen", Vogel-Verlag, Würzburg, FRG.

5. Mersmann, A., 1980, "Thermische Verfahrenstechnik", Springer-Verlag, Berlin, Heidelberg, New York.

6. Hunt, C.D'A., Hanson, D. N., Wilke, R., 1955, AIChE J. 4, 441-451.

7. Hughmark, G.A., O'Connell, H.E., 1953, Chem. Eng. Prog. 3, 127-132.

8. Kolodzie, P.A., van Winkle, M., 1957, AIChE J. 3, 305-312.

9. Smith, P.L., van Winkle, M., 1958, AIChE J. 3, 266-268.

TABLE 1 Geometric parameters of the test column

Column diameter	m	0.072
Column height	m	6.4
Active height	m	4.3
Settler diameter	m	0.15
Volume	m^3	0.049
Plate distance	m	0.1
Number of sieve-plates	-	43

TABLE 2 Physical properties of the test systems

Liquid	Density kg/m^3	Viscosity $10^{-3} \cdot Pa \cdot s$	Surface tension $10^{-3} \cdot N/m$	Interfacial tension in Water $10^{-3} \cdot N/m$
Water	998.2	1.003	72.75	0.00
Toluene	866.7	0.586	28.40	35.40
Butylacetate	881.5	0.730	24.90	14.10
Butanol	810.0	2.948	24.60	1.75

TABLE 3 Sieve-plate parameters

Hole diameter $10^{-3} \cdot m$	Pitch $10^{-3} \cdot m$	Thickness $10^{-3} \cdot m$	Fractional free area
2.0	4.0	1.0	0.23
4.0	6.0	1.5	0.39
6.0	8.0	1.5	0.49

TABLE 4 Specifications of the pressure transmitters

Typ	Pressure range $10^{-2} \cdot Pa$	Response time s	Accuracy
1	-5 ÷ 20	0.5	±0.2% of calibrated span
2	-34 ÷ 34	0.03÷0.05	±0.15% of full span

TABLE 5 Comparison of calculeted and measured drag coefficients

Drag coefficient	Fractional free area		
	49%	39%	23%
measured in present work	3.2	5.1	21.7
calculated from Zelfel [1]	2.1	4.7	21.9
calculated from Hughmark [7] with fig. 3	6.3	9.9	30.7

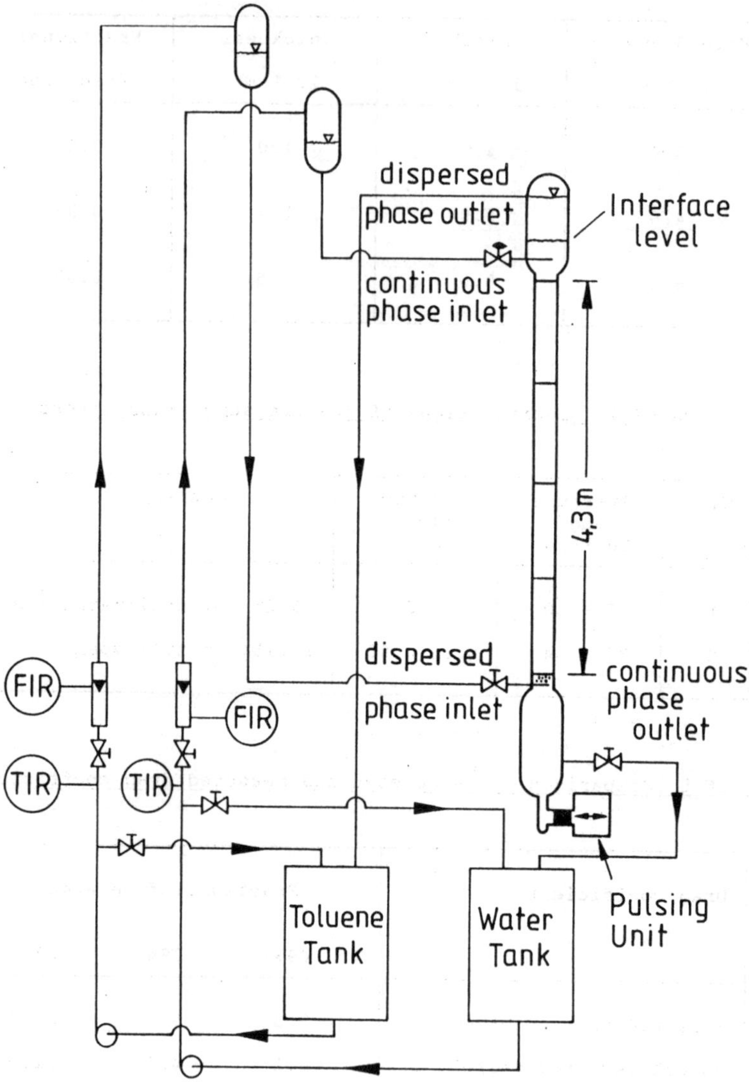

Fig. 1 Experimental set-up

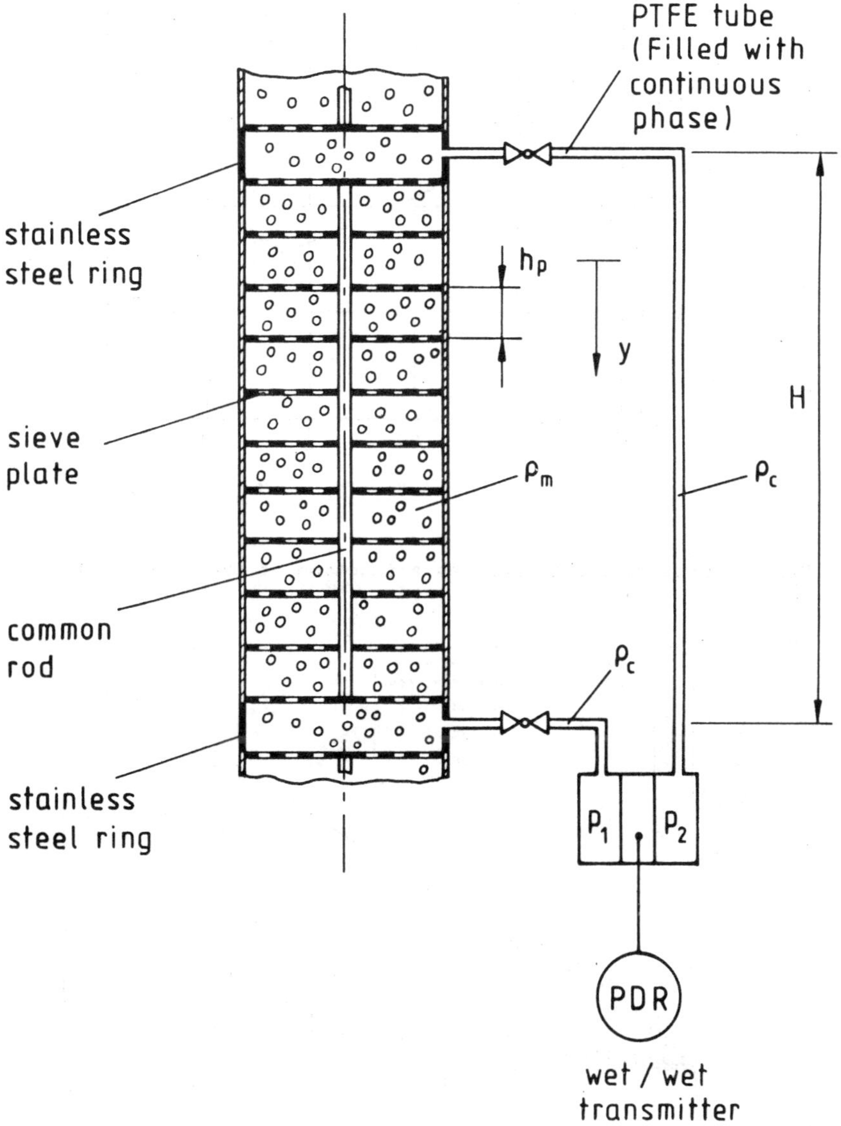

Fig. 2 Scheme of pressure drop measurement technique

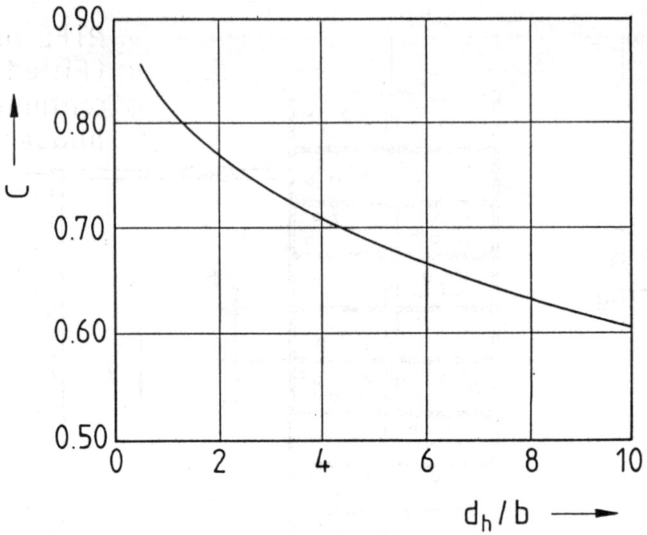

Fig. 3 Orifice coefficient as a function of the ratio d_h/b (7)

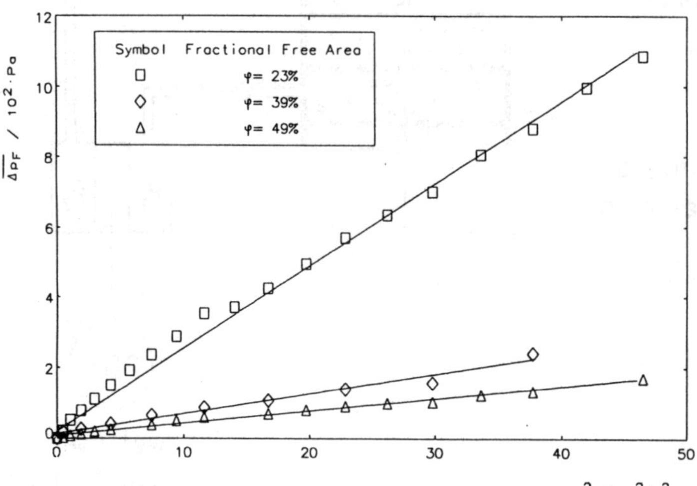

Fig. 4 Pressure drop due to friction for steady-state flow

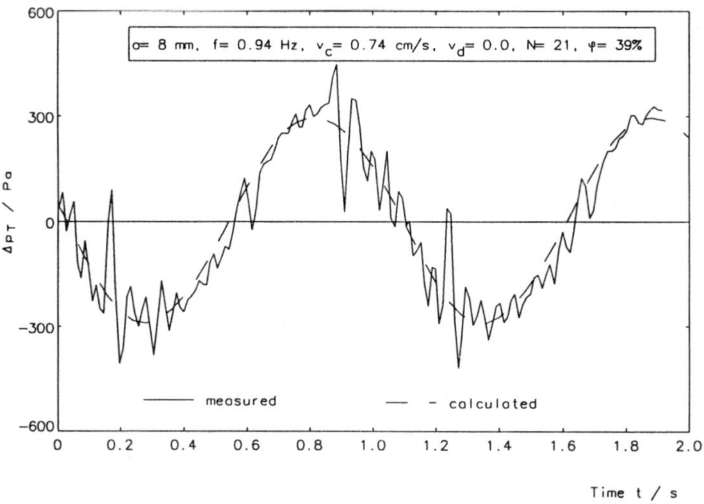

Fig. 5 Pressure drop signals for single-phase operation (low pulsation intensity)

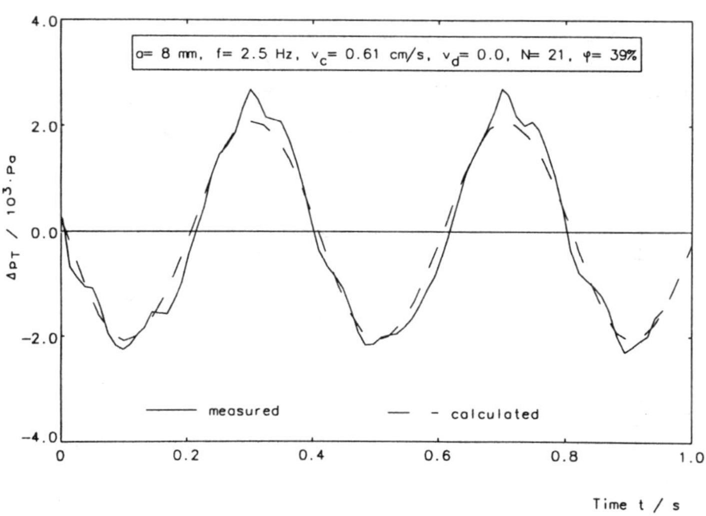

Fig. 6 Pressure drop signals for single-phase operation (high pulsation intensity)

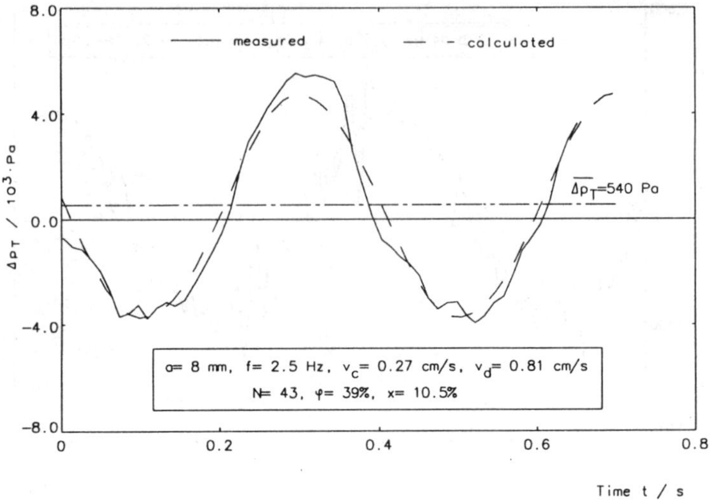

Fig. 7 Pressure drop signals for two-phase operation

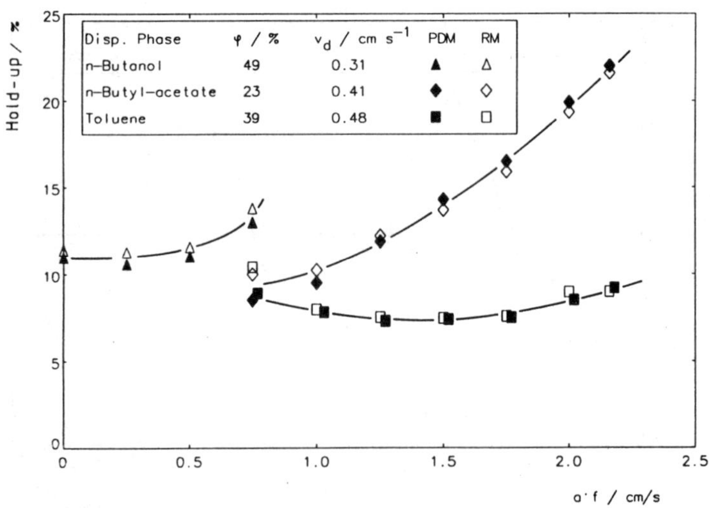

Fig. 8 Comparison of hold-ups calculated from pressure drop measurements and measured with reference method (a=8 mm, v_d/v_c=1, PDM=pressure drop method, RM=reference method)

INTERFACIAL INSTABILITY AT LIQUID/LIQUID EXTRACTION WITH CHEMICAL REACTION

L.M.Rabinovitch* and I.A.Belikova*

Linear interfacial instability analysis is carried out for liquid/liquid system with a finite depth of the phases. The following cases are considered: (i) one-component extraction;(ii) extraction accompanied by chemical reaction at the interface; (iii) interfacial reaction, with the reagents dissolved in contacting phases. New regimes of instability are obtained, which are determined by the ratio of the depths of phases, the ratio of diffusion coefficients as well as the reaction rate constant and by surface activities of the two components changing surface tension.

INTRODUCTION

A lot of extraction processes in liquid/liquid systems are accompanied by interfacial instability phenomena, which substantially effect the properties of an interface and mass transfer rate. These phenomena are caused by interaction of hydrodynamic, diffusion and chemical processes under the Marangoni effect. While the surface-active substance is transfering through the interface and taking part in interfacial chemical reaction, the surface concentration is changing continuously, even new interfacial structures may be formed. That leads to local surface tension gradients, which may cause a loss of hydrodynamic stability of the interface and thus a substantial change of hydrodynamics near the interface; this in turn may lead to essential change of mass transfer kinetics (1,2). As a result of such a coupling of different processes in the system spontaneous convection may occur in the form of spatially ordered convective circulating cells or interfacial turbulence which is characterized by violent chaotic motion of an interface, ruptures and eruptions. Such phenomena may lead to a sharp increase of mass transfer rate and thus the rate of extraction (3-5).

The investigation of mass transfer and interfacial reaction effect upon the conditions for the convective instability to occur becomes rather interesting for the study of hydrodynamic interactions and microkinetic regularities of diffusion controlled reactions in liquid phase. The question under investigation seems to be also of considerable practical interest in connection with the use of the regimes with interfacial instability

* Karpov Institute of Physical Chemistry, Obukha 10, Moscow 103064, USSR

for intensification of a number of gas-liquid chemical processes as well as separation processes, liquid-liquid extraction in the first instance. Instability criteria obtained enable one to find optimal physico-chemical parameters for highly intensive extraction regimes as well as to look for effective extractants and solvents, whose physico-chemical properties may bring about interfacial instability.

The investigation of the conditions for the Marangoni instability to occur in liquid-liquid systems with chemical reaction has been attracting vast attention hitherto (6-8). In the paper presented the linear analysis of convective instability of the system consisting of two liquid layers is carried out for three cases: one-component extraction, extraction accompanied by chemical reaction at the interface and interfacial reaction with reagents dissolved in different phases. New regimes of non-oscillatory instability determined by depths of liquid layers ratio, the ratio of diffusion coefficients as well as chemical reaction rate constant are found. It is found also that the presence of two substances with surface-active properties substantially changes the conditions for the interfacial instability to occur compared with the situation when there is only one surface active component in the system.

PHYSICAL MODELS AND STABILITY ANALYSIS

In the coordinate system (x, y, z) two horisontal plane motionless layers infinite in x and y directions, divided by undeformable interface $y=0$ and bounded by rigid surfaces $y=h_1$ and $y=-h_2$ are considered (Fig. 1). Case (a) corresponds to transfer of component **A** from phase 2 into phase 1. In the case (b) irreversible chemical reaction of the second order between component **A** and component **B** dissoluble in phase 1 only takes place. The rate of chemical reaction is comparable with the rate of diffusion of components up to the interface. Component **A** change interfacial tension while the product of the reaction is considered to have no surface active properties. In the case (c) the reagents dissolved in phases 1 and 2 take part in irreversible chemical reaction $A+B \rightarrow C$ and completely convert into the product which goes out of the interface into phase 1. Reagent **A** and product **C** change interfacial tension. The rate of chemical reaction is described by the expression $K a_s b_s$, where a_s and b_s - the concentrations of reagents in the vicinity of the inter face, K - reaction rate constant.

Introducing stream function equations of motion of viscous incompressible fluid in the first and in the second phases (j = 1,2) may be written down as follows:

$$\left(\frac{\partial}{\partial t} + \frac{\partial \psi_j}{\partial y}\frac{\partial}{\partial x} - \frac{\partial \psi_j}{\partial x}\frac{\partial}{\partial y}\right)\Delta\psi_j = \nu_j \Delta^2 \psi_j \qquad (1)$$

where $\Delta = \partial^2/\partial x^2 + \partial^2/\partial y^2$, t - time, ν - kinematic visco sity.

Distribution of concentrations of **A**, **B** and **C** in the first and in the second phase is described by convective diffusion equations

$$\frac{\partial c_n}{\partial t} + \frac{\partial \Psi_j}{\partial y}\frac{\partial c_n}{\partial x} - \frac{\partial \Psi_j}{\partial x}\frac{\partial c_n}{\partial y} = D_n \Delta c_n \quad (n = 1, 2, 3) \tag{2}$$

where c_1 - concentration of A or C in the first phase, c_2 - the concentration of A in the second phase, c_3 - concentration of B in the first phase, D_n - diffusion coefficients, $j = 1$ at $n = 1;3$, $j = 2$ at $n = 2$.

The boundary conditions at the interface $y = 0$ and at the rigid surfaces are as follows:

$$\Psi_j = \partial\Psi_j/\partial x = 0, \quad \partial\Psi_1/\partial y = \partial\Psi_2/\partial y, \qquad y=0 \tag{3}$$

$$(\partial^2/\partial y^2 - \partial^2/\partial x^2)(\mu_2\Psi_2 - \mu_1\Psi_1) = \partial\sigma/\partial x, \qquad y=0 \tag{4}$$

$$\sigma = \sigma_0 - \alpha_1 c_1 - \alpha_2 c_2, \qquad y=0 \tag{5}$$

$$D_1 \partial c_1/\partial y = D_2 \partial c_2/\partial y, \qquad y=0 \tag{6a}$$

$$D_1\frac{\partial c_1}{\partial y} - D_2\frac{\partial c_2}{\partial y} = D_3\frac{\partial c_3}{\partial y} = K c_2 c_3, \quad c_1 = \beta c_2, \qquad y=0 \tag{6b}$$

$$D_1\frac{\partial c_1}{\partial y} = D_2\frac{\partial c_2}{\partial y} = -D_3\frac{\partial c_3}{\partial y} = -K c_2 c_3, \qquad y=0 \tag{6c}$$

$$\partial\Psi_1/\partial x = \partial\Psi_1/\partial y = 0, \quad c_1 = 0, \quad c_3 = b_0, \qquad y = h_1 \tag{7}$$

$$\partial\Psi_2/\partial x = \partial\Psi_2/\partial y = 0, \quad c_2 = a_0, \qquad y = -h_2 \tag{8}$$

Boundary condition (4) expresses the tangential stress balance at the interface, conditions (6a), (6c) describe the balance of matter at the interface in the cases (a), (b) and (c) presented in Fig. 1. The normal stress balance at the interface is not considered because the surface tension coefficient is supposed to be rather large so that the interface may be considered as the plane one even when unstable. In the problem formulated the following quantities are used: h_1, h_1^2/D_1, D_1, a_0 in order to make dimensionless the variables x, y, t, Ψ_j, c_n. The following dimensionless parameters are introduced

$$\mu = \mu_1/\mu_2, \quad h = h_2/h_1, \quad d = D_1/D_2, \quad d_1 = D_2/D_3$$

$$Ma_1 = \frac{\alpha_1 b_0 h_1}{\mu_1 D_3}, \quad Ma_2 = \frac{\alpha_2 a_0 h_2}{\mu_2 D_2}, \quad Ti = \frac{K a_0 h_1}{D_3} \tag{9}$$

Here Ma_1, Ma_2, Ti - Marangoni numbers and Tile number.

Eq. (1) - (8) have the trivial stationary solution:

$$\psi_j^{(s)} = 0, \quad c_n^{(s)} = a_n y + b_n \tag{10}$$

Because of the nonlinearity of the boundary conditions (6b) and (6c) there exists a multiplicity of steady states of (10) and the constants a_n and b_n are chosen so that the concentrations at the interface will be positive.

Let us treat the stability of the steady states described by (10) with respect to small disturbances of stream function and concentrations. We shall seek these disturbances in the form of separate Fourier-components characterised by dimensionless wave number k and complex decrement ω

$$\psi_j = \Psi_j(y)\, e^{ikx+\omega t}, \quad c_n - c_n^{(s)} = C_n(y)\, e^{ikx+\omega t} \tag{11}$$

After substituting (11) into (1) and (2) and linearizing the equation obtained one may get the set of equations for the amplitudes of the disturbances $\Psi_j(y)$ and $C_n(y)$. These equations have a nontrivial solution only at the certain value of spectral parameter ω . Decrements ω must be found as eigenvalues of the boundary problem, the appropriate eigenfunctions Ψ_j and C_n determining the structure of characteristic disturbances of fluid velocity and concentrations.

Now we shall investigate the marginal stability of the system at different values of the physico-chemical parameters.

After substituting the solutions for Ψ_j and C_n into boundary conditions (3) - (8) one obtains a system of algebraic equations for the constants, 10 equations in the case (a), 14 - in the cases (b) and (c). These uniform sets of equations have nontrivial solutions only if their characteristic determinants equal zero.

The appropriate characteristic equations may be put in the following form

$$\mathcal{F}\,(k, Ma_1, Ma_2, Ti, h, \mu, d, \dots) = 0 \tag{12}$$

These equations establish the connection between wave number k and all other parameters of the system in question. Numerical solution of the equation obtained make it possible to study the dependence of the Marangoni numbers upon wave number; also the Marangoni number - Tile number dependence at different values of k. In order to find neutral stability curves dividing stable ($Re\,\omega < 0$) and unstable ($Re\,\omega > 0$) regions one has to determine the roots of complex function $\mathfrak{F}$, given in the determinant form. The method of parabols has been used for that, the determinant values having been found by means of a standard program; the double precision calculations have been carried out on an ES computer.

In Fig.2-12 presented are the neutral stability curves at different values of the parameters pointed out in (9). These curves enable one to study the stability bound dependence upon physico-chemical and technological parameters as well as to find critical Marangoni and Tile numbers. They also make it possible to obtain the wavelengths of the disturbances which grow most quiquely with enhancement of the Marangoni number and thus determining to a greate extent the convective dissipative structure being formed as a result of convective instability and further stabilization of the disturbances.

In Fig. 2-5 are presented the plots of Ma against k and Ma against Ti ($Ma = Ma_1(a_0/b_0)$, $Ma_2 = 0$) for the case of extraction with chemical reaction (b in Fig. 1) at μ =0,5 , $d_1 = 5/6$, $\beta = 5$, $b_0/a_0 = 1.3$. Analysis of the curves presented leads one to the conclusion that the stability of the system strongly depends upon the ratio of diffusion coefficients of the transfering substance which change the surface tension. Depending on the ratio d the system may become unstable for the transfer of substances lowing the surface tension (surface active substances, SAS) and for the transfer of substances raising it (surface inactive substances, SIS) as well. While reaction rate and the Tile number increase the absolute values of the critical Marangoni numbers corresponding to the extremums of the curves. This indicates that chemical reaction may stabilize the system.

There exist certain values of d at which Tile number enhancement causes the principal change in the stability characteristics. For example Fig. 3 shows that for the case d = 0.7 only mass transfer of SAS causes interfacial instability at Ti = 0: 1, while at Ti =1.6, 3 - only mass transfer of SIS makes the system unstable. At the same time if Ti = 1.45, 1.5 the instability occurs if either SAS or SIS are transfered. Thus, the variation of the Tile number and particularly the reaction rate constant leads to a change in the stability for a substance with a given surface activity.

The analysis of the neutral stability curves presented in Fig. 4 leads to the following conclusion: at different thickness of the contacting layers there exist three intervals of diffusion coefficients ratio d corresponding to different regimes of instability. One of the regimes occus when instability appears at the transfer of SAS, the other at the transfer of SIS only. The third regime is characterized by instability occurance at the transfer of components with different of surface activity, i.e. at the fixed value of d and given mass transfer direction the instability may be brought about both by the transfer of SAS and by the transfer of SIS. The regimes pointed out are realized in the absence of chemical reaction ($Ti = 0$) and in its presence ($Ti = 1;3$) as well. Comparison of the curves in Fig. 4 enables one to illustrate the effect of the Tile number on the threshold of convective instability at $h > 1$. An increase in the reaction rate constant leads to a rise in the absolute values of the critical Marangoni numbers for each of the three regimes.

In Fig. 5 are presented the curves characterizing the correlation between the numbers Ma and Ti corresponding to the stability boundary ($\omega = 0$). The regions above the curves at $Ma > 0$ and below them at $Ma < 0$ correspond to unstable states. Vertical asymptotes existing at $d = 0.2, 0.3$ testify to the possibility of interfacial instability both for the transfer of SAS and for the transfer of SIS.

In Fig. 6-12 are presented the neutral stability curves for the case of interfacial reaction where the reagent A and product C change the surface tension (c in Fig. 1). The calculations have been carried out at $b_0 / a_0 = 1.3$, $\mu = 1.8$, $d_1 = 5/6$ ($Ma_A = Ma_2$, $Ma_C = Ma_1$, $Ti = K b_0 h_1 / D_3$).

The behaviour of the curves in Fig. 6 testifies to the fact that the stability characteristics of the system in question substantially depend upon the ratio of the diffusion coefficients of components A and C possessing surface active properties.

The sets of neutral stability curves in Fig. 7, 8 show the effect of the surface activity of the second component and the corresponding Marangoni number Ma_C on the stability of the system for the case when the reagent diffusion coefficient is larger than the diffusion coefficient of the product ($d = 0.8$), as well as for the case when the diffusion coefficients ratio becomes more than one ($d = 2$). At $d = 0.8$ the product decreasing surface tension stabilizes the system while the product increasing surface tension destabilizes it. At $d = 2$ the situation becomes opposite, i.e. the surface active product promotes instability development while the inactive product suppresses it. The curves in Fig. 7-9 show that an increase of reaction rate constant leads to an increase of critical Marangoni number and thus stabilizes the system.

It follows from the behaviour of the curves in Fig. 10 , that there exist certain values of the ration of the depths of phases at which the system becomes unstable both in the presence of surface active reagents and in the presence of inactive ones. The conclusion about the influence of parameter h on the conditions for interfacial convection to occur and on its intensity is in accordance with the results of experiments in (5) .

In Fig. 11 are presented curves of Ma_A against Ti which devide stability regions (below the curves) from regions of instability (above the curves) at different values of d and Ma_C. These curves are not monotonous and they have minima which indicate the stabilizing effect of chemical reaction on convective stability for the system at low and high Tile numbers. Such a conclusion makes lucid physical sence because at $K \rightarrow 0$ surface diffusion fluxes also stream to zero according to (6c) and the feed of reagents to reaction surface stops. At $K \to \infty$ concentration gradients along the interface become more smooth thus leading to the decreasing of surface tension gradients and there fore to produce convective instability one hase to increase the surface activity of the components and the corresponding Marangoni numbers.

In Fig. 12 are presented curves shows the correlation between critical Marangoni numbers Ma_A^{cr} and Ma_C^{cr} at different values of d and h.

The presented analysis of the behaviour of neutral stability curves demonstrates the variety of unstable regimes in a system with two components possessing surface active properties and taking part in a chemical reaction.

CONCLUSIONS

The mechanism of hydrodynamic interaction in liquid-liquid system is described and convective instability caused by interfacial mass transfer and heterogeneous chemical reaction is investigated. As a result of the analysis made the following basic qualitative conclusions may be drawn:

- convective instability of an interface may develop in the system without transient flux of matter through the interface in the case when interfacial chemical reaction occurs;
- dissipative processes connected with chemical reaction may enlarge the instability threshold and thus stabilize the system;
- the presence of two components which change the surface tension substantially changes the instability criteria in comparison with the case when there is only one component in the system;
- there exist two basic instability regimes associated with the sign (positive or negative) of the surface activity of the reagent or the sign of the substance transfered through the interface;
- Altering the ratio of the diffusion coefficients of the surface active components as well as the ration of the depths of the phases may lead to the principal change of the system stability character. The critical values of these parameters for the system with heterogeneous chemical reaction depends upon the reaction rate constant and correspondingly upon the Tile number;
- hydrochemical instability generation is possible also for chemical reactions with more complex kinetics.

In conclusion, we note that the use of interfacial instability phenomena seems to be one of the most universal methods of intensification of mass transfer in liquid-liquid extraction. This method deals with the utilization of extractants, reagents, solvents and surface-active microcomponents whose physico-chemical properties cause interfacial instability. Highly intensive regimes of extraction processes may be also realized by choosing concentration parameters which promote interfacial convection.

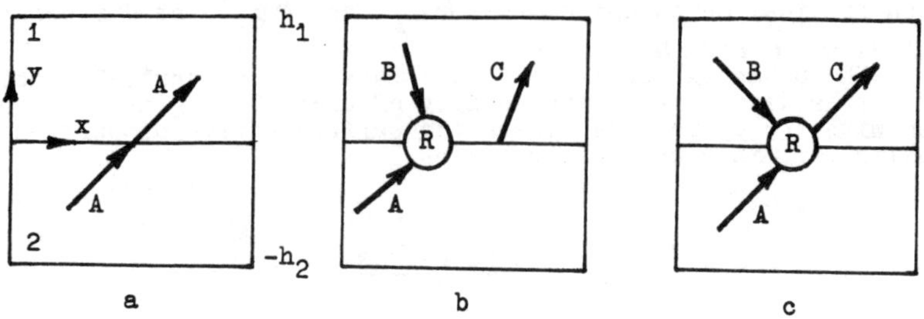

Figure I Models of two-layer systems

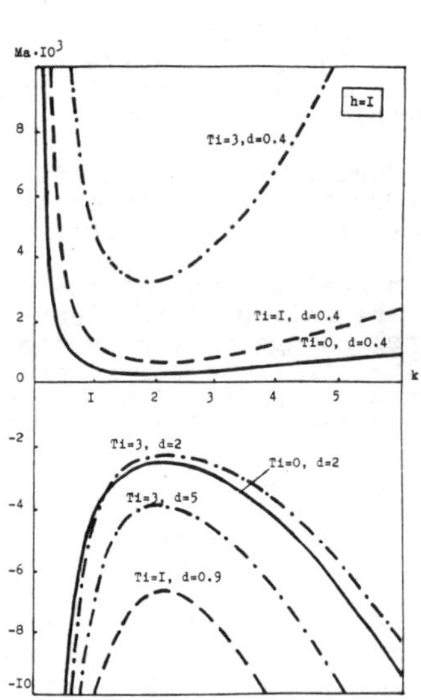

Figure 2

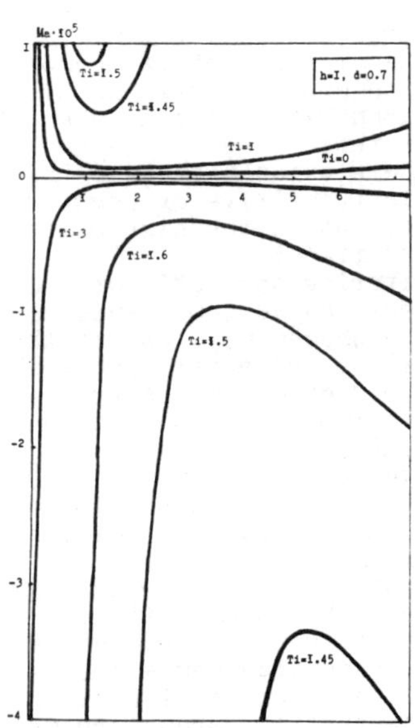

Figure 3

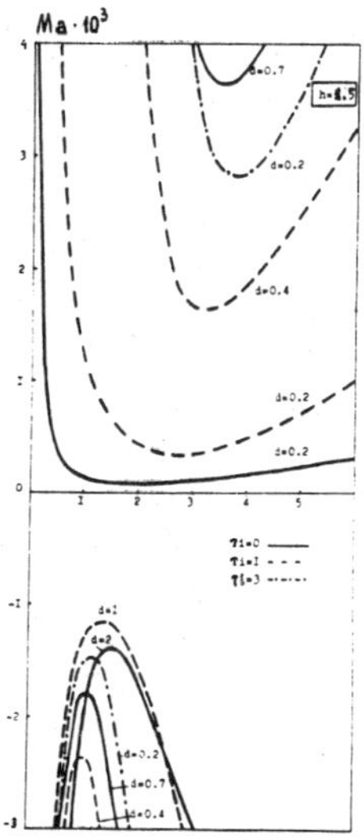

Figure 4

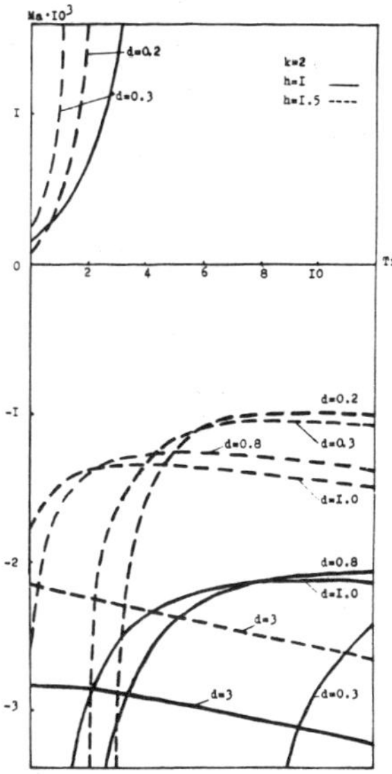

Figure 5

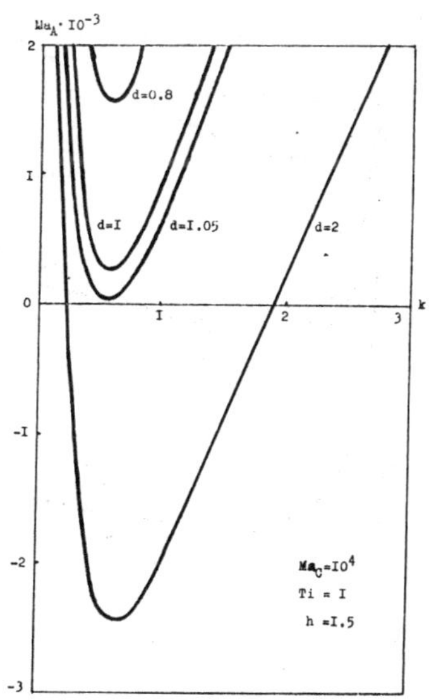

Figure 6

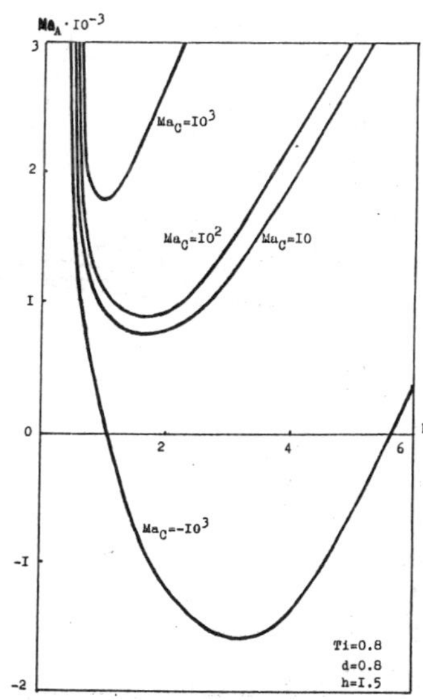

Figure 7

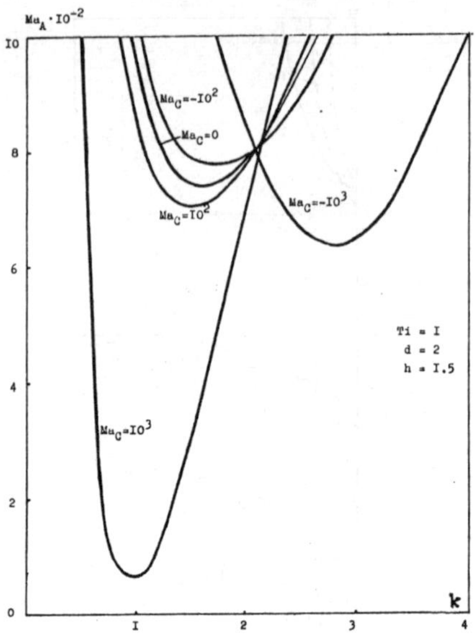

Figure 8

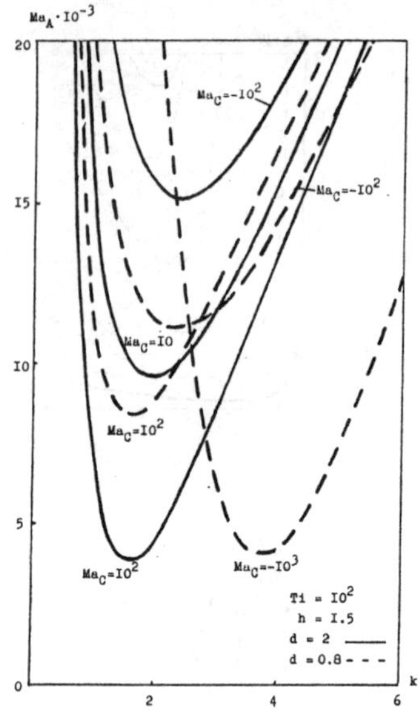

Figure 9

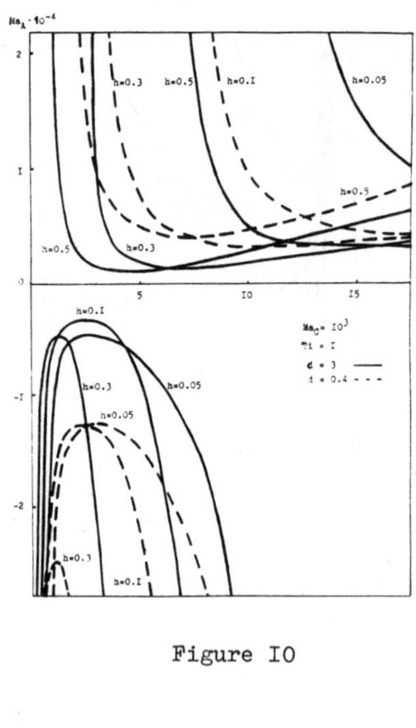

Figure IO

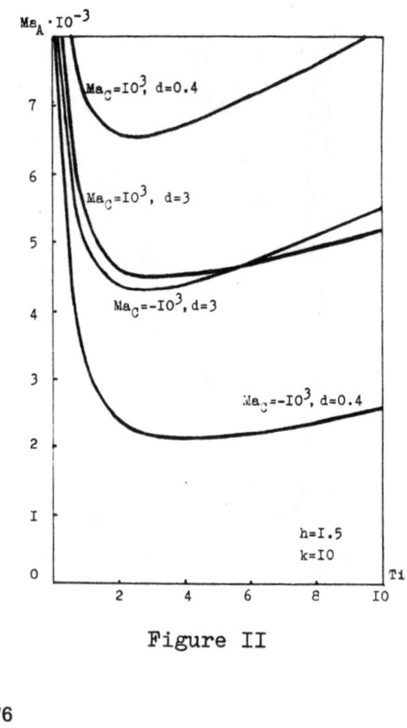

Figure II

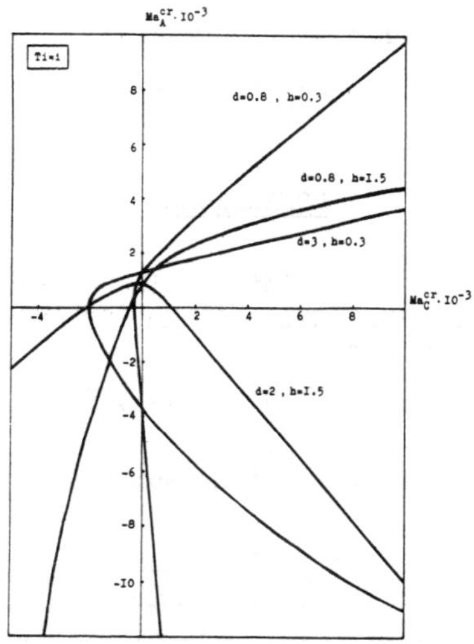

Figure I2

SYMBOLS USED

c = concentration

D = diffusion coefficient

h = liquid layer depth

K = reaction rate constant

k = wave number of disturbances

Ma = Marangoni number

Ti = Tile number

α = surface activity of the substance

β = local equilibrium constant

μ = dynamic viscosity

ν = kinematic viscosity

ψ = stream function

σ = interfacial tension

ω = complex decrement of disturbances

REFERENCES

1. Sawistowski,H., 1971, "Interfacial phenomena", In"Recent Advances in Liquid-Liquid Extraction"(ed. C.Hanson), Pergamon Press, Oxford.

2. Rabinovitch,L.M.,and Slin'ko,M.G., 1988, ISEC'88, Conf.Papers, Moscow, 2, 91.

3. Thompson,P.J.,Batey,W.,and Watson,R.J., 1984,I.Chem.E.Symp.Ser., 88, 231.

4. Ermakov,A.A.,Rabinovitch,L.M.,and Slin'ko,M.G., 1988,Proc.Acad.Sci. USSR, Phys.Chem.Section., 303, 429 (in Russian).

5. Pichugin,A.A.,et.al., 1988, ISEC'88,Conf.Papers, Moscow, 2, 101.

6. Dalle Vedove,W.,and Sanfeld,A., 1981, J.Coll.Interface Sci., 84, 318.

7. Perez de Ortiz,E.S.,and Mendes Tatsis,M.A., 1988, ISEC'88, Conf.Papers,Moscow, 2, 104.

8. Golovin,A.A.,and Rabinovitch,L.M., 1988, ISEC'88, Conf.Papers, Moscow, 2, 97.

MECHANISM FOR THE REMOVAL OF ALCOHOL FROM BEER BY EMULSION LIQUID MEMBRANES

B. R. Etuk* and K. R. Murray*

A diffusion mass transfer model has been proposed for the removal of alcohol from beer by emulsion liquid membrane technique on a batch process during the period in which the membrane is stable. The model is based on the immobilised globule model and assumes interfacial equilibrium between the beer and membrane phases. The model predicts accurately the total alcohol separation and is in good agreement with experimental data obtained at longer mixing times.

INTRODUCTION

The use of emulsion liquid membrane (ELM) systems for liquid phase separation was first proposed by Li (1 - 3) about two decades ago. Since its development, the technique has been introduced to various separation processes singly or in combination to achieve the desired separation and is now generally regarded as a technology for encapsulation of materials and separation of mixtures (4 - 7). The technique has found potential applications in the separation of a mixture of hydrocarbons (3, 8 - 12), weak acids (e.g., phenol and cresol) and weak bases (e.g., ammonia and aromatic amines) from wastewater (4, 13 - 20). Other potential areas of application include the recovery of metal ions such as Cu^{2+} (14, 21 - 25), Cr^{6+} (14, 26, 27), Ni^{2+} (28), Co^{2+} (29, 30), Cd^{2+} (14), Hg^{2+} (14, 30) and U^{6+} (31, 32); anions such as phosphates, nitrates, cyanides and sulphides (4); the separation and concentration of amino acids and organic acids from fermentation broths (16, 20, 33 - 35); and the separation of alcohol from normally fermented beer (36).

Most investigations on liquid membrane transfer have been concentrated on facilitated transport within the emulsion globule (i.e., permeation with chemical reaction or carrier transport with reaction), hence several mathematical models have been proposed based on these to predict mass transfer rate and separation efficiency. In general the ELM separation systems have been modelled based on different configuration of the

*Department of Chemical & Process Engineering
Heriot-Watt University, Riccarton, Edinburgh, EH14 4AS.

emulsion globule (Figure 1) and these have been reviewed (37 - 39) and are summarised in Table 1 below:

TABLE 1 - Proposed Models for facilitated transport within the Emulsion Globule of Liquid Membrane systems.

Globule model	Description and major assumptions	Reference	Comments
Uniform flat-sheet (planar geometry)	Fine encapsulated droplets within the emulsion globule coalesced into a single droplet and mass transfer of the solute is by diffusion across a flat membrane of constant thickness into internal interface where it reacts irreversibly with reagent. Mass transfer rate is proportional to average concentration difference between external continuous and internal phases.	13,40-42	Model is a rough approximation useful only as a preliminary approach to engineering design.
Hollow-sphere model	Similar to the uniform-flat sheet model except that the single droplet is surrounded by a spherical shell of the membrane.	23,27, 29,43, 44	Model is not satisfactory as it does not reflect the emulsion globule heterogeneity that results from the presence of encapsulated internal droplets.
Hollow-sphere -advancing model	Basically similar to the hollow-sphere model except that the encapsulated internal droplets do not coalesce but are immobilised and uniformly distributed within the emulsion globule. Also the reaction of the solute with the internal reagent is irreversible	18,45 -48	Model usually overpredicts the removal rate of the solute.

	and takes place at a reaction surface which advances into the droplet centre as reagent is being consumed.		
Immobilised globule model	Solute diffuses through the external boundary layer and the reacted globule to the internal encapsulated phase where it reacts reversibly with the reagent. Local equilibrium between the internal and membrane phases exists.	17,38, 49	Locating the reaction front becomes unnecessary since satisfying reaction equilibrium turns the reaction on and off. Model usually predicts more closely the experimental results than does the advancing front theory.

The process mechanism utilised in this study is molecular diffusion of the solute (alcohol) through the liquid membrane with no chemical reaction in the internal droplet. This process has been used for the separation of hydrocarbons (2, 3, 50) and the study of membrane stability (51), and mathematical models have been proposed for the process mechanism though analysis has not been developed that is adequate for data correlation and comparative study. For instance, Li (2, 3) analysed single drop experimental data for hydrocarbon separation at steady state by the model which assumed that the transfer rate of the hydrocarbon is directly proportional to the average concentration difference of the solute between the external continuous and internal reagent phases. The liquid membrane film was assumed to be a flat sheet with constant thickness (Figure 1.a) which is quite remote from the actual geometrical configuration of the emulsion globule.

Alessi et al (52) considered the liquid membrane as a particle in the separation of the hydrocarbon mixture and assumed the volume fraction of the hydrocarbon mixture to be constant up to the surface of the liquid membrane. Also the feed and aqueous solution are assumed to have attained equilibrium during primary stirring. This model which was proposed to attempt to explain the effect on permeation rate of the different operative techniques observed in the experimental results was not very satisfactory for data analysis.

In their study, Casamatta et al (50) proposed a model of mass transfer within an emulsion globule for spherical geometry (Figure 1.e) for hydrocarbon separation. The model assumed that the oil-in-water (O/W) emulsion globule is rigid and surrounded by a peripheral water layer of infinitesimal thickness. An unsteady state diffusion process was considered. The model identifies the peripheral water layer of the O/W emulsion globule as the controlling mass transfer resistance. However, the assumption of a closely packed internal droplets in the emulsion globule is not realistic, therefore the assumption of solute diffusion across each droplet may be untenable due to large spaces which may exist between the droplets in real systems.

Kopp et al (48) proposed a model to describe mass transfer process in liquid membranes where the liquid membrane emulsion globule is considered as a sphere with a great number of immobile reaction sites embedded. The model is based on the theory of reversible sorption at the immobile sites with a finite reaction rate (53). However, any model based on this theory for diffusion through a liquid membrane is flawed because the reaction is hypothetical and the rate constant for sorption of the diffusing solute is a function of physical rather than chemical factors.

Kremesec (12) and Kremesec and Slattery (54) proposed a model based on planar geometry and integral mass balance technique for approximate analysis of experimental data obtained from a batch liquid membrane separation system. In their method the geometric and internal circulation effects are lumped into an overall mass transfer coefficient. Like other planar models proposed, this model is not realistic as it does not consider the true geometric configuration of a liquid membrane emulsion globule.

In this paper, a mass transfer model is presented based on the immobilised globule model (Figure 1.d) which assumes interfacial equilibrium between beer and membrane phases. The experimental data are analysed using measurable physical parameters on the basis of the model, and only batch processes are considered during the period of assumed emulsion stability.

Theory

The model is developed based on a typical liquid membrane emulsion globule schematically shown in Figure 2. This consists of internal aqueous droplets suspended in the external continuous aqueous phase (beer). The solute from the external continuous phase diffuses through the oil membrane of the emulsion globule into the encapsulated aqueous phase where it is dissolved. This transfer process follows the step sequence:-

(1) Transfer of solute (alcohol) from the external continuous phase to the external membrane phase interface.
(2) Diffusion of alcohol across this external interface into the oil membrane phase.
(3) Diffusion of alcohol through the oil membrane to the internal interface.
(4) Diffusion of alcohol across the internal interface into the internal aqueous phase.
(5) Dissolution of the solute in the internal aqueous phase.

As pointed out by Li (55), the membrane phase and its two interfaces offer higher mass transfer resistances than the two aqueous phases. This is attributed to the membrane viscosity, diffusivity, thickness (which is related to membrane-to-external-phase ratio), internal circulation, if any, of the internal phase droplets within a globule, and temperature. In the interfaces, the higher mass transfer resistances is attributed to surfactant types and concentrations, the surfactant molecular packing at the interface and temperature.

For mathematical treatment, the following assumptions are proposed.

(1) Only batch processes are considered.
(2) Membrane and external continuous phases as well as membrane and internal phases are totally immiscible.
(3) No chemical reaction occurs.
(4) The membrane is dispersed in the external continuous phase as rigid globules consisting of internal aqueous droplets and transport is merely by diffusion.
(5) Only the transport of alcohol is considered; the other beer constituents are considered insoluble in the internal aqueous phase.
(6) Local phase equilibrium exists at each interface, therefore at any given radial position inside the globule, the concentration of the solute in the membrane dictates the concentration of the solute in the internal phase at the same position.
(7) No internal circulation within globules, i.e., the internal droplet is immobilised.
(8) Globule size variations can be lumped into a single effective mean diameter.
(9) The system is well agitated, therefore the external mass transfer resistance can be considered negligible and the external continuous phase is of uniform solute concentration.
(10) Concentration of the solute within the internal droplets does not depend on position.
(11) Membrane rupture is neglected. This implies that the transfer process is considered only when the emulsion remains stable.

The diffusion equation of the solute in an emulsion globule of the type shown in Figure 2 is given by Fick's law of diffusion (57) as

$$(1-\phi)\frac{\partial C_m}{\partial t} + \phi\frac{\partial C_i}{\partial t} = D_{eff}\left(\frac{\partial^2 C_m}{\partial r^2} + \frac{2}{r}\frac{\partial C_m}{\partial r}\right) \qquad (1)$$

where an effective diffusivity, D_{eff} is used, and with the initial and boundary conditions as follows:

$$\text{I.C.:} \quad t = 0, \quad C_m = 0 \quad (R > r \geq 0) \qquad (2)$$

$$\text{B.C.1:} \quad r = R, \quad C_m = K_{mb}C_b \quad (t \geq 0) \qquad (3)$$

B.C.2: $r = o, \quad \frac{\partial C_m}{\partial r} = 0 \quad (t \geq 0)$ (4)

the initial condition expressing the fact that the globule is initially free from the solute.

According to assumption 9, if the system is well stirred then the concentration of the solute in the external continuous phase depends only on time, and is determined essentially by the condition that the rate at which the solute leaves the external solution is always equal to that at which it enters the surface of the emulsion globule (56). This is expressed mathematically as follows:

$$-V_b \frac{\partial C_b}{\partial t} = \frac{3V_e}{R} D_{eff} \frac{\partial C_m}{\partial r} \Big/_{r = R} \tag{5}$$

with initial condition

$$t = 0, \quad C_b = C_{bo} \tag{6}$$

where $3V_e/R$ is the total surface area of the emulsion globule, C_{bo} is the initial concentration of alcohol in the external continuous phase, V_b is the total volume of the external continuous phase. According to assumption 6,

$$C_m = K_{mi} C_i \tag{7}$$

where K_{mi} is the equilibrium constant of alcohol between the internal aqueous phase and the membrane phase. Combining Equations 1 and 7 gives

$$(1-\phi + \phi/K_{mi}) \frac{\partial C_m}{\partial t} = D_{eff} \left(\frac{\partial^2 C_m}{\partial r^2} + \frac{2}{r} \frac{\partial C_m}{\partial r} \right) \tag{8}$$

or,

$$\frac{\partial C_m}{\partial t} = D^*_{eff} \left(\frac{\partial^2 C_m}{\partial r^2} + \frac{2}{r} \frac{\partial C_m}{\partial r} \right) \tag{9}$$

where

$$D^*_{eff} = \frac{D_{eff}}{1-\phi + \phi/K_{mi}} \tag{10}$$

The solution to equations 5 and 9 with initial and boundary conditions given by equations 2 to 4 and 6 is given by Crank (58) as follows:

$$\frac{M_t}{M_\infty} = \frac{C_{bo} - C_b}{C_{bo} - C_{beq}} = 1 - \sum_{n=1}^{\infty} \frac{6\alpha(1+\alpha)\exp(-D^*_{eff}\, q_n^2\, t/R^2)}{9(1+\alpha) + q_n^2\, \alpha^2} \tag{11}$$

where, q_n's are the non-zero roots of

$$\tan q_n = \frac{3q_n}{3 + q_n^2} \tag{12}$$

$$\alpha = \frac{V_b}{V_e K} \tag{13}$$

$$M_\infty = V_b\, C_{bo}/(1 + \alpha) \tag{14}$$

M_t and M_∞ are the total amounts of alcohol in the globule after time t and infinite time respectively. Here it is assumed that $K_{mb} = K_{mi} = K$, which is reasonable for the system used in this work as the internal phase and external continuous phase are basically the same, i.e., water. Thus when the membrane phase concentration profile is known, that in the internal aqueous phase can be calculated from equilibrium assumption at the interface, i.e.,

$$C_m = K\, C_i \quad \text{(at equilibrium)} \tag{15}$$

Mass Balance

In the final equilibrium state, the total amount of solute in solution, membrane and internal droplets was originally in the solution of concentration C_{bo}. In mathematical form

$$V_m C_{meq} + V_i C_{ieq} + V_b C_{beq} = V_b C_{bo} \tag{16}$$

$$(1-\phi)\, V_e C_{meq} + \phi\, C_{ieq} + V_b C_{beq} = V_b C_{bo} \tag{17}$$

Combining Equations 15 and 17 we have

$$(1-\phi)\, V_e C_{meq} + \frac{\phi}{K} C_{meq} + V_b C_{beq} = V_b C_{bo} \tag{18}$$

Dividing Equation 18 through by $V_e C_{beq}$ and rearranging we have

$$C_{beq} = \frac{V_b C_{bo}}{V_e\ [(1-\phi)K + \phi + V_b/V_e]} = \frac{V_b C_{bo}}{[(1-\phi)K + \phi]V_e + V_b} \tag{19}$$

Recall that from Equation 11

$$\frac{M_t}{M_\infty} = \frac{C_{bo} - C_b}{C_{bo} - C_{beq}}$$

Substituting for C_{beq} from Equation 19 we have

$$\frac{M_t}{M_\infty} = \frac{C_{bo} - C_b}{C_{bo}\left(1 - \frac{V_b}{[(1-\phi)K + \phi]V_e + V_b}\right)} \tag{20}$$

Combining Equations 11 and 20 we have

$$\frac{C_{bo}-C_b}{C_{bo}} = \left(1 - \frac{V_b}{[(1-\phi)K + \phi]V_e + V_b}\right)\left(1- \sum_{n=1}^{\infty} \frac{6\alpha(1+\alpha)\exp(-D^*_{eff}q_n^2 t/R^2)}{9(1+\alpha) + q_n^2 \alpha^2}\right) \tag{21}$$

or,

$$1 - \frac{C_b}{C_{bo}} = (1 - V^*_b)\Sigma^* \tag{22}$$

where

$$V^*_b = \frac{V_b}{[(1-\phi)K + \phi]V_e + V_b} \tag{23}$$

and

$$\frac{M_t}{M_\infty} = \Sigma' = 1 - \sum_{n=1}^{\infty} \frac{6\alpha(1+\alpha)\exp(-D^*_{eff}q_n^2 t/R^2}{9(1+\alpha) + q_n^2 \alpha^2} \tag{24}$$

From Equation 22,

$$\frac{C_b}{C_{bo}} = 1 - (1 - V^*_b)\Sigma^* \tag{25}$$

or

$$\frac{C_b}{C_{bo}} = 1 - (1 - V^*_b) \frac{M_t}{M_\infty} \tag{26}$$

which gives the solute concentration profile in the external continuous phase.

To determine the final fractional uptake we recall the definition:

$$M_\infty = V_b \; (C_{bo} - C_{beq}) \tag{27}$$

Combining Equations 19 and 27 gives

$$\frac{M_\infty}{V_b C_{bo}} = 1 - \frac{V_b}{[(1-\phi)K + \phi]V_e + V_b} \tag{28}$$

$$= 1 - V'_b \tag{29}$$

or

$$\frac{M_\infty}{V_b C_{bo}} = 1 - \frac{\alpha}{1 + \alpha + \phi/K - \phi} \tag{30}$$

It should be noted that for a sphere (not emulsion)

$$\phi = 0$$

Therefore, the final fractional uptake becomes (58)

$$\frac{M_\infty}{V_b C_{bo}} = \frac{1}{1 + \alpha} \tag{31}$$

Thus knowing the parameters ϕ, K, D^*_{eff}, C_{bo} and the emulsion size and globule size distribution, the concentration profile of the solute in the external continuous phase can be calculated from Equation 26 while that in the membrane phase can be calculated from equilibrium assumption at the interface.

EXPERIMENTAL

(a) Emulsion Liquid Membrane Preparation and Procedure

The experimental apparatus and procedure for the ELM used

in this study were the same as reported in the previous paper (36). The membrane (oil) phase was made of light mineral oil (SIGMA, Cat. No. M3516) and a surfactant, 3% wt lecithin (with 10 - 20% phosphatidycholine content based on choline determination - SIGMA TYPE III-S). Measured volume of distilled water which formed the dispersed phase of the emulsion was added slowly to a known quantity of the oil solution contained in a 500 ml beaker immersed in a water bath maintained at 10° - 14°C. The mixture was then stirred at 8000 - 9000 rpm for 60 min using a high-speed Silverson[R] turbine-type laboratory mixer with a stationary emulsor screen covering the four impellers. The resulting water-in-oil (W/O) emulsions whose temperatures rose to 45° - 50°C were allowed to cool to 20° - 25°C before they were used in the liquid membrane separation process.

For the separation process, a measured quantity of the W/O emulsion was added to 150 ml of normally fermented decarbonated beer sample of known alcohol content in a 1 litre unbaffled beaker immersed in a water bath through which water at 15° - 20°C was circulated. The mixture was stirred at 230±10 rpm using a gear-driven Citenco[R] propeller laboratory mixer which has three 6.5 cm diameter impellers. The dispersed emulsion globules decreased steadily in size until a steady state was reached. After an agreed period of contact, the stirring was stopped and the beer allowed to separate from the emulsion phase by gravity for 10 min. The separated beer was then collected, its volume measured, filtered twice through WHATMAN Nos. 2 and 5 filter paper in a covered funnel and the specific gravity and alcohol content determined (59 - 61). The quantity of the alcohol transferred into the emulsion globules in the different time intervals was evaluated and the fractional extraction calculated by the difference in the alcohol content of the beer before and after mixing over the alcohol content before mixing.

(b) Estimation of Mass Transfer Model Parameters

In order to compare the model predictions with the experimental results, the equilibrium partition coefficient, K, of alcohol between the external continuous (beer) and membrane (oil) phases, the effective diffusivity, D_{eff}, of alcohol in the emulsion phase, and the mean diameter (Sauter mean diameter, d_{32}) of the emulsion globules are required. The estimation of these parameters is discussed below.

Equilibrium partition coefficient. The partition coefficient, K, of alcohol between the membrane phase (oil and surfactant) and beer was measured by adding 100 ml of the oil phase to 150 ml of decarbonated beer of known alcohol content, and mixing at 180±10 rpm with the propeller mixer for 30 min at 20°C. (The method of component addition, mixer speed and time were chosen after many trials, to avoid the emulsification of the beer and allow the system to attain equilibrium). On settling, the oil phase and the beer were separated under gravity, filtered through a dry filter paper (WHATMAN No.2) in a covered funnel, and the alcohol content of the beer determined at 20°C/20°C (59 - 61). The concentration of the alcohol in the oil phase was then determined by mass balance. The value of K was calculated from the ratio of the equilibrium concentration of alcohol in the oil to beer phases.

Effective diffusivity. The effective diffusivity, D_{eff}, of alcohol in the emulsion phase was estimated from the Jefferson-Witzell-Sibbett equation for diffusion through a composite medium (62, 63) based on a concentration driving force defined in terms of the membrane phase concentration, C_m (45). For this purpose, the molecular diffusivity of alcohol in both the oil and aqueous phases of the emulsion are required. These were estimated by the widely used Wilke-Chang correlation (64, 65) using measured or computed values of the viscosity and molecular weight of the oil and aqueous phases as well as the molar volume of the solute (alcohol) at its normal boiling point. These latter parameters were estimated as briefly described below.

Viscosity of the oil and aqueous phases. The viscosity of the oil phase (oil and surfactant) was estimated with the aid of an Ostwald viscometer (Size D) (66, 67), at 20°C using 70% wt aqueous glycerol which has a specific gravity and viscosity at 20°C of 1.18125 and 2.29 x 10^{-4} Ns/m^2 respectively (68) as the standard solution. The specific gravity of the oil was given as 0.85 and this was assumed to be the same for the oil phase used which contained 3% wt lecithin. The viscosity of the aqueous phase was taken as that for water.

Molecular weight of the oil and aqueous phases. The average molecular weight of the oil phase containing 3% wt lecithin was determined by osmometry methods using chloroform as the solvent (66, 69). The average molecular weight of the aqueous phase was taken as that for water.

Molar volume of alcohol. The molar volume of alcohol at its normal boiling point was estimated from the ratio of the relative molecular mass of alcohol to the density of alcohol at normal boiling point (66).

Mean diameter of emulsion globule. A strip of paper 1 mm wide was attached to the inside wall of a 1 litre beaker and 150 ml of water added to it. The water was stirred with the propeller mixer used for the separation process at 230±10 rpm and a measured volume of the W/O emulsion prepared as described previously added to it. After about 5 min, photographs of the resulting emulsion globules and the strip were taken under adequate lighting, using a fast film with 1600 ASA and a shutter speed of 1/500 to 1/1000 seconds. The emulsion globule size distribution, as seen on the printed photographs, were then determined using the Kontron Mop Video Plan image analyser with the 1 mm strip of paper as the scale, and the results analysed by the Kontron 64 (Version 5.30) Standard Software computer programme which calculated the droplet distribution by number, equivalent circular diameter, standard deviation and standard error of the mean of the measurements. The mean diameter (Sauter mean diameter, d_{32}) of the emulsion globules was then calculated (70, 71).

RESULTS AND DISCUSSIONS

(a) Model Parameters Estimation

The results obtained for the model parameters and the pertinent operating variables are shown in Tables 2 and 3. From an earlier report (72) it has been shown that the presence of lecithin up to 3% wt in the oil phase had little or no effect on the value of the partition coefficient of alcohol between the oil and beer phases. It should be noted that most of these parameters of the model are estimated or measured independently of the separation experiments and does not require any curve fitting.

TABLE 2 - Values of the Parameters used for Model Calculations of the Alcohol Separation from Beer by Liquid Membrane technique.

K	[-]	0.04
T	K	293
μ_w	Ns/m^2	0.998×10^{-3}
μ_m	Ns/m^2	27.60×10^{-3}
M_w	kg/kg-mol	18
M_m	kg/kg-mol	339
V_{ETOH}	m^3/mol	62.6×10^{-3}
D_w	m^2/s	1.242×10^{-9}
D_m	m^2/s	1.204×10^{-10}
D_{eff}	m^2/s	4.671×10^{-10}
ϕ	[-]	0.383
D^*_{eff}	m^2/s	4.587×10^{-11}

TABLE 3 - W/O Emulsion Globule size Analysis

Emulsion : V_m/V_i = 1.61; S_L = 3.0% wt; N_P = 8000 rpm
N_t = 60 min; V_F = 150 ml

Mixer speed (rpm)	Volume of emulsion (ml)	
	60	100
	Mean diameter, d_{32} (mm)	
220 - 240	0.664	0.838
$\alpha = \dfrac{V_b}{V_e K}$	62.50	37.50

(b) Comparison of Model Prediction with Experimental Results

The final fractional uptake of alcohol by the emulsion

phase calculated from Equation 28 are given in Table 3. Using these values, the results of the preceding section and the solution to Equation 11 given by Crank (44), the model prediction data were computed from Equation 26.

Figure 3 shows a comparison of the proposed model predictions with the experimental results whereby the dimensionless concentration of alcohol in the liquid membrane treated beer (LMTB) defined as the ratio of alcohol content in the LMTB to the untreated beer sample, is plotted as a function of the mixing time of beer and emulsion phases during the liquid membrane separation process. The proposed model overpredicts the separation rate of alcohol from beer at mixing times below 20 min, but gives good agreement with the experimental results at longer mixing times. The differences in the theoretical and experimental results may be due to errors in the estimation of the model parameters; the assumption of a monodisperse character of the dispersed emulsion phase (the use of a single globule size, the Sauter mean diameter); membrane break-up; and the none attainment of equilibrium between the beer and emulsion phases during the liquid membrane separation process. It is possible, however, to accurately predict the final alcohol separation if the partition coefficient of the alcohol in the membrane phase is determined accurately, since this involves only parameters which can be estimated independently of the separation measurements themselves (Table 4).

TABLE 4 - Computed final fractional Alcohol Separation from Beer by Liquid Membrane technique

V_e (ml)	V_b (ml)	V^*_b [-]	$M_\infty/V_b C_{bo} = 1 - V^*_b$
60	150	0.8599	0.1401
100	150	0.7864	0.2136

CONCLUSIONS

The parameters of a diffusion mass transfer model for the separation of alcohol from beer by liquid membrane technique proposed were estimated, and the model predictions compared with the experimental results. At an early stage of the separation process (below 20 min) there were differences between the theoretical and experimental data, but good agreement was obtained at longer mixing times. The model, however, predicts accurately the final alcohol separation from beer by liquid membrane technique.

SYMBOLS USED

C_b = alcohol concentration in external continuous phase (beer) (% vol or g/l).
C_{bo} = initial alcohol concentration in beer (% vol or g/l).
C_i = alcohol concentration in internal aqueous droplet (% vol or g/l).
C_m = alcohol concentration in membrane phase (% vol or g/l).
D_m = diffusivity of alcohol in membrane phase (m^2/s).
D_w = diffusivity of alcohol in aqueous phase (m^2/s).
D_{eff} = effective alcohol diffusivity in emulsion mixture (m^2/s).
D^*_{eff} = $D_{eff}/(1-\phi + \phi/K)$, effective alcohol diffusivity in emulsion mixture (m^2/s).
d_{32} = Sauter mean diameter of W/O emulsion globule (m).
K = partition coefficient of alcohol between the aqueous and membrane phases [-].
K_{mb} = partition coefficient of alcohol between the external continuous and membrane phases [-].
K_{mi} = partition coefficient of alcohol between the internal aqueous and membrane phases [-].
M_t, M_∞ = total amount of alcohol in the emulsion phase after time t, infinite time (kg).
N_p, N_t = emulsification speed (rpm), time (min).
n_p = mixing speed of emulsion and external continuous phase (rpm).
r = radial coordinate (m).
R = W/O emulsion globule radius (m).
R_f = reaction front position (m).
R^- = radius of inner core of an emulsion globule (m).
rpm = revolutions per minute.
S_L = lecithin concentration in the membrane phase (% wt).
T = absolute tempearture (K).
t = time (s).
V_{ETOH} = molar volume of alcohol at normal boiling point ($m^3/kmol$).
V_b, V_F = external continuous phase volume (m^3).
V^*_b = $\alpha/[1+\alpha+\phi(1/K - 1)]$.
V_e = total volume of W/O emulsion (m^3).
V_i = total volume of internal aqueous phase in emulsion (m^3).
V_m = total volume of membrane phase in emulsion (m^3).

Greek Letters

α = V_b/V_eK, the ratio of the volume of the external continuous and emulsion phases [-].
δ = membrane thickness (m).
ε = thickness of the peripheral water layer (m).
μ = solvent viscosity (Ns/m^2).
ϕ = $V_i/(V_i + V_m)$, volume fraction of the internal droplet in W/O emulsion [-].
Σ^* = M_t/M_∞, ratio of the total amount of alcohol in emulsion phase after time t to that after infinite time [-].

Subscripts

b,F = external continuous phase (beer, water).
e = emulsion phase.
eff = effective.
eq = equilibrium value.
ETOH = ethanol (alcohol).
i = internal aqueous phase.
m = membrane phase.
o = initial value.
w = aqueous phase.

REFERENCES

1. Li, N. N., 1968, U. S. Patent, 3,410,794.
2. Li, N. N., 1971, AIChE J., 17, 459.
3. Li, N. N., 1971, Ind. Eng. Chem. Proc. Des. Dev., 10, 215.
4. Li. N. N. and Shrier, A. L., 1972, Rec. Dev. Sep. Sci., 1, Li, N. N. (ed.), CRC, Cleveland, Ohio, p. 163.
5. Li, N. N. 1981, TCE, 7, 325.
6. Draxler, J. and Marr, R., 1986, Chem. Eng. Process., 20, 319.
7. Gu, Z. M., Wasan, D. T. and Li, N. N., 1988, in "Surfactants in Chemical/Process Engineering", Wasan, D. T., Ginn, M. E. and Shah, D. O. (eds.), Marcel Dekker, New York & Basel, Chap. 4.
8. Shah, N. D. and Owens, R. C., 1972, Ind. Eng. Chem. Prod. Res. Dev., 11, 58.
9. Cahn, R. P. and Li, N. N., 1976, in " Membrane Separation Proceses", Meares, P. (ed.), Elsevier, Amsterdam & New York, Chap. 9.
10. Alessi, P., Kikic, I. and Mirella, O. V., 1980, Chem. Eng. J., 19, 221.
11. Halwachs, W., Flaschel, E. and Schugerl, K., 1980, J. Membr. Sci., 6, 33.
12. Kremesec, V. J., 1981, Sep. Purif. Methods, 10, 117.
13. Cahn, R. P. and Li, N. N., 1974, Sep. Sci., 9, 505.
14. Kitagawa, T. Nishikawa, Y., Frankenfeld, J. W. and Li, N. N., 1977, Envir. Sci. Tech., 11, 602.
15. Cahn, R. P., Li, N. N. and Minday, R. M., 1978, Envir. Sci. Tech., 12, 1051.
16. Terry, R. E., Li, N. N. and Ho, W. S., 1982, J. Membr. Sci., 10, 305.
17. Teramoto, M., Takihana, H., Shibutani, M., Yuasa, T., Miyake, Y. and Hara, N., 1983, Sep. Sci. Tech., 18, 397.
18. Kim, K., Choi, S. and Ihm, S., 1983, Ind. Eng. Chem. Fund., 22, 167.
19. Hatton, T. A., Lightfoot, E. N., Cahn, R. P. and Li, N. N., 1983, Ind. Eng. Chem. Fund., 22, 27.
20. Boey, S. C., Garcia Del cerro, M. C. and Pyle, D. L., 1987, Chem. Eng. Res. Des., 65, 218.
21. Martin, T. P. and Davies, G. A., 1976/77, Hydrometall., 2, 315.
22. Strzelbicki, J. and Charewicz, W. A., 1978, J. Inorg. Nucl. Chem., 40, 1415.
23. Kondo, K., Kita, K., Koida, I., Irie, J. and Nakashio, F., 1979, J. Chem. Eng. Japan, 12, 203.
24. Frankenfeld, J. W., Cahn, R. P. and Li, N. N., 1981, Sep.

Sci. Tech., 16, 385.
25. Cahn, R. P., Frankenfeld, J. W., Li, N. N., Naden, D. and Subramanian, K. N., 1981, Rec. Dev. Sep. Sci., 6, Li, N. N. (ed.), Boca Raton, Florida, Chap. 4.
26. Schiffer, D. F., Cussler, E. L., Evans, D. F. and Hochhauser, A. M., 1974 , Nature, 250, 484.
27. Hochhauser, A. M. and Cussler, E. L., 1975, AIChE Symp. Ser., 152, (71), 136.
28. Draxler, J. and Marr, R., 1986, Chem. Eng. Process., 20, 319.
29. Stzelbicki, J. and Charewicz, W. A., 1978, Sep. Sci. Tech., 13, 141.
30. Wasan, D. T., Gu, Z. M. and Li, N. N., 1984, Faraday Disc. Chem. Soc., 77, 67.
31. Hayworth, H. C., Ho, W. S. and Burns, W. A., 1983, Sep. Sci. Tech.,18, 493.
32. Akiba, K., Takeuchi, T. and Kanno, T., 1984, Bull. Chem. Soc. Japan, 57, 2618.
33. Scheper, Th., Halwachs, W. and Schugerl, K., 1984, Chem. Eng. J., 29, B31.
34. Makryaleas, K., Scheper, Th., Schugerl, K. and Kula, M., 1985, Germ. Chem. Eng., 8, 345.
35. Thien, M. P., Hatton, T. A. and Wang, D. I. C., 1986, in "Separation, recovery, and Purification in Biotechnology", Asenjo, J. A. and Hong, J. (eds.), ACS Symp. Ser. 314, p. 67.
36. Etuk, B. R. and Murray, K. R., 1990, Process Biochem., 25, 24.
37. Chan, C.C. and Lee, C.J., 1984, J. Membr. Sci., 20, 1.
38. Bunge, A.L. and Noble, R.D., 1984. J. Membr. Sci., 21, 55.
39. Way, J., Noble, R.D., Flynn, T.M. and Sloan, E.D., 1982, J. Membr. Sci., 12, 239.
40. Cussler, E.L., 1971, AIChE J., 17, 1300.
41. Carraciolo, F., Evans, D.F. and Cussler, E.L., 1975, AIChE J.,21, 160.
42. Boyadzhiev, L., Sapundzhiev, T. and Benzenshck, E., 1977, Sep. Sci., 12, 541.
43. Matulevicius, E. S. and Li, N. N., 1975, Sep. Purif. Methods, 4, 73.
44. Gladek, L., Stelmaszek, J. and Szust, J., 1982, J. Membr. Sci., 12, 153.
45. Ho, W. S., Hatton, T. A., Lightfoot, E.N. and Li, N. N., 1982, AIChE J., 28, 662.
46. Stroeve, P. and Varanasi, P.P., 1984, AIChE J., 30, 1007.
47. Fales, J.L. and Stroeve, P., 1984, J. Membr. Sci., 21, 35.
48. Kopp, A., Marr, R. J. and Moser, F. E., 1978, I. Chem. Eng. Symp. Ser., 54, 279.
49. Chan, C.C. and Lee, C.J., 1987, Chem. Eng. Sci., 42, 83.
50. Casamatta, G., Chavarie, C. and Angelino, H., 1978, AIChE J. 24, 946.
51. Kopp, A., 1978, Dissertation, Tech. Univ. Graz.
52. Alessi, P., Kikic, I., Canepa, B. and Costa, P., 1978, Sep. Sci. Tech., 13, 613.
53. Crank, J., 1975, "The Mathematics of Diffusion", 2nd ed., Claredon Press, Oxford, Chap. 14.
54. Kremesec, V. J. and Slattery, J. C., 1982, AIChE J., 28, 492.
55. Li, N. N., 1980, in "Wastewater Treatment and Metal

Recovery Conference", University of Manchester Institute of Science & Technology, (Manchester), p. 9.
56. Crank, J., 1975, "The Mathematics of Diffusion", 2nd ed., Claredon Press, Oxford, Chap. 4.
57. Ficks, A., 1855, Ann. Phys. Lpz., 170, 59.
58. Crank, J., 1975, "The Mathematics of Diffusion", 2nd ed., Claredon Press, Oxford, Chap. 6.
59. Recommended Methods of Analysis, 1977, The Institute of Brewing, London.
60. Methods of Analysis, 1976, 7th ed., American Society of Brewing, St. Paul, Minesota.
61. European Brewing Convention: Analytica-EBC, 1975, 3rd ed., Schweizer Brauerei-Rundschau, Zurich.
62. Crank, J., 1975, "The Mathematics of Diffusion", 2nd ed., Claredon Press, Oxford, Chap. 12.
63. Jefferson, T.B., Witzell, O.W. and Sibbett, W.L., 1958, Ind. Eng. Chem., 50, 1589.
64. Wilke, C.R. and Chang, P., 1955, AIChE J., 1, 264.
65. Reid, R. C., Prausnitz, J. M. and Sherwood, T. K., 1977, "The Properties of Gases and Liquids", 3rd ed., McGraw-Hill, New York, Chap. 11.
66. James, A. M. and Prichard, F. E., 1974, "Practical Physical Chemistry", 3rd ed., Longman, Harlow, England, Chaps. 2 & 3.
67. McGoury, T.E. and Mark, H., 1949, in "Physical methods of Organic Chemistry", Weissberger, A. (ed.), 2nd ed., Vol. 1, Part 1, Interscience, New York, Chap. 8.
68. Lang, N.A., 1956, "Handbook of Chemistry", 10th ed., McGraw-Hill, New York, pp. 1139 & 1681.
69. Wagner, R.H., 1949, in "Physical methods of Organic Chemistry", Weissberger, A. (ed.), 2nd ed., Vol. 1, Part 1, Interscience, New York, Chap. 11.
70. Beddow, J. K., 1980, "Particulate Science and Technology", Chemical Publ. Co., New York, Chap. 5.
71. Drinker, P. and Hatch, T., 1954, "Industrial Dust", 2nd ed., McGraw-Hill, New York.
72. Etuk, B. R., 1988, PhD Thesis, Heriot-Watt University, Edinburgh.

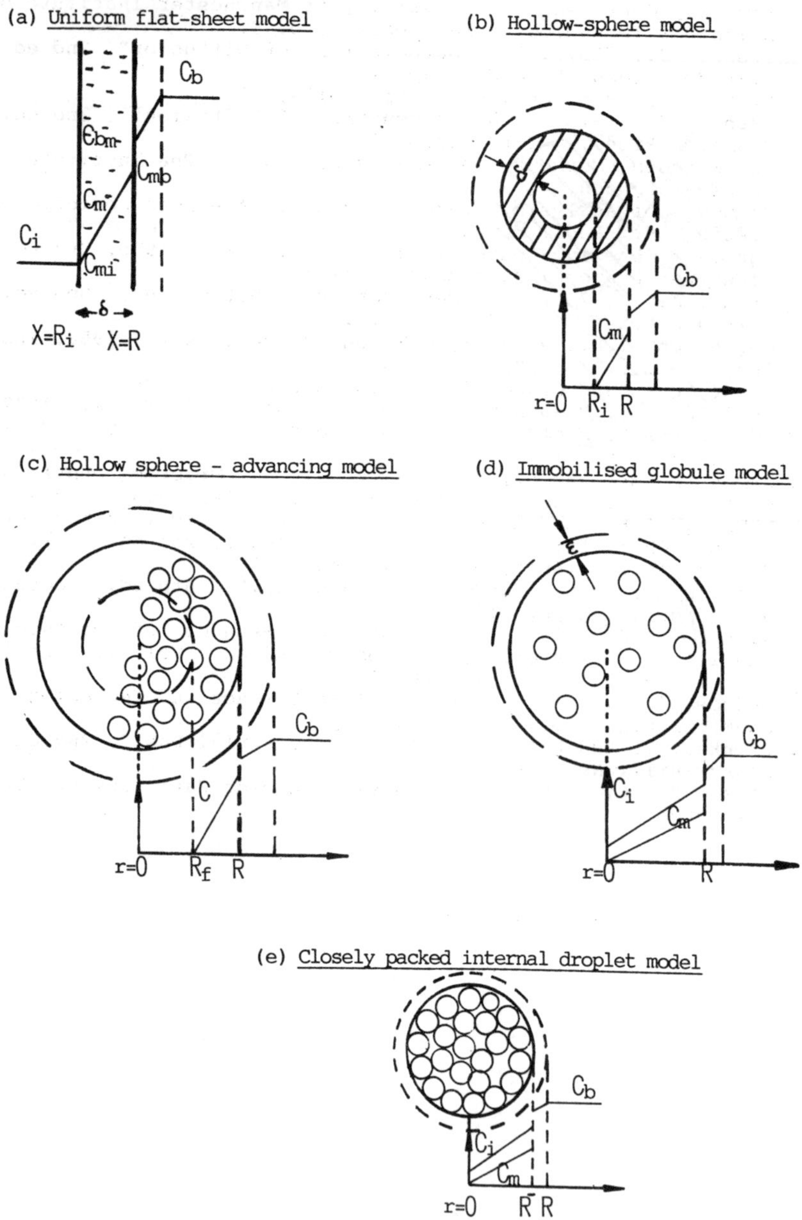

Figure 1. Schematic diagrams of emulsion globule models in a liquid surfactant membrane separation system.

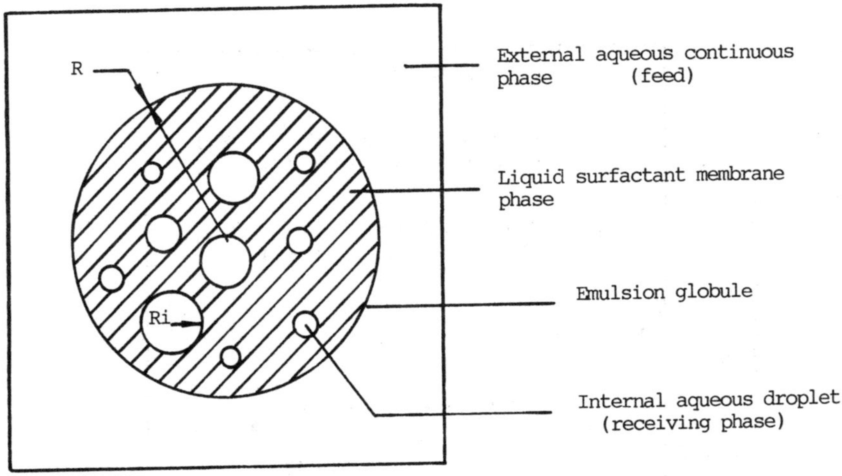

Figure 2. Schematic diagram of liquid surfactant membrane system (emulsion globule)

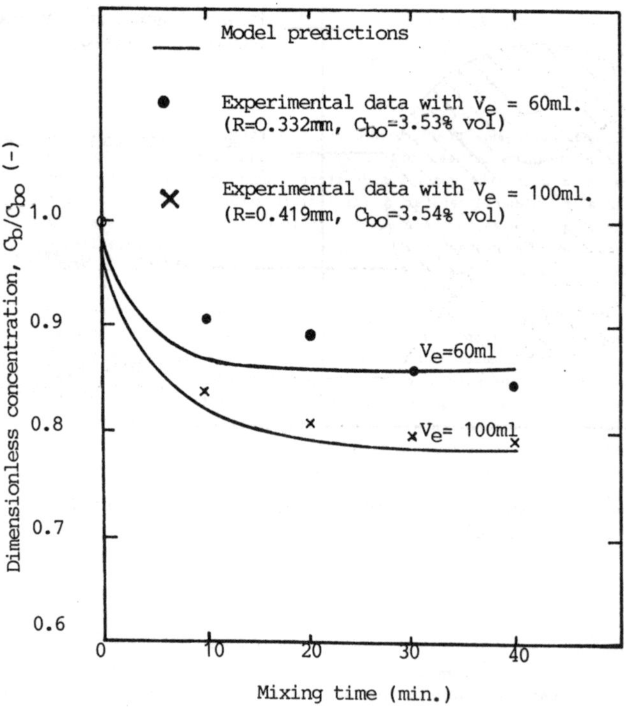

Figure 3. Comparison of model predictions with experimentat data for the liquid membrane separation of alcohol from beer (V_m/V_i=1.61, S_L=3.0 wt %, N_p =8000rpm, N_t=60min, n_p=230±10 rpm, V_b =150ml.).

FURTHER STUDIES ON THE TIME DEPENDENCE OF THE MASS TRANSFER COEFFICIENTS FOR THE SYSTEM URANYL NITRATE - NITRIC ACID - TBP

D. Rogers and P.J. Thompson

Overall mass transfer coefficients as a function of time for the transfer of uranium between nitric acid and tri-butyl phosphate have been measured using the quasi-steady state droplet technique. The time dependence of the mass transfer coefficient is known to be a result of a corresponding decrease in the frequency of interfacial turbulence as the interface ages. The present study reports several supporting experiments carried out in an attempt to gain a better understanding of this phenomenon.

The experimental conditions investigated concentrated on the extraction of uranyl nitrate into and from tri-n-butylphosphate, and included measurements of the mass transfer coefficients to examine the effect of solute loading of the solvent continuous phase, the effect of drop size, and the effect of a change of solvent phase diluent, A study of the backwashing of uranyl nitrate, in which interfacial turbulence is not observed, from the solvent phase to an aqueous one was made.

An explanation for the cause of the dampening of interfacial turbulence is proposed and discussed.

INTRODUCTION

It is well established (1) that the presence of interfacial turbulence in a liquid-liquid system has a profound effect on the mass transfer coefficient, increasing it several times over the value predicted by theory. A factor connected with interfacial turbulence is the physical age of the interface, such that the turbulence decreases as the interface ages. Over the past few years, several studies have been carried out to examine the dependence of the mass transfer coefficients on the age of the interface for systems exhibiting interfacial turbulence (2, 3, 4, 5, 6). For a freshly formed drop the mass transfer coefficient has been shown to be initially high, decreasing to a steady state value as the interface ages, as illustrated by Figure 1. Depending on the system studied, there can be up to an order of magnitude change in the mass transfer coefficient. These studies have shown without a doubt that the time dependence is accompanied by a corresponding decrease in the frequency of interfacial turbulence. Consideration of this phenomenon may be of significance in the design of liquid-liquid extraction contactors; a knowledge of the appropriate mass transfer coefficients is a fundamental requirement for their accurate design and efficient operation.

The experimental technique used to measure the time dependence of the mass transfer coefficient was the quasi-steady state droplet technique (2, 3, 4,

AEA Fuel Services, AEA Technology, Dounreay, Thurso, Caithness KW14 7TZ, Scotland

5, 6). This involved the formation of a suspended quasi-steady state hemispherical drop of the aqueous phase in a stagnant solvent continuous phase. The quasi-steady state droplet technique is unique in its ability to measure the change in mass transfer coefficient over an extended period of time whilst maintaining a constant drop volume, surface area and driving force for mass transfer.

The reason for the decay of interfacial turbulence, and hence the mass transfer coefficient, with time is unresolved. Several theories have been put forward by previous workers. Sawistowski and Goltz (7) considered that the turbulence decayed as a result of increased interfacial tension brought about by liquid eddies of the bulk phase arriving at the interface. Javed et al (6) pointed out that for the present experimental technique this was not likely to be a feasible explanation with regard to the time-scale and the state of rapid circulation within the droplet. Ruckenstein (8) suggested that the liquid elements taking part in the roll cells of the interfacial activity became progressively saturated with solute as contact time increased thereby reducing concentration gradients which would cause the turbulence to decay. An interfacial fluid model was used by Nakaike et al (9) to describe the decay of individual roll cells in terms of a reduction in the interfacial tension gradient as solute was transferred across the interface. However, the degree of interfacial instability differs from system to system, and it is only in some cases that the observation of roll cells is a characteristic. These instabilities are typically associated with liquids at rest. Transient, or disordered interfacial convection may be associated with turbulence or forced convection within the bulk liquid, as in the present experimental technique, for which eruptions of interfacial turbulence are typical (10).

Mas et al (11) calculated the mass fluxes arising from a consideration of kinetic and diffusion control in the extraction of plutonium. The fluxes were found to be large and of opposing sign at the beginning of interphase transport, which was considered to be a possible explanation for the occurrence of interfacial instabilities. They also demonstrated that as equilibrium was approached, the sum of the opposing fluxes tended to zero. This was calculated to occur after about 300 seconds, the time approximately corresponding to the steady state situation (Figure 1) in systems studied by Javed (3, 4, 6) and Rogers (2, 5).

Javed (3) suggested that the decay in interfacial turbulence was a result of a build-up of contaminants at the interface causing a reduction in the interfacial tension and an increase in interfacial viscosity. Javed et al (6) pointed out that whilst this proposal was attractive, it was all too easy to attribute the decay to this cause. Along this line of thought however, Rogers et al (2) postulated that a build up of an intermediate or complexed species may be responsible for the dampening of interfacial activity. This paper continues in the attempt to gain a better understanding of this phenomenon.

URANIUM MASS TRANSFER

Liquid-liquid extraction is the process used in the separation and purification of uranyl nitrate and plutonium nitrate during the reprocessing of irradiated nuclear fuels. As part of the Underlying Research programme of the UKAEA, work has been in progress at the Dounreay Nuclear Power Development Establishment to measure the mass transfer coefficients for systems of interest in the reprocessing of nuclear fuels, namely the transfer of uranyl nitrate and plutonium nitrate between nitric acid and tri-n-butylphosphate (TBP).

The extraction of uranyl nitrate by TBP is accompanied by interfacial turbulence. The relevance of interfacial turbulence to the systems used in nuclear fuel reprocessing has been discussed (12). A previous study (2) reported the effect of concentration on the time dependence of the mass transfer coefficient for the extraction of uranyl nitrate from 3M nitric acid into 20% TBP in odourless kerosene (20% TBP/OK), and this paper reports further work.

Firstly, the effect of solvent loading on the frequency of interfacial turbulence, and hence the mass transfer coefficient, was considered. A high proportion of droplets in a contactor are dispersed amongst a continuous phase with a significant solute concentration, especially in a counter-current contactor. Secondly, the back-extraction, or backwashing reaction was examined. Backwashing is an essential step in the fuel cycle if the extracted solutes are to be recovered. The transfer of solute from the organic phase back into the aqueous phase is achieved by changing the acidity of the aqueous phase in order to reverse the equilibrium of the reaction

$$UO_2^{++} + 2NO_3^- + 2TBP \rightleftharpoons UO_2(NO_3)_2.2TBP$$

A distribution of drop sizes inevitably arises within any liquid-liquid extraction contactor, and for this reason it was considered necessary to investigate the effect of changing the nozzle size to produce a drop of different surface area and Reynolds number. A Reynolds number can be applied to a droplet forming at a nozzle at a flow of F mm³/s to give a nozzle Reynolds number, Re_N, based on the inlet orifice diameter d_N.

$$Re_N = \frac{4 F \rho}{\pi \mu d_N}$$

Burkhart et al (13) described the importance of the nozzle Reynolds number in relation to the amount of circulation within a drop, and demonstrated the effect of different degrees of internal circulation on the mass transfer rates in liquid drops. They showed that the maximum internal circulation occurred for nozzle Reynolds numbers $Re_N > 30$.

Finally, the effect of a change in solvent phase diluent was examined. The diluent generally chosen as most suitable for TBP in its application in nuclear fuel reprocessing is odourless kerosene, a mixture of hydrocarbons with n-dodecane as the main constituent. However, a wide variety of diluents are available, ranging from normal, branched and cyclo-paraffins to aromatics and halogenated compounds. The choice of diluent may have important effects on density, interfacial tension, viscosity and even on the kinetics of extraction (14).

EXPERIMENTAL

The feature which allows the quasi-steady state droplet experiment to be unique in its ability to maintain a constant driving force for mass transfer is the nozzle used to form and suspend the droplet. The nozzle was constructed from two concentric glass tubes. The flow of liquid through the nozzle was arranged such that fresh aqueous liquid was fed through the inner capillary tube into the drop, and withdrawn at the same flowrate through the nozzle annulus thereby maintaining the pendant drop at a constant volume over an extended period of time. The experimental technique was such that the effects of mass transfer during drop formation could be disregarded, and has been described in detail elsewhere (5).

It has been shown, using this apparatus, that

- i surface renewal due to hydrodynamic turbulence does not occur in the absence of mass transfer
- ii circulation of the continuous phase in order to eliminate the possibility of concentration gradients had no effect on the progressive decrease of the mass transfer coefficients with time (6).

The residence time for liquid within the drop was estimated to be one second. All experiments were performed at 25°C.

The depleted uranyl nitrate hexahydrate and nitric acid used were of "Analar" quality. The nitric acid solutions were prepared by dilution of 16M nitric acid with demineralised water. The aqueous uranium solutions were prepared by dissolving depleted uranyl nitrate hexahydrate crystals in 3M nitric acid and filtering. For each extraction condition studied, the aqueous phase consisted of a 50g/l solution of uranium, as uranyl nitrate, in 3M nitric acid.

The 20% v/v TBP/OK solutions were made up from industrial stock supplies of TBP and OK, purchased to standard UKAEA specifications. Feeds were taken from the same batch throughout in order to eliminate any effects of fluctuation in batch quality. The solvent was purified before use by washing with a 2.5% w/v sodium hydroxide / 2.5% w/v sodium carbonate solution to remove any degradation products (mono- and di-butyl phosphate) and impurities. This was followed by a pre-equilibration of the solvent with 3M nitric acid in order to eliminate any effects due to transfer of nitric acid between the two phases during extraction. The solvent was then filtered through phase separating paper to remove any suspended aqueous droplets.

Solutions of uranyl nitrate in solvent were prepared by dissolving the required quantity of uranyl nitrate hexahydrate crystals in 3M HNO_3, and then shaking with 20% TBP/OK, which extracted part of the uranium. The layers were separated; the solvent layer was put to one side and the aqueous layer repeatedly shaken with solvent until the uranium was extracted such and the aqueous solution appeared colourless. The solvent layers were combined, filtered and made up to the required volume.

The Effect of a Solute-Loaded Continuous Phase

20% TBP/OK solutions over a range of uranium concentrations served as the extracting media, thus reducing the net driving force for extraction from the aqueous drop. The solvent phase uranium concentrations were chosen to represent (i) a maximum driving force, (ii) a partially loaded continuous phase ,and (iii) one closely approaching the equilibrium distribution concentration corresponding to the aqueous feed. Therefore the extraction was from 50g/l uranium into a solvent phase containing (i) no uranium, (ii) 37g/l uranium, and (iii) 60g/l uranium.

Backwashing Conditions

The backwashing experiments studied were from a uranium loaded solvent continuous phase into an aqueous drop of 0.02M nitric acid. The solvent uranium concentrations examined were 50g/l, 20g/l and 18g/l.

The Effect of Drop Size

In the majority of experiments the nozzle was constructed from an outer tube of 9mm od, 8mm id with a concentric inner tube of 7mm od, 2.5mm id. To study the effect of drop size a nozzle with approximately a third the usual diameter was employed. The "mini-nozzle" had an outer tube of 3.3mm od, 2.4mm i.d, with an inner tube of 1.7mm od, 1.1mm id.

The Effect of Solvent Phase Diluent

Hexane was chosen as an alternative to odourless kerosene for the TBP diluent as it provided a solvent phase with a considerably different range of physical properties. The 20% TBP/hexane solution was prepared in a similar manner to the 20% TBP/OK solutions.

RESULTS

The quasi-steady state droplet technique allowed the mass transfer rate from the droplet to be measured as a function of time by following the uranium concentration in the aqueous stream leaving the droplet through the nozzle annulus. Therefore the samples collected were analysed for uranium and the results plotted as concentration versus time.

In order to determine mass transfer coefficients from the concentration-time curves, the following mathematical procedures were adopted.

Dispersed to Continuous Phase Transfer

An unsteady state mass balance for a hemispherical droplet with the flow conditions described in Figure 2 can be written

$$FC^\circ = FC + KA\,(C-C^*) + V\frac{dC}{dt}$$

where C^* is the equilibrium concentration of the aqueous phase, $(C-C^*)$ is the overall driving force based on the dispersed phase, and thus K is the overall mass transfer coefficient based on the dispersed phase driving force.

Assuming the solute concentration in the continuous phase is negligible, ie approaches zero, at all times then the equilibrium concentration in the aqueous droplet, C^*, will also tend to zero.

$$K = \frac{F(C^\circ - C)}{AC} - \frac{V}{AC} \cdot \frac{dC}{dt}$$

Since K also depends on time, a finite difference technique is adopted to solve the equation. K may be considered constant over a small time interval Δt during which there is a corresponding concentration change $\Delta C'$. Therefore,

$$K = \frac{F(C^\circ - C^m)}{AC^m} - \frac{V}{AC^m} \cdot \frac{\Delta C'}{\Delta t}$$

where C^m is the mean concentration in the drop during the time interval Δt.

Dispersed to Solute-Loaded Continuous Phase Transfer

The same flow conditions apply as in the above situation, but in this case the solute concentration in the continuous phase is not negligible. Therefore the equation becomes

$$K = \frac{F(C^\circ - C^m)}{A(C^m - C^*)} - \frac{V \Delta C'}{\Delta t} \cdot \frac{1}{A(C^m - C^*)}$$

Continuous to Dispersed Phase Transfer

An unsteady state mass balance for a droplet with the flow conditions described in Figure 3 can be written

$$FC^\circ + KA(C^* - C) = FC + V\frac{dC}{dt}$$

The initial aqueous feed concentration to the drop is barren of solute, ie $C^\circ = 0$. Therefore the equation can be rearranged, using the finite difference technique, to

$$K = \frac{V}{A(C^* - C^m)} \frac{\Delta C'}{\Delta t} + \frac{FC^m}{A(C^* - C^m)}$$

The graphs depicting the overall mass transfer coefficient versus time are shown for each of the conditions studied in Figures 4 to 7.

DISCUSSION

The main advantage of the quasi-steady state droplet experiment over other methods for measuring the mass transfer coefficient is in the maintenance of a constant driving force throughout the lifetime of the drop. In other methods, for example the Lewis and single drop methods, the driving force decreases with the time of contact.

For ease of discussion, each set of conditions studied are considered separately.

The Effect of a Solute-Loaded Continuous Phase

Solute loading of the continuous phase was investigated using uranium concentrations in TBP/OK of 0g/l, 37g/l and 60g/l, which provided different driving forces for extraction. The mass transfer coefficient is plotted as a function of time in Figure 4. It can be seen that the coefficient curve for the partially loaded (37g/l) continuous phase was considerably reduced compared to that for the barren continuous phase. A continuous phase concentration approaching the equilibrium value (60g/l) removed any detectable time dependence and reduced the mass transfer coefficient to a minimal level.

These results are consistent with observations of the intensity of interfacial turbulence. In the case of the partially loaded continuous phase, interfacial turbulence was considerably reduced compared with that observed during extraction into a barren solvent. No interfacial turbulence was observed with the 60g/l solution.

The effect of a different interfacial tension was reflected in the shape of the droplet as it was suspended from the nozzle, compared with a barren solvent.

Burger (15) studied the effect of a solvent loading using a Lewis cell, and reported that the reaction rate slowed down greatly as equilibrium was approached. He proposed that a different mechanism of extraction became important at this stage, and was a function of the degree of equilibrium approached rather than a function of elapsed time. His observations corresponded well with the theory put forward by Sawistowski and Goltz (7), who proposed that the degree of interfacial turbulence, and hence the rate of mass transfer, was dependent upon a critical driving force or change in interfacial tension.

Backwashing Conditions

In a comprehensive visual study (12) interfacial turbulence was shown to occur during the transfer of uranium from a nitric acid aqueous phase into a TBP/diluent phase. However the transfer of solute from the organic to the aqueous phase was shown not to be accompanied by any visible turbulence.

In the present study, Figure 5 shows a time independent mass transfer coefficient for the backwashing system, corresponding to an absence of interfacial turbulence. Again, the effect of a different interfacial tension on the size and shape of the suspended droplet was observed.

These results may be used to support arguments against Javed's proposal (3) that the decay of interfacial turbulence is caused by a build up of contaminants at the interface. Consider for one moment that a build up of contaminants is the reason for the time dependence of the mass transfer coefficient for extraction, then it would be expected that a similar time dependence would be apparent for the backwash condition, regardless of interfacial turbulence. However the mass transfer coefficient proved to be independent of time, an indication surely that the decay of turbulence during extraction is not usually caused by an accumulation of contaminants.

The Effect of Drop Size

A reduction in the nozzle size, as indicated, had the effect of increasing the nozzle Reynolds number from 46 for the larger nozzle, to 97 for the mini-nozzle. The influence of Reynolds number and drop size is illustrated by Figure 6. The results are indicative of a dramatic enhancement in the mass transfer coefficient as t approaches zero. More importantly, however, there appears to be little difference in the mass transfer coefficient at short contact times, when compared with the same system operated with the normal nozzle.

It is well established (13) that an increase in Re_N is associated with an increase in mass transfer coefficient. The results of this study therefore suggest that with freshly formed interfaces and at short contact times, the value of the mass transfer coefficient, for the drop size range studied, is controlled more by interfacial turbulence than by Reynolds number and internal circulation.

The Effect of Solvent Phase Diluent

The choice of n-hexane as an alternative to odourless kerosene initially arose from a consideration of third-phase formation as a possible cause for the decay of interfacial turbulence as the interface ages. Third phase formation refers to the phenomenon which occurs when limiting solute solubility causes the original organic phase to split into two immiscible organic phases: an upper diluent-rich layer and a lower TBP-solute layer. Although it has been shown (16, 17) that no formation of a third phase occurs with uranyl nitrate systems below 9M aqueous phase acidity, an experiment using n-hexane as diluent was chosen as TBP-hexane systems are less likely to yield a third phase under any "normal" condition (14).

However, n-hexane was ultimately chosen as an appropriate alternative diluent for TBP as a 20% TBP/n-hexane solution has a much lower kinematic viscosity than the corresponding TBP/OK solution, 0.69cS compared to 2.28cS for 20% TBP/OK. This fact had been considered in previous studies (12, 18) and was therefore considered necessary to examine the effect of a change in viscosity on the time dependence phenomenon.

The first mathematical study of the onset of interfacial turbulence was by Sternling and Scriven (19). Their model was based on two semi-infinite, quiescent fluid phases in contact along a plane interface which were in thermal, but not in chemical equilibrium. Application of the model resulted in the Sternling-Scriven criterion for instability, which stated that interfacial turbulence was promoted by solute transfer out of the phase of higher kinematic viscosity and lower molecular diffusivity for systems where $\partial\sigma/\partial C < 0$. For systems in which $\partial\sigma/\partial C > 0$ an instability is possible (10) if mass transfer is into the phase of higher kinematic viscosity and lower molecular diffusivity. It has been shown (12) that the TBP/OK system followed the Sternling-Scriven criterion; when n-hexane was used as diluent the kinematic viscosity ratio between the two phases was such that the criterion did not apply, yet there were still strong interfacial instabilities (16).

This fact was clearly demonstrated in the present study. The intensity of interfacial turbulence was in fact enhanced compared to the TBP/OK system. As can be seen in Figure 7, this corresponded with an increase in the relative value of the mass transfer coefficient. The TBP/n-hexane system still showed a time dependence which coincided with a dampening of interfacial turbulence; it is interesting to note the similarity in the shapes of the curves.

The implications of these results were directed towards the proposal previously put forward (2) that the dampening of interfacial turbulence could be due to a build up of an intermediate or complexed species at the interface. The use of n-hexane as a diluent for TBP had the effect of significantly lowering the viscosity of the solvent phase. If a build up of a species on the solvent side of the interface was the cause of the time dependence, it would be expected that this change of diluent would reduce the time dependent effect considerably, as the species would be able to diffuse away from the interface more easily. As Figure 7 shows, this is not the case as the shapes of the curves are similar. This therefore leads to the proposal that any build up of species would be on the aqueous side of the interface.

CONCLUSIONS

For systems which exhibit interfacial turbulence, the value of the mass transfer coefficient has been shown to depend on the time of contact, and hence the age of the interface. The results of the present study confirm that the time dependence of the coefficient is associated with a corresponding decay in the frequency of interfacial turbulence.

The reasons for the dampening of interfacial turbulence have been discussed. The present study has suggested that a build up of contaminants at the interface is not responsible for the decay

An increase in drop Reynolds Number has the effect of increasing the mass transfer coefficient at long contact times; however at short contact times the magnitude of the mass transfer coefficient is dominated by the effect of the interfacial turbulence.

The mass transfer coefficient for transfer to a partially loaded continuous phase was reduced, and therefore the effect of continuous phase loading is a significant parameter to be considered in the design of liquid-liquid contactors.

ACKNOWLEDGEMENT

The authors thank the Chief Executive, AEA Fuel Services, for permission to publish this work.

NOMENCLATURE

A	- interfacial area (mm^2)
C	- concentration of outlet phase (g/l)
C^m	- mean concentration (g/l)
C^o	- concentration of inlet phase (g/l)
C^*	- equilibrium concentration of aqueous phase (g/l)
d_N	- nozzle inlet orifice diameter (mm)
F	- flowrate (mm^3/s)
K	- overall mass transfer coefficient (mm/s)
Re_N	- nozzle Reynolds number
V	- drop volume (m^3)
$\Delta C'$	- concentration change (g/l)
Δt	- time interval (s)
ρ	- density (Kg/m^3)
σ	- interfacial tension (N/m)
μ	- dynamic viscosity (m^2/s)

REFERENCES

1 Lewis J B: Chem Eng Sci, **8**, 295 (1958)

2 Rogers D, Thompson PJ and Thornton JD: I ChemE Symp Ser 103 "Extraction 87" (1987)

3 Javed KH: Study of Surface Phenomena in Liqiud-Liquid Extraction, PhD Thesis, University of Newcastle upon Tyne (1982)

4 Javed KH and Thornton JD: I Chem E Symp Ser 88 "Extraction 84" (1984)

5 Rogers D: A Study of Factors Determining Mass Transfer Rates in Solvent Extraction, M Phil Thesis, The University of Aston in Birmingham (1988)

6 Javed KH, Thornton JD and Anderson T: AIChJ **35(7)** 1125 (1989)

7 Sawistowski H and Goltz GE: Trans Inst Chem Eng **41** 174 (1963)

8 Ruckenstein E: Int J Heat Mass Transfer **11** 1753 (1968)

9 Nakaike Y, Tadenuma Y, Sato T and Fujinawa K: Int J Heat Mass Transfer **14** 1951 (1971)

10 Sawistowski H: Recent Advances in Liquid-Liquid Extraction, Ed Hanson C, Ch 9, (1971)

11 Mas C, Knittel G and Von Ammon R: Proc ISEC 83, 1 I 223 (1983)

12 Thompson PJ, Batey W and Watson R: I Chem E Symp Ser 88 "Extraction 84" 231 (1984)

13 Burkhart L, Weathers PW and Sharer PC: AIChEJ 22 1090 (1976)

14 Vandegrift GF: The Science and Technology of Tri-butyl Phosphate 1 Ch 4 (1984)

15 Burger LL: HW-62087 (1959)

16 Mills AL and Logan WR: Solv Extr Chem, N Holland, 322 (1967)

17 Boukis N and Kanellakopulos B: KfK-3352 (1983)

18 Perez de Ortiz ES, Thompson PJ and MacDonald CN: Proc PROGEC Summer School on Extraction, Toulouse, September 1987

19 Sternling CV and Scriven LE: AIChJ 5 514 (1959)

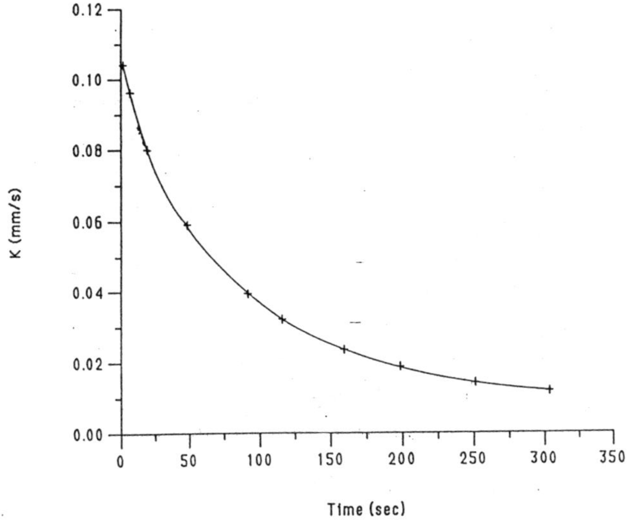

Figure 1

Overall mass transfer coefficient for extraction of 50g/l U as a function of time

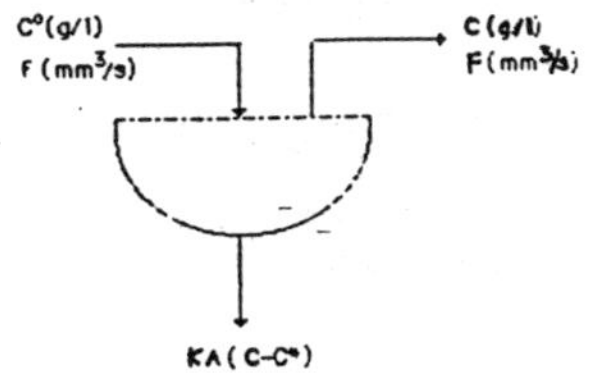

Figure 2

Flow conditions for extraction

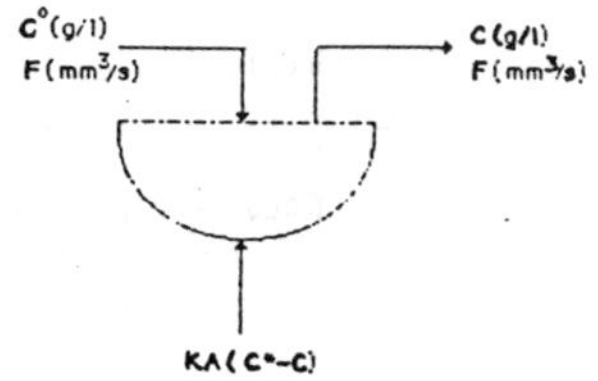

Figure 3

Flow conditions for backwash

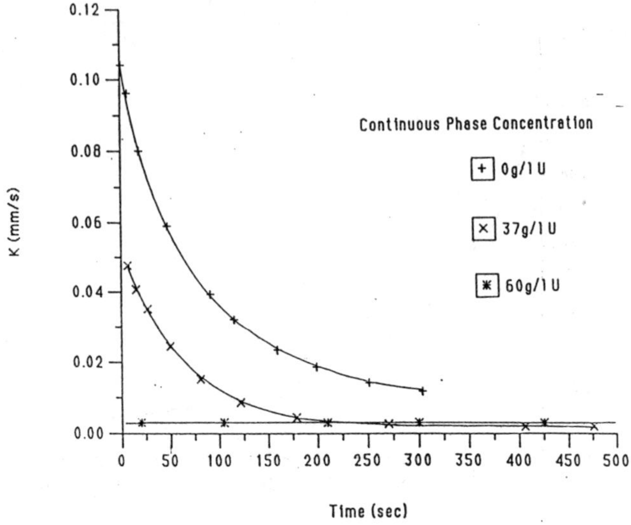

Figure 4

Variation of overall mass transfer coefficient with time and continuous phase loading

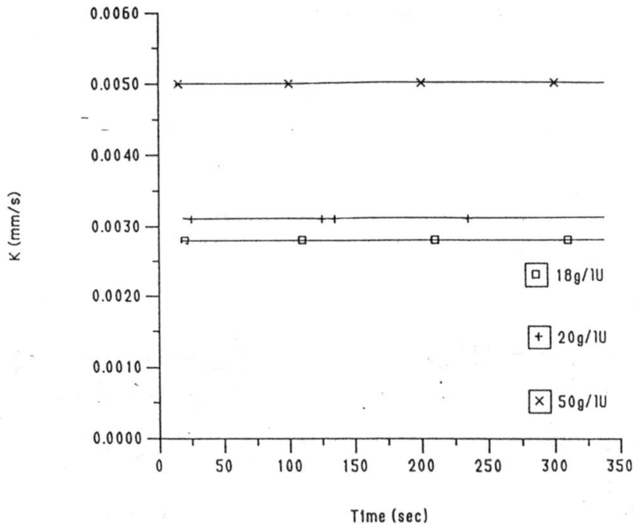

Figure 5

Variation of overall mass transfer coefficient with time under backwash conditions

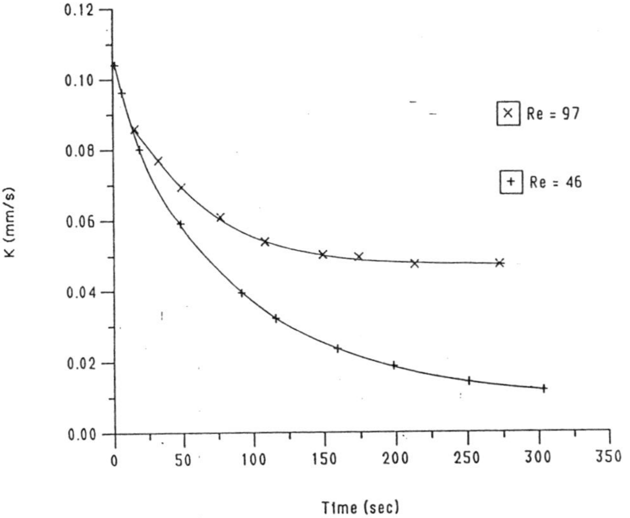

Figure 6

Variation of overall mass transfer coefficient with time and drop size

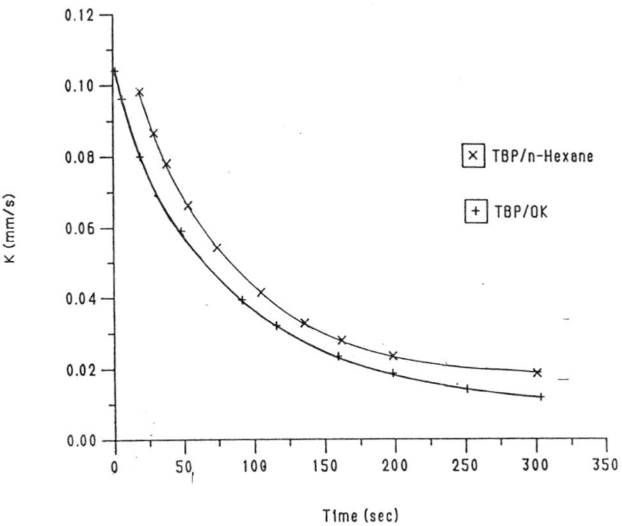

Figure 7

Variation of overall mass transfer coefficient with time for two diluents

INTERFACIAL TURBULENCE AND MASS TRANSFER AT LIQUID/LIQUID EXTRACTION

A.A.Golovin*

A semiempirical mass transfer model for liquid/liquid extraction accompanied by interfacial turbulence is presented. In the framework of the model the linear dependence of mass transfer coefficient on the driving force is obtained which is confirmed by a number of experiments. Methods are presented which allow mass transfer rates to be predicted for extraction accompanied by interfacial turbulence in growing as well as in moving droplets. A method is proposed for calculating the parameters of a countercurrent extraction column under the conditions of interfacial turbulence.

INTRODUCTION

Interfacial turbulence, or spontaneous interfacial convection, is known to increase substantially the mass transfer rates in the course of liquid/liquid extraction and seems to be a good way to enhance mass transfer in industrial processes. To succeed in the industrial use of interfacial turbulence one has to know the conditions for interfacial instability to occur as well as the kinetics of mass transfer for extraction accompanied by interfacial turbulence. A large number of paper have been devoted to the appearance of interfacial instability. However, mass transfer kinetics under conditions which give interfacial instability have been much less investigated hitherto. The results of experiments in liquid/liquid extraction allow us to point out the following characteristic features of mass transfer accompanied by interfacial convection:

- mass transfer coefficients highly exceed their values in the diffusion regime (without interfacial convection);
- mass transfer coefficients in stirred systems (Lewis cells, moving droplets and films) are time dependent whereas in the diffusion regime they are constant;
- the time dependence of mass transfer coefficients in systems with stagnant phases does not agree with the penetration theory;
- mass transfer coefficients depend upon the driving force and the properties of an interface, interfacial tension in partic-

* Karpov Institute of Physical Chemistry, Moscow, USSR

ular;

- there exists a crirical value of the driving force such that after it is reached mass transfer begins to deviate from that found in the diffusion regime;

- there exist several regimes of mass transfer accompanied by interfacial convection which are distinguished by different mass transfer coefficient dependence upon physico-chemical parameters (i.e. different kinetics).

So far there have been obtained several empirical correlations for mass transfer coefficients for liquid/liquid extraction under the conditions of interfacial turbulence (Sawistowski and Goltz (1), Sawistowski and Austin (2), Maroudas and Sawistowski (3), Olander and Reddi (4), Ostrovskii et al (5), Ostrovskii and Golyakova (6), Ermakov et al (7,8), Konshin and Ermakov (9)). Most of these correlations establish linear dependence of mass transfer coefficient on the driving force. The same dependence is predicted by a number of works in which nonlinear analysis of hydrodynamical stability of an interface is treated (Lebon and Cloot (10), Ho and Chang (11)) and also when a numerical analysis of the whole problem is carried out ((11), Elenin et al (12)). Unfortunately neither empirical correlations nor the results of nonlinear stability analysis may be suitable for chemical engineering as they are restricted to a certain class of extraction systems and to theoretically ideal cases.

More promissing seem to be semiempirical models of mass transfer for liquid/liquid extraction accompanied by interfacial turbulence which reflect the main kinetic regularities and contain the minimum number of empirical parameters. A number of semiempirical models have been proposed in order to account for mass transfer enhancement and concentration and time dependence of mass transfer coefficients when interfacial turbulence is present (Bakker et al (13), Linde and Winkler (14), Imaishi et al (15), Ruckenstein (16), Sethy and Cullinan (17), Rabinovich and Slinko (18), Rehakova and Rod (19), Golovin and Rabinovich (20, 21), Golovin et al (22)). In the present work a new semiempirical model of mass transfer for extraction accompanied by interfacial turbulence manifested first in (20) and developed further in (21),(22) is applied to extraction in drops as well as to the calculation of the parameters of a countercurrent extraction column.

The Scope of the Model

Consider the interfacial instability to be a system of convective cells spartially ordered or pulsating in a chaotic manner and occupying a part of the contact area. An interfacial stress balance at the interface of the cell

$$\eta_1 \frac{\partial v_1}{\partial y} - \eta_2 \frac{\partial v_2}{\partial y} = \frac{\partial \sigma}{\partial x} \qquad (1)$$

leads to the following evaluation:

$$(\eta_1 + \eta_2) \frac{v_s}{\delta_\nu} \sim \frac{\partial \sigma}{\partial c} \frac{\Delta_{\parallel} C}{l} \tag{2}$$

Taking note of $\delta_\nu \sim \sqrt{\nu \tau}$ and considering the interfacial concentration difference along the cell $\Delta_{\parallel} C$ connected with interfacial mass flux j_0 as follows

$$\Delta_{\parallel} C \sim j_0 \tau / \delta_D, \qquad \delta_D \sim \sqrt{D\tau} \tag{3}$$

one may obtain

$$v_s^2 \sim \frac{1}{\eta_1 + \eta_2} \frac{\partial \sigma}{\partial c} Sc^{1/2} j_0 \tag{4}$$

Eq. [4] gives a general correlation between the circulation velocity in the convective cell and interfacial mass flux.

It is easy to show that the concentration difference along the cell $\Delta_{\parallel} C$ is equal to that between the interface and the depth of diffusion boundary layer $\Delta C'$, the latter being proportional to the whole driving force $\Delta C = C_1 - C_2/K_e$

$$\Delta C' \sim b \, \Delta C$$

Here b is an empirical parameter characterizing the extent of the surface renewal in the convective cell (21,22). Considering the characteristic scale of the cell l to be of the same order of magnitude as the characteristic depth of momentum penetration δ_ν as well as supposing the fraction of the interface occupied by convective cells ε to be dependent upon the deviation of the concentration driving force ΔC from its critical value ΔC_{cr}: $\varepsilon = f(\Delta C - \Delta C_{cr})$, one may obtain the following correlations between the total interfacial flux and the driving force (as well as the correlations for mass transfer coefficient and the Sherwood number):

$$j = K_D \Delta C + K_{IT} \Delta C (\Delta C - \Delta C_{cr}),$$

$$K_M = K_D + K_{IT}(\Delta C - \Delta C_{cr}), \qquad Ma = \frac{L}{D(\eta_1 + \eta_2)} \frac{\partial \sigma}{\partial c} \Delta C \tag{5}$$

$$Sh = Sh_D + B\, Sc^{-1/2} (Ma - Ma_{cr}), \qquad K_{IT} = \frac{B}{\eta_1 + \eta_2} \frac{\partial \sigma}{\partial c} Sc^{-1/2}$$

Formulae [5] are similar to empirical correlations obtained in experiments (7,8). They are valid for systems with both stagnant and stirred phases (Lewis cells, moving droplets and films, etc.).

Time Dependent Mass Transfer Coefficient

Correlation [5] leads to time dependence of mass transfer coefficient in stirred systems. In particular, for extraction accompanied by interfacial turbulence in a stirred diffusion cell with constant contact area one may write for the concentration in the phase limiting mass transfer

$$\frac{dC_i}{dt} = - S_* j = - S_* \Delta C(K_D + K_{IT}(\Delta C - \Delta C_{cr})) \qquad (6)$$

(in case mass transfer is directed out of the phase).

Time dependence of mass transfer coefficient is hence easy to obtain

$$K_M = \frac{K_0(K_{IT}\Delta C_{cr} - K_D)\exp(-\ae t)}{K_{IT}\Delta C_{cr} - K_D + K_0(1 - \exp(-\ae t))}, \qquad (7)$$

$$\ae = S_*(K_{IT}\Delta C_{cr} - K_D),$$

$$K_0 = K_D + K_{IT}(\Delta C_0 - \Delta C_{cr})$$

The plot of characteristic $K_M(t)$ dependence is presented in Fig.1.

Usually $\ae$ is rather large (23) and dependence [7] seems to be nearly exponential. This fact is in accord with Javed et al (24) experimental results. The initial mass transfer coefficient K_0 increases when increasing the initial driving force ΔC_0 which is also experimentally confirmed (24). If $K_{IT} = 0$ (no interfacial turbulence) $K_M = K_D = \text{const}$.

Interfacial Instability During Drop Formation

Applying eq. [5] to mass transfer in a growing droplet one may simplify mass balance equation by considering the term $K_{IT}(\Delta C)^2$ in [5] to prevail because the conditions under which the drop is formed are usually far from the critical ones and interfacial turbulence is highly intensive.

The mass balance equation for a growing droplet is

$$\frac{dC}{dt} = \frac{d}{dt}\left(\frac{M}{V}\right) = \frac{1}{V}\frac{dM}{dt} - \frac{M}{V^2}\frac{dV}{dt} = j\frac{S}{V} + \frac{Q}{V}(C_0 - C) \qquad (8)$$

Taking note of

$$j = \frac{B}{\eta_1 + \eta_2} Sc^{-1/2}(\Delta C)^2, \qquad \Delta C = C_1 - C(t)/K_e$$

$$V(t) = Qt, \; S(t) = 3(4\pi)^{1/3}V^{2/3}(t)$$

one may obtain for the degree of extraction $A_* = (C_* - C_0)/(C_e - C_0)$ to the moment of drop formation (20), Golovin et al (25)

$$A_* = 1 - \frac{3}{m^2}(m\,\mathrm{cth}(m) - 1) \qquad (9)$$

$$m = 3\sqrt{3}B\left(\frac{1}{\eta_1 + \eta_2} \cdot \frac{\partial \sigma}{\partial C}\frac{1}{K_e}(C_1 - C_0/K_e)\frac{t_*}{R_*}\right)^{1/2} Sc^{-1/4}$$

(the case of transfer into the drop is considered; in the opposite case one should replace in [9] $m \rightarrow m\sqrt{K_e}$).

Formula [9] has been proved by experiments carried out by N.I.Polomarchuk and A.A.Ermakov (25) and the following correlation for parameter B for the case of a forming drop has been obtained (25):

$$B = 3.2 \; 10^{-6} \; Pe_f^{0.5} \qquad (10)$$

Figure 2 presents experimental data (dots) together with [9] , [10] (contineous line). The mean deviation is less than 11% (25).

Interfacial Turbulence in a Moving Droplet

When a droplet is set in motion after it has been formed interfacial turbulence becomes less intensive and decays while extraction is going on. Thus eq. [5] in the complete form should be applied when describing mass transfer kinetics in a moving drop. One may obtain the following formular for the time dependence of the degree of extraction in a moving droplet:

$$A = A_* + \left(A_{cr} - A_* + \frac{K_D}{K_{IT}\Delta C_0}\right)\frac{(1-A_*)(\exp(\beta(t-t_*)) - 1)}{(1-A_*)\exp(\beta(t-t_*)) - \left(A_{cr} - A_0 + \frac{K_D}{K_{IT}\Delta C_0}\right)} \quad (11)$$

$$\beta = \frac{3}{R}K_e^{-1}(K_{IT}\Delta C_0(1 - A_{cr}) - K_D), \quad K_{IT} = \frac{B}{\eta_1+\eta_2}\frac{\partial\sigma}{\partial C}Sc^{-1/2}, \quad t > t_*$$

Fig.3 presents the results of the experiments made by Polomarchuk and Ermakov (see Golovin et al (26)) for different extraction systems together with the curves obtained from [11] (contineous lines). Dotted lines correspond to the diffusion regimes. Experimental kinetics is in a good agreement with the model predictions. It should be noted that as time goes on interfacial turbulence becomes weaker and decays (when A becomes less than A_{cr}) and the slopes of the contineous lines become equal to those of the dotted ones.

It should be emphasized that not only mass transfer coefficient in the diffusion regime K_D but also the parameters B and A_{cr} characterizing interfacial turbulence substantially depend upon the drop motion. The latter may effect the fraction of the drop surface occupied by interfacial turbulence and may influence the critical conditions (e.g. the critical driving force) for interfacial instability to appear.

For practical use formula [11] requires the empirical correlations for parameter B and for the critical degree of extraction A_{cr}.

Effect of Interfacial Turbulence on Countercurrent Extraction Column Parameters

We consider here a countercurrent extraction column with complete stirring in both dispersed and contineous phases. According to eq. [5] partial mass transfer coefficients K_c and K_d should be linear dependent upon partial driving forces

$$K_c = K_c^0\left(1 + (K_{IT}^c/K_c^0)((y_s - y) - (y_s - y)_{cr})\right)$$

$$K_d = K_d^0\left(1 + (K_{IT}^d/K_d^0)((x - x_s) - (x - x_s)_{cr})\right) \quad (12)$$

$$K_{IT}^{c} = \frac{B}{\eta_1 + \eta_2} \frac{\partial \sigma}{\partial y} Sc^{-1/2}, \quad K_{IT}^{d} = \frac{B}{\eta_1 + \eta_2} \frac{\partial \sigma}{\partial x} Sc^{-1/2},$$

$$(K_{IT}^{c} = K_e^{-1} K_{IT}^{d})$$

Consider the dispersed phase to be the limiting one , thus $x_s = y/K_e$, $K_d = K_{od}$ (K_{od} and K_{oc} are the total mass transfer coefficients) and neglect the effect of interfacial turbulence on coalescence and breakup thus considering the contact area to be constant.

Solutions of the differential equations of mass transfer for an element of a column height as well as of material balance equations yield two expressions for the dimensionless column height Z needed for the given degree of extraction φ_{d2}, which corresponds to the different cases:

1) Interfacial turbulence takes place along the whole column height: $(1-\varphi_{d2} > \omega_{cr})$

$$Z = \frac{1}{1-\varkappa_d \omega_{cr}} \frac{G}{G-1} \ln \frac{[1+\varkappa_d(1-\varphi_{d2}-\omega_{cr})](1-\varphi_{d2}/G)}{[1+\varkappa_d(1-\varphi_{d2}/G-\omega_{cr})](1-\varphi_{d2})} \quad (13)$$

2) Interfacial turbulence decays at a certain height shorter than the needed one: $(1-\varphi_{d2} < \omega_{cr} < 1-\varphi_{d2}/G)$

$$Z = \frac{1}{1-\varkappa_d \omega_{cr}} \cdot \frac{G}{G-1} \ln \frac{1-\varphi_{d2}/G}{[1+\varkappa_d(1-\varphi_{d2}/G-\omega_{cr})]\omega_{cr}} + \frac{G}{G-1} \ln \frac{\omega_{cr}}{1-\varphi_{d2}} \quad (14)$$

$$Z = (K_d^0 S_* \Sigma / V_d)H, \quad G = LK_e, \quad L = V_c/V_d, \quad \varkappa_d = \frac{K_{IT}^{d}(x_1 - x_{2e})}{K_d^0},$$

$$\varphi_d = \frac{x_1 - x}{x_1 - x_{2e}}, \quad \omega_{cr} = \frac{(x - x_e)_{cr}}{x_1 - x_{2e}}$$

(the case of $G > 1$ is considered).

Here ω_{cr} is the dimensionless critical driving force at which interfacial turbulence appears, æ_d characterizes its intensity. At $\text{æ}_d \to 0$ both (13) and (14) are reduced to

$$Z_0 = \frac{G}{G-1} \ln \frac{1 - \varphi_{d2}/G}{1 - \varphi_{d2}}$$

corresponding to extraction without interfacial turbulence (Brounstein et al (27)).

Fig.4 presents the relative column height $\zeta = Z/Z_0$ dependence upon æ_d - parameter. A considerable decrease of a column height is demonstrated. It is worth noting that ζ is nothing but the ratio of the mass transfer unit height in the presence of interfacial turbulence to that without it.

CONCLUSIONS

1. A semiempirical model is proposed describing mass transfer kinetics for extraction accompanied by interfacial turbulence. The mass transfer coefficient is shown to be linearly dependent upon the driving force thus leading to its time dependence.

2. The model proposed is applied to mass transfer calculations for extraction in growing drops as well as in moving droplets. The formulae are obtained for the degree of extraction in both cases which give fair agreement with experimental results.

3. The developed model is used to calculate extraction column parameters. Formulae are obtained for the column height needed for a given degree of extraction. A considerable decrease of the column height due to interfacial turbulence is demonstrated.

SYMBOLS USED

A = degree of extraction

b, B = empirical parameters

C = concentration (kg mole/m^3)

Δ C = concentration driving force (kg mole/m^3)

D = diffusivity (m^2/s)

H = extraction column height (m)

j = total mass flux (kg $m^{-2}s^{-1}$)

j_0 = interfacial mass flux through the cell interface (kg mole/ /(m^{2}s)

K_D = mass transfer coefficient in the diffusion regime (m/s)

K_M = the total mass transfer coefficient (m/s)

K_0 = initial mass transfer coefficient (m/s)

K_e = phase equilibrium constant

K_{IT} = defined in [5] ($m^4 (kg\ mole)^{-1} s^{-1}$)

L = characteristic scale of mass transfer (m)

Ma = Marangoni number

M = the whole mass of transferring substance in a drop (kg)

Pe_f = Peclet number in the forming droplet

R = drop radius (m)

S = drop surface (m^2)

S_* = specific area of mass transfer (m^{-1})

Sc = ν/D = Schmidt number

Sh = $L\ K_M/D$ = Sherwood number

t = time (s)

t_* = the time of a drop formation (s)

V_c = volume rate of contineous phase feeding (m^3/s)

V_d = volume rate of dispersed phase feeding (m^3/s)

V = volume of a droplet (m^3)

v = velocity (m/s)

x = tangential coordinate (m) (concentration in the dispersed phase in extraction column, (kg mole/m^3), 1 - inlet, 2 - outlet)

y = normal coordinate (m) (concentration in the contineous phase in extraction column (kg mole/m^3)

Z = dimensionless extraction column height

δ_D = diffusion penetration depth (m)

δ_ν = momentum penetration depth (m)

ε = the fraction of the contact area occupied by convective cells

η = viscosity (kg $m^{-1}s^{-1}$)

ν = kinematic viscosity (m^2/s)

Σ = cross section area of an extraction column (m^2)

σ = interfacial tension (J/m^2)

Indice

cr = critical value

i = phase number (i=1 - contineous phase, i=2 - dispersed phase)

0 = initial value

s = surface

* = formation

e = equilibrium

REFERENCES

1. Sawistowski,H.,and Goltz,G.E.,1963,Trans.Inst.Chem.Eng. 41, 174.
2. Sawistowski,H.,and Austin,L.I.,1967,Chem.Ing.Techn. 39, 224.
3. Maroudas,N.G.,and Sawistowski,H.,1964,Chem.Eng.Sci. 19, 919.
4. Olander,D.R.,and Reddi, L.B.,1964,Chem.Eng.Sci. 19, 67.
5. Ostrovskii,M.V.,Barsukov,I.I. and Abramzon,A.A.,1973, Izv.Vuzov.Khim. i Khim.Technol. 16, 955.
6. Ostrovski,M.V.,and Golyakova,I.V.,1975,Theor.Osnovi Khim. Technol. 9, 643.
7. Ermakov,A.A.,Konshin,Yu.A.and Nazarov,V.I.,1977,Zh.Phys. Chem. 51, 2151.
8. Ermakov,A.A.,Rabinovich,L.M.and Slinko,M.G.,1988,Dokl. Acad.Nauk SSSR 303, 429.
9. Konshin,Yu.A.,and Ermakov,A.A.,1986,J.Appl.Chem (USSR) 59, 2222.
10. Lebon,G.and Cloot,A.,1982,Acta Mechan. 43, 141.
11. Ho,K.L.,and Chang,H.C.,1988,AIChE J. 34, 705.

12. Elenin,G.G.,Kalachinskaya,I.G.,and Rabinovich,L.M.,1987, Izvestiya Acad.Nauk SSSR, Mech.Zhidk. i Gaza N6, 4.

13. Bakker,C.A.P.,van Buytenen,P.M.,and Beek,W.J.,1966,Chem. Eng.Sci. 21, 1039.

14. Linde,H.,and Winkler,K.,1965,Z.Phys.Chem. 230, 207.

15. Imaishi,N.,Hozawa,M.,and Fujinawa,K.,1976,J.Chem.Eng.Jap. 9, 499.

16. Ruckenstein,E.,1968,Int.J.Heat Mass Transfer 11, 1753.

17. Sethy,A.,and Cullinan,H.T.,1975,AIChE J. 21, 575.

18. Rabinovich,L.M.,and Slinko,M.G.,1988, "The Problems of Mass Transfer Intensification and Mathematical Methods of Interfacial Self-Organization at Liquid/Liquid Extraction", ISEC-88 Conference Papers, Moscow, USSR.

19. Rehakova,M.,and Rod,V.,1979,Coll.Czech.Chem.Commun. 44, 2780.

20. Golovin,A.A.,and Rabinovich,L.M.,1988, "Interfacial Convection and Mass Transfer at Extraction in Drops", ISEC-88 Conference Papers, Moscow, USSR.

21. Golovin,A.A.,and Rabinovich,L.M.,1990,Theor.Osnovi Khim. Technol. 24, 592.

22. Golovin,A.A.,Ermakov,A.A.,and Rabinovich,L.M.,1989,Dokl. Acad.Nauk SSSR 305, 921.

23. Golovin,A.A.,1989, Dissertation, Karpov Inst. of Phys. Chem., Moscow, USSR.

24. Javed,K.H.,Thornton,J.D.,and Anderson,T.J.,1989,AIChE J. 35, 1125

25. Golovin,A.A.,Polomarchuk,N.I.,and Ermakov,A.A.,1988, Khimicheskaya Promyshlennost N12, 740.

26. Golovin,A.A.,Polomarchuk,N.I.,and Ermakov,A.A.,1990, Theor.Osnovi Khim.Technol. 24, 450.

27. Brounstein,B.I.,and Shögolev,V.V.,1988, "Hydrodynamics, heat and mass transfer in column apparatus", Khimiya, Leningrad, USSR.

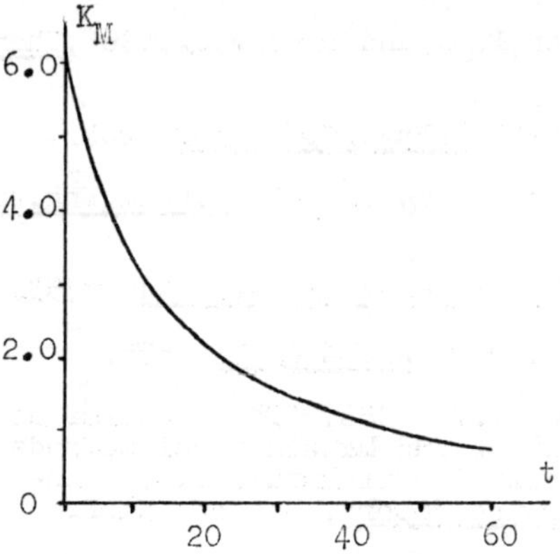

K_M - mass transfer coefficient (m/s 10^5)
t - contact time (min)

Figure 1 Time dependence of mass transfer coefficient

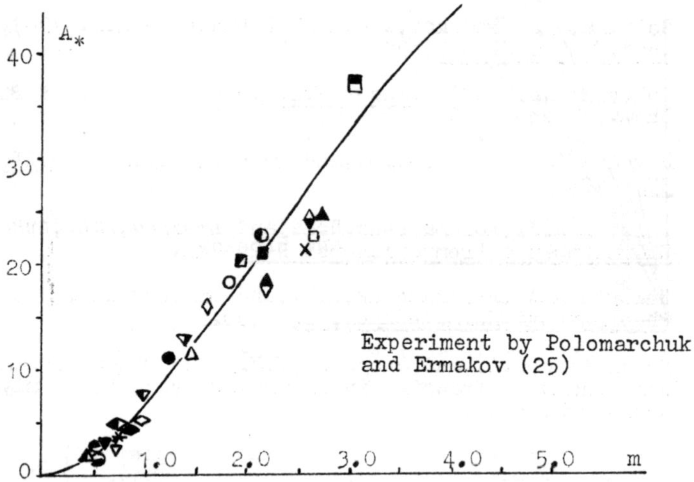

A_* - the degree of extraction to the moment of drop formation (%)

m - defined in [9]

Figure 2 The dependence of the degree of extraction A_* on the parameter m

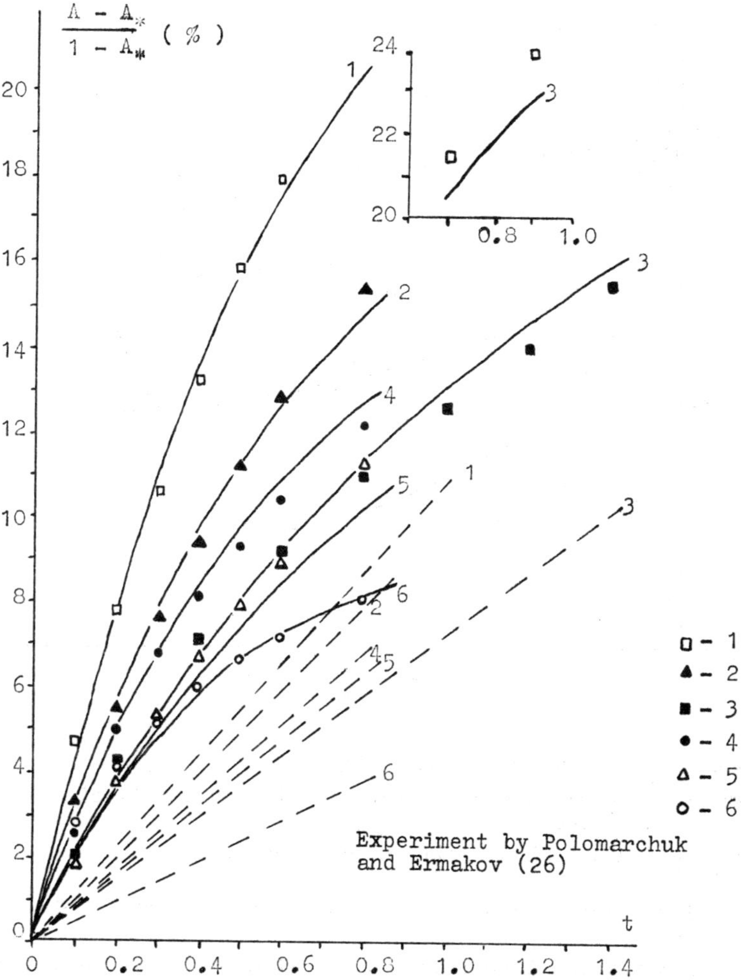

A - the degree of extraction in the moving droplet (%)

A_* - the degree of extraction at the beginnig of the drop motion (%)

t - contact time (s)

Figure 3 Mass transfer kinetics in a moving droplet during extraction accompanied by interfacial turbulence

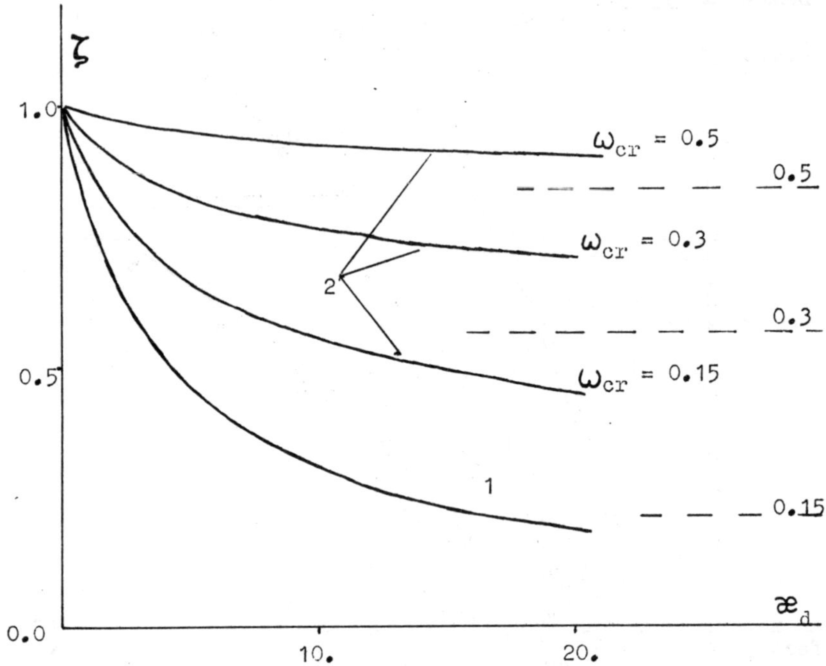

$\zeta = Z/Z_0$ - relative column height

$æ_d$ - the parameter characterizing the intensity of interfacial turbulence

ω_{cr} - dimensionless critical driving force

1 - interfacial turbulence takes place along the whole column height

2 - interfacial turbulence decays at a certain height shorter than the needed one

Figure 4 The relative column height ζ dependence upon the parameter $æ_d$

THE ONSET OF THE INTERFACIAL INSTABILITY OF DROPS DURING EXTRACTION

A.A.Golovin*

The onset of interfacial instability induced by the Marangoni effect accompanying mass transfer during liquid/liquid extraction to or from a steady drop is theoretically investigated. The non-stationary concentration profiles are considered. The finite rate of the adsorption of transferring substance at the interface is taken into account. It is shown that in case of diffusion controlled extraction in drops there exist two universal critical values of diffusion coefficients ratio. restricting the interval where interfacial instability may occur regardless of mass transfer direction.

INTRODUCTION

Liquid/liquid extraction is well known to be often accompanied by interfacial turbulence caused by hydrodynamic instability of an interface during mass transfer according to the Marangoni effect. The conditions for interfacial instability to occur have been attracting vast attention since the pioneering work of Sternling and Scriven (1). However few papers are devoted to the onset of interfacial instability of a spherical interface (Haydon (2), Davies and Haydon (3), Deyhimi and Sanfeld (4), Sörensen et al (5), Sörensen and Hennenberg (6), Sörensen (7), Sörensen and Castillo (8), Sörensen (9), Patzer and Homsy (10), Golovin and Rabinovich (11,12,13)) though it seems to be of considerable interest from the chemical engineering point of view.

The most comprehensive analysis has been carried out in (6). The authors took into account an arbitrary number of transferring substances as well as arbitrary kinetics of surface chemical reactions, surface rheology and deformation, and the concentration profiles in the bulks of phases. One of the most important conclusions made is that the critical ratio of the diffusion coefficients in the phases (in the case of one transferring substance) which determins the direction of mass transfer at which interfacial instability may occur depends upon the harmonic perturbation number and in the case of parabolic concentration profiles is less than one. This fact may account for the large number of experimental data which do not accord with Sternling and Scriven theory.

* Karpov Institute of Physical Chemistry, Moscow, USSR.

The analysis of the conditions for interfacial instability to occur during extraction in drops (6) has been generalized in (11,12,13) for the case of a finite adsorption rate of the transferring substance at the interface and more general shape of bulk concentration profiles. It has been found that finite adsorption rate (i.e. nonequilibriumness between phases and an interface) plays an important role. It has been shown also that there exist two critical values of diffusion coefficients ratio in phases, depending upon the adsorption kinetics as well as the shape of concentration profiles. These two values restrict the interval out of which interfacial instability may occur if mass transfer goes in one direction only and within which it is possible for interfacial instability to appear independently of mass transfer direction.

In the papers mentioned somewhat artificial stationary concentration profiles in the phases were used. In real extraction systems interfacial instability usually occurs at the very beginning of the extraction process when diffusion penetration depth is much less than drop radius and concentration gradients are localized in the vicinity of an interface. At the initial stage of extraction without chemical reaction there is no difference between the curved and the plane interface, so one may use the self-similar concentration distributions of the plane problem in order to approximate concentration distributions at the beginning of extraction in a drop. Therefore it seems interesting to analise interfacial instability of a spherical interface with such concentration profiles.

Marginal Stability Condition

In the present work the following concentration profiles approximating concentration distribution in diffusion layer at the beginning of extraction process in a drop are used:

$$C_i^{(0)} = C_{i0} + (C_{is} - C_{i0})\, \mathrm{erfc}\left[(-1)^i (R - r)\sqrt{\pi}/(2\delta_i)\right] \qquad (1)$$

where $\delta_i = \sqrt{\pi D_i t_0}$, t_0 - is the time at which interfacial instability begins to develop. Usually $R \sim 10^{-3}$m, $D \sim 10^{-9}\mathrm{m}^2\mathrm{s}^{-1}$, $t_0 \sim 0.1$ s, thus $\delta_i/R \sim 0.01$ and we have to assume further $\delta_i/R \ll 1$.

Interfacial instability of the system in question is investigated by means of linear stability analysis of Navier-Stokes, continuity and convective diffusion equations as well as the equation of surface active substance conservation at the interface with finite adsorption rate being taken into account. The well known Marangoni boundary condition derived by Scriven (14) for a spherical interface is used. The analysis is similar to that carried out in (6). This leads to marginal stability condition in the following form (the surface is considered undeformable):

$$\frac{\partial\sigma}{\partial\Gamma}\frac{\partial C_2^{(0)}}{\partial r}\bigg|_{r=1}\cdot R = H_\ell(\eta,\varkappa,\varepsilon)\,\frac{K_1^\Gamma\dfrac{D_1(\ell+1)}{RK_1^c + D_1(\ell+1)} + K_2^\Gamma\dfrac{D_2\ell}{RK_2^c + D_2\ell} + \dfrac{D_s}{R^2}\ell(\ell+1)}{\dfrac{RK_2^c\varphi_{\ell 2}}{RK_2^c + D_2\ell} - \dfrac{D_2}{D_1}\cdot\dfrac{RK_1^c\varphi_{\ell 1}}{RK_1^c + D_1(\ell+1)} + \dfrac{\Gamma_0\ell(\ell+1)}{\dfrac{\partial C_2^{(0)}}{\partial r}\Big|_{r=1}\cdot R^2}} \qquad (2)$$

$$H_\ell = \frac{1}{\ell(\ell+1)}\left[(\eta_1+\eta_2)(2\ell+1) + \frac{(\varkappa+\varepsilon)}{R}\ell(\ell+1) - \frac{2\varepsilon}{R}\right]$$

$$K_i^c = \frac{\partial J_i(C_i,\Gamma)}{\partial C_i}, \quad K_i^\Gamma = -\frac{\partial J_i(C_i,\Gamma)}{\partial \Gamma}$$

$$\varphi_{\ell 1} = \frac{1}{2}\int_0^\infty z(z+2)(z+1)^{-2\ell-1}\exp\left(-\frac{\pi R^2 z^2}{4\delta_1^2}\right)dz\ ; \quad \varphi_{\ell 2} = \frac{1}{2}\int_0^1 z(2-z)(1-z)^{2\ell+1}\exp\left(-\frac{\pi R^2 z^2}{4\delta_2^2}\right)dz$$

Eq. [2] generalizes the one obtained in (6) for the case of a finite adsorption rate and more real "erfc"-concentration profiles.

Equation [2] has two limit forms: in the case of diffusion controlled extraction and in the case of adsorption controlled one.

Diffusion controlled extraction: $D_i \ll k_i^c \delta_i$. In that case one has the following boundary conditions at the drop surface in the unpertubed state:

$$D_1\frac{\partial c_1^{(0)}}{\partial r} = D_2\frac{\partial c_2^{(0)}}{\partial r}, \quad c_1^{(0)} = K_e c_2^{(0)}, \quad r = R,$$

which lead to the following formulae for C_{is}:

$$C_{2s} = \frac{C_{10} + \sqrt{D_2/D_1}\ C_{20}}{\sqrt{D_2/D_1} + K_e}, \quad C_{1s} = K_e C_{2s} \tag{3}$$

If the adsorption Γ_0 in the unpertubed state and surface diffusion D_s may be neglected eqs. [2] and [3] after taking the asymptotics of φ_{11} and φ_{12}

$$\varphi_{11} \sim (2/\pi)\left[(\delta_1/R)^2 - (2l + 1/2)(\delta_1/R)^3\right] + O((\delta_1/R)^4)$$

$$\varphi_{12} \sim (2/\pi)\left[(\delta_2/R)^2 - (2l + 3/2)(\delta_2/R)^3\right] + O((\delta_2/R)^4) \tag{4}$$

lead to the following expression for dimensionless critical Marangoni number:

$$Ma_{diff} = \eta_*(\varepsilon_1, \varepsilon_2, l)\frac{[K_e(l+1) + D_* l](\sqrt{D_*}+K_e)}{l(l+1)\sqrt{D_*}K_e[(4l+1) - \sqrt{D_*}(4l+3)]} \tag{5}$$

$$Ma_{diff} = \left(-\frac{\partial\sigma}{\partial C_2}\right)(\Delta C)_0 \frac{t_0}{R(\eta_1+\eta_2)}, \quad (\Delta C)_0 = C_{20} - C_{10}/K_e \text{ - initial}$$

driving force, $D_* = D_2/D_1$, $\eta_*(\varepsilon_1, \varepsilon_2, l) = 2l+1+\varepsilon_1 l(l+1) - 2\varepsilon_2$,

$$\varepsilon_1 = \frac{æ+\varepsilon}{R(\eta_1+\eta_2)}, \quad \varepsilon_2 = \frac{\varepsilon}{R(\eta_1+\eta_2)}$$

The direction of mass transfer at which interfacial instability may occur is determined by D_*. It follows from [5] that in the case of diffusion controlled extraction the critical value of D_* depends upon the harmonic perturbation number only: $D_{*cr} = [(4l+1)/(4l+3)]^2$, and there exist two universal critical values of D_*

$$D_{*1} = 25/49 \approx 0.51 \text{ and } D_{*2} = 1 \tag{6}$$

restricting the interval out of which the interfacial instability (without oscillations) may occur if mass transfer goes in one direction only and within which it is possible for interfacial instability to appear regardless of mass transfer direction.

It also follows from [5] that interfacial instability in a drop may appear only a certain period of time t_0 after the phases have been put in contact, namely, if

$$t_0 \geqslant t_{cr} = \frac{R(\eta_1+\eta_2)}{\left(-\frac{\partial\sigma}{\partial C_2}(\Delta C)_0\right)} f(\varepsilon_i, K_e, D_*, l) \tag{7}$$

This time is determined by the radius of a drop and the viscosities of the phases: the larger is the drop and the more is the viscosity the greater is the critical time. It is also determined by the "driving force of interfacial instability" $(\partial\sigma/\partial C)\Delta C$: less time is necessary for interfacial instability to occur in extraction system with the larger driving force.

The critical time depends also upon the harmonic perturbation number l. It may be shown, however, that there exists the minimum critical time t_* depending only upon the physico-chemical parameters of an extraction system and the direction of mass transfer.

It is noteworthy that if the contact time t_0 is very small eqs [5] and [7] are no more valid as the diffusion controll condition $D_i \ll k_i^c \delta_i$ is not true. If this is the case one may consider adsorption controlled extraction.

<u>Adsorption controlled extraction: $D_i \gg k_i^c \delta_i$</u>. In this case one has the following boundary conditions at the drop surface in the unpertubed state:

$$-\frac{1}{R} D_2 \frac{\partial C_2^{(0)}}{\partial r} = k_2^c C_2^{(0)} - k_2^{\Gamma} \Gamma_0$$

$$\frac{1}{R} D_1 \frac{\partial C_1^{(0)}}{\partial r} = k_1^c C_1^{(0)} - k_1 \Gamma_0$$

$$D_1 \frac{\partial C_1^{(0)}}{\partial r} = D_2 \frac{\partial C_2^{(0)}}{\partial r}$$

which lead to the algebraic system for C_{is}, Γ_0. Using the solution of that system and the asymptotics [4] one may obtain the expression for the critical Marangoni number

$$Ma_{ads} = \frac{D_*}{\chi_1} \eta_*(\varepsilon_1, \varepsilon_2, l) \frac{\frac{l+1}{\chi_1+l+1} + E\frac{1}{\chi_2+l}(\chi_2+1)(\chi_1+l+1)}{l^2(l+1)(K(l+1)/l - D_*)} \quad (8)$$

$$Ma_{ads} = \left(-\frac{\partial\sigma}{\partial\Gamma}\right)(\Delta\Gamma)_0 \frac{t_0}{R(\eta_1+\eta_2)}, \quad K = k_2^c/k_1^c, \quad E = k_2^\Gamma/k_1^\Gamma,$$

$$\chi_i = Rk_i^c/D_i, \quad (\Delta\Gamma)_0 = C_{20}k_2^c \frac{k_1^\Gamma}{k_1^\Gamma+k_2^\Gamma} - C_{10}k_1^c \frac{k_2^\Gamma}{k_1^\Gamma+k_2^\Gamma}$$

(as in the case of diffusion controlled extraction we neglect adsorption Γ_0 in the denominator of [8] as well as the surface diffusion).

It follows from [8] that in case of interface adsorption controlled extraction the critical diffusion coefficients ratio is $D_{*cr} = K(l+1)/l$. There also exist two critical values of D_* depending upon the adsorption constants ratio

$$D_{*1} = K, \quad D_{*2} = 2K \quad (K = k_2^c/k_1^c) \quad (9)$$

restricting the interval out of which interfacial instability (without oscillations) may appear if mass transfer runs in one direction only and within which it is possible for interfacial instability to occur independently of mass transfer direction. It is worth noting that in case of adsorption controlled extraction D_{*1} and D_{*2} may be more than one and may account for the certain experimental results (2) on "kicking droplets", adduced in (6).

Obviously in case of adsorption controlled extraction there also exists the critical contact time after which interfacial instability is possible to occur. Now it depends also upon the adsorption kinetics.

The existence of the critical contact time for Marangoni instability during extraction in drops seems to have the following qualitative physical explanation. Interfacial instability of the harmonic number l may not appear if the contact time is less than the characteristic circulation time $\tau \sim \lambda/\upsilon_s$ in the convective cell of the scale $\lambda \sim R/l$ having the circulation velocity $\upsilon_s \sim \sim \eta^{-1}(\partial\sigma/\partial c)\Delta_{\parallel}C$ where $\Delta_{\parallel}C$ is the concentration difference along the cell. It is natural to suppose that in case of diffusion controlled extraction $\Delta_{\parallel}C \sim (\Delta C)_0/f(K_e, D_*, l)$, i.e. the concentration difference along the cell is proportional to the whole driving force and determined by diffusion coefficients ratio and phase equilibrium constant. Thus it is easy to obtain from [5] $t_0 \sim \tau$. The same evaluations may be made for adsorption controlled extraction.

In conclusion it should be noted that the asymptotics [4] and the critical conditions [5], [8] for diffusion and adsorption controlled extraction are valid if the harmonic perturbation number is not too large. Contrariwise one is obleged to take the final adsorption rate into account as well as the surface effects (surface viscosities, diffusion and adsorption) and to use eq.[2] for interfacial instability analysis.

CONCLUSIONS

1. The onset of interfacial instability during extraction in drops is highly effected by the finite rate of adsorption of the substance transferred.

2. Unlike a plane interface, for a spherical one there are two critical values of the ratio of the diffusion coefficients in phases: D_{*1} and D_{*2}. They restrict the interval within which interfacial instability may occur if mass transfer goes in one direction only and out of which it is possible for the interfacial instability to appear regardless of the direction of mass transfer.

3. In the case of diffusion controlled extraction at the beginning of mass transfer one has $D_{*1} = 25/49$, $D_{*2} = 1$ and in the case of adsorption controlled extraction $D_{*1} = K^2$, $D_{*2} = 2K$. The latter may be greater than one, thus interfacial instability may occur if mass transfer goes into the phase with a smaller diffusion coefficient.

4. Interfacial instability in a drop may appear only a certain period of time after the beginning of the phase contact, both in the cases of diffusion and adsorption controlled extraction. This time is determined by the critical Marangoni numbers defined by eqs. [5] and [8].

SYMBOLS USED

C_i = concentration in i-phase (kg/m^3)

D_i = diffusion coefficient in i-phase (m^2/s)

$D_* = D_2/D_1$

D_s = surface diffusivity (m^2/s)

$J_i(C_i, \Gamma)$ = adsorption-desorption flux in i-phase ($kg\ m^{-2}s^{-1}$)

$K = k_2^c/k_1^c$

K_e = phase equilibrium constant

l = harmonic perturbation number

Ma_{diff} = Marangoni number for diffusion controlled extraction

Ma_{ads} = Marangoni number for adsorption controlled extraction

r = radial coordinate

R = radius of a drop

t_0= the time at which interfacial instability begins to develop

Γ= adsorption of transferring substance (kg/m^2)

δ_i = diffusion penetration depth (m)

ε= surface shear viscosity (kg/s)

η_i = bulk shear viscosity of i-phase

$\varkappa$= surface dilatational viscosity (kg/s)

σ= interfacial tension (J/m^2)

Indice:

e = equilibrium state

i = phase number, i=1 - contineous phase, i=2 - dispersed phase

l = harmonic perturbation number

0 = steady state

s = surface

REFERENCES

1. Sternling, C.V., and Scriven, L.E., 1959, AIChE J. 5, 514.

2. Haydon, D.A., 1958, Proc.Roy.Soc. A243, 483.

3. Davies, T.V., and Haydon, D.A., 1958, ibid, 492.

4. Deyhimi, F., and Sanfeld, A., 1974, C.R.Acad.Sci. D279, 437.

5. Sörensen, T.S., Hennenberg, M., Steinchen, A., and Sanfeld, A., 1976, J.Coll.Int.Sci. 56, 191.

6. Sörensen,T.S.,and Hennenberg,M.,1979,Lecture Notes in Physics,N105, Springer Verlag, Berlin, p.276.

7. Sörensen,T.S.,1980,J.Chem.Soc.Farad.Trans.II 76, 1170.

8. Sörensen,T.S.,and Castillo,J.L.,1980,J.Coll.Int.Sci. 76, 399.

9. Sorensen,T.S.,1982,J.Act.Chem.Scand. A36, 141.

10. Patzer,J.F.II,and Homsy,G.M.,1981,Phys.Fluids 24, 567.

11. Golovin,A.A.,and Rabinovich,L.M.,1988,Izv.Acad.Nauk SSSR, Mech.Zhidk. i Gaza N2, 137.

12. Golovin,A.A.,and Rabinovich,L.M.,1988,"Interfacial Convection and Mass Transfer at Extraction in Drops", ISEC-88 Conference Papers, Moscow, USSR

13. Golovin,A.A.,and Rabinovich,L.M.,1988,J.Prikl.Mech. i Tech Phys. N5. 101.

Extraction '90

(10–14 September 1990)

IChemE Symposium Series No. 119

Discussion of Papers

Paper 1
The breakup of a jet of chlorobenzene in water
E.A. Elhag (*Industrial Research and Consulting Centre, Sudan*) and R.M. Edge (*University of Strathclyde, UK*)

QUESTION from Dr. L.R. Weatherley (*Heriot-Watt University, UK*)
What do you think would be the effect of high viscosity in the drop upon the detachment mechanism?

ANSWER from Dr. R.M. Edge
Some work has been done by other authors in which the dispersed phase is viscous and non-Newtonian. In this case they have observed segments joined by filaments.

QUESTION from Dr. P.J. Thompson (*AEA Technology, UK*)
Have you done any work to demonstrate whether similar breakup mechanisms occur in the presence of mass transfer?

ANSWER from Dr. R.M. Edge
So far we have only studied systems in which mass transfer is absent.

QUESTION from Dr. P.J. Thompson (*AEA Technology, UK*)
Could the measurement of the critical velocity form the basis of a simple method for determination of physical properties such as interfacial tension?

ANSWER from Dr. R.M. Edge
In a subsequent paper we will be examining the critical velocity in more detail. As the results which we have are for one system only I cannot give a definite answer to this question. However, it is not easy to get an accurate measure of the critical nozzle velocity and I do not foresee it being used to estimate the physical properties of the system.

Paper 2
Terminal velocity relationships of two-phase drops
M.A. Hashim and M. Balasubramaniam (*University of Malaya, Malaysia*)

QUESTION from Dr. P.J. Thompson (*AEA Technology, UK*)
Have you any information on how the presence of a surfactant will affect the behaviour of the two-phase drop system?

ANSWER from Dr. M.A. Hashim
We do not have any experimental data with regards to the effect of surfactants on two-phase drops but we realise that it is difficult to eliminate all surface-active contaminants from whatever system is employed. Based on the available evidence, I can suggest that surfactants have the greatest influence on terminal velocity near the point of transition from rectilinear to oscillating motion. This is presumably because internal circulation can drastically alter the wake structure of a fluid particle leading to delayed boundary layer separation, smaller wakes and delayed vortex shedding. Surfactants play a particularly important role in high interfacial tension systems since surface tension reductions are largest for these systems.

<u>Paper 3</u>

The visualisation of droplet motion, breakdown and coalescence in a packed bed

R.M. Edge (*University of Strathclyde, UK*)

<u>QUESTION</u> from Dr. I.S. Denniss (*BNFL, UK*)
Does the need to match the refractive indices of the phases make it difficult to duplicate all physical properties of the aqueous phase?

<u>ANSWER</u>
It is easy to vary the viscosity of the two liquid phases. To change the densities of the two liquids is more difficult and we are currently attempting to reduce the density difference between the phases. The density of the aqueous phase is a problem as at present it is rather high.

<u>QUESTION</u> from J. Jenkins (*AEA Technology, UK*)
In view of the usefulness of this technique in looking at overall flow patterns, channelling etc in columns, how wide a slice of column can be viewed?

<u>ANSWER</u>
The thickness of the slice examined by the laser is about 1 mm.

<u>QUESTION</u> from Dr. P.J. Thompson (*AEA Technology, UK*)
What were your criteria for selection of packing size?

<u>ANSWER</u>
Whilst developing the optical method we used a packing which was readily available. The larger the size of the packing the greater the path length in the column that can be examined without problems arising due to scattering of the laser beam.

<u>QUESTION</u> from Professor B. Waldie (*Heriot-Watt University, UK*)
The drops shown look very sharp. Sharpness presumably depends on relative velocity of drops and laser scan. What were the drop velocities in the examples shown?

<u>ANSWER</u>
The laser beam was scanned in the opposite direction to the direction of the fall of the drops. This gives a clear image of a moving drop but there will be some distortion of the drop shape in the direction of scan. This distortion increases with increase in the velocity of the drops and decreases with increase in the velocity of scan. At present we are scanning 45 cm in about 10 s. We hope to reduce this to between 1 and 2 s. During the development programme we did not take successive scans and therefore we do not know the effective drop velocities in the direction of scan. In future work we will be able to obtain measurements.

Paper 4

Further studies on the time dependence nature of the mass transfer coefficients for the system uranyl nitrate - nitric acid - TBP
D. Rogers and P.J. Thompson (*AEA Technology, UK*)

QUESTION from Dr. A.A. Golovin (*Karpov Institute of Physical Chemistry, USSR*)
Which is the cause of interfacial instability in the system under investigation; the concentration dependence of interfacial tension or its temperature dependence?

ANSWER from Dr. P.J. Thompson
The interfacial tension of the uranyl nitrate - nitric acid - TBP/OK system increases with uranium concentration, but I believe that the variation of interfacial tension with temperature is a dominating factor for interfacial instability in this system; the heat of reaction for 50 g/l uranium is of the order -21.2 kJ/mole. This effect was reported by Ortiz and Thompson at the PROGEC Summer School on Extraction at Toulouse 7/11 September 1987, and showed that a stability analysis predicted interfacial instability for both OK and hexane diluents.

Paper 6

Interfacial instability at liquid/liquid extraction with chemical reaction
L. M. Rabinovitch and I. A. Belikova (*Karpov Institute of Physical Chemistry, USSR*)

QUESTION from Dr. P.J. Thompson (*AEA Technology, UK*)
Do you consider the general published correlations for diffusion coefficients accurate, or is it necessary to measure them experimentally?

ANSWER from Dr. L.M. Rabinovitch
I consider the general published correlations for diffusion coefficients may be used for evaluation of the conditions for interfacial instability to appear. However, there are a lot of physio-chemical factors which may influence the diffusion coefficients and it may be that you will have to measure the diffusion coefficients in order to get the most correct information. I should say that you have to measure the diffusion coefficients after you have found that the use of general correlations leads to the contradiction between the theory and the experiment.

QUESTION from Dr. P.J. Thompson (*AEA Technology, UK*)
What measurements are necessary to enable design of a liquid/liquid extraction contactor?

ANSWER from Dr. L.M. Rabinovitch
Several steps are necessary to apply interfacial instability theory to contactor design. Firstly one should measure physico-chemical parameters which influence the conditions for interfacial instability to appear. Then the critical parameters have to be measured, such as the critical driving force (or the critical Marangoni number), the critical Reynolds number in the particular system (in case of moving droplets and films) and the critical contact time at which interfacial instability appears. After the critical parameters have been measured the empirical parameters

of the semiempirical model are to be obtained either from the already existing correlations or from the experiments.

QUESTION from E. Brejza (*Imperial College, UK*)
What sort of system and apparatus have you based your models on? Is each system tested on different types of apparatus and then a model based on the results obtained overall?

ANSWER from Dr. L.M. Rabinovitch
The models developed are rather general and may be applied to the analysis of mass transfer under interfacial instability in any systems and apparatus. When applying the obtained criteria for the determination of the conditions for interfacial instability to appear it is necessary to use physico-chemical characteristics of contacting fluids, kinetic constants and geometrical parameters of dispersed phases and apparatus.

Paper 8
Phenol recovery with supported liquid membranes
A.M. Urtiaga, M.I. Ortiz and A. Irabien (*University of Pais Vasco, Spain*)

QUESTION from J. Blyton (*University of Strathclyde, UK*)
How specific is the continuous process for phenol recovery? For example, is it possible to separate and recover different isomers of phenolic derivatives using appropriate solvents for high throughputs?

ANSWER from Dr. M.I.Ortiz
If it is possible to have a membrane formulation so that solubilities and diffusivities of the phenolic derivatives and different enough the answer is yes. In general terms, when the LM technology is to be applied and there is a need for selectivity of a given solute the most common approach is to use facilitated transport LM.

QUESTION from Dr. R. Bush (*AEA Technology, UK*)
Is the feed solution pre-equilibrated with MIBK, as well as the strip solution?

ANSWER from Dr. M.I. Ortiz
Yes, both aqueous phases were pre-equilibrated in order to avoid losses of MIBK in either of the two phases.

QUESTION from E. Brejza (*Imperial College, UK*)
Are you planning to use any rate enhancers eg. microemulsions to increase the extraction rate?

ANSWER from Dr. M.I. Ortiz
At the moment we are waiting for some promising results from a well known research group at Imperial College. If these results check the enhancement of solute fluxes through the LM we will probably start adding rate enhancers to the LM.

QUESTION from R. Klaassen (*T N O, The Netherlands*)
Is there data available about the lifetime of the SLM?

ANSWER from Dr. M.I. Ortiz
At the beginning of our research we performed longtime experiments. After three months we measured the flux of phenol and we observed that there was no significant change (c10%) when compared with the initial flux.

QUESTION from Professor B. Waldie (*Heriot-Watt University, UK*)
In batch laboratory experiments the aqueous phases were saturated with MIBK for experimental purposes. In a real continuous process how would loss of MIBK to the aqueous streams be dealt with?

ANSWER from Dr. M.I. Ortiz
If a mixed membrane containing MIBK is used in an industrial process, after the extraction step a further separation operation has to be performed in order to eliminate MIBK from the aqueous phases.

QUESTION from Dr. D.J. Gunn (*University College of Swansea, UK*)
With reference to the internal or boundary layer mass transfer coefficient inside the fibre, a theoretical estimate for the mass transfer coefficient may be obtained from the classical 19th century analysis of Graetz who obtained a solution to the equation for flow in a round tube:

$$\frac{D}{R}\frac{\partial}{\partial R}\left(R\frac{\partial c}{\partial R}\right) - \frac{u_o}{z}\left(1-r^2\right)\frac{\partial c}{\partial z} - \frac{\partial c}{\partial t} = 0$$

under a mass transfer driving force between the fluid and the wall. His series solution may be expressed as a mass transfer coefficient that may be used to check the authors' experimental estimate.

ANSWER from Dr. M.I. Ortiz
The 19th century analysis of Graetz for flow in a round tube under a mass transfer driving force between the fluid and the wall gives good estimates of the internal mass transfer coefficients in the case when the concentration of the solute along the wall is kept constant. Average values of the wall concentration of PhOH can be considered for low extraction yields, that is when the difference between PhoH at the entrance of the fibre and the end is less than 10%.

One of the main objectives of this work is to reach high extraction yields (>90%) in order to reduce PhOH concentration in the feed solution. The authors applied the Graetz solution obtaining very poor estimates under the experimental range. This result was not surprising if one takes into account that the integration of the system of differential equations for a variable PhOH concentration at the wall of the fibre has been recently published (1989, Stroeve et al).

Paper 9
New extractants for the treatment of nuclear waste solutions
C. Cuillerdier, C. Musikas and P. Hoel (*Commissariat a L'Energie Atomique, France*)

QUESTION from H. Schmieder (*Kernforschungszentrum Karlsruhe GmbH, West Germany*)
You have applied a long residence time of 15 minutes in the settler. Is this really necessary?

ANSWER from C. Cuillerdier
Perhaps it is not necessary, but we took this precaution as the behaviour of the solvent in a mixer-settler was not known because it was the first experiment.

QUESTION from H. Schmieder (*Kernforschungszentrum Karlsruhe GmbH, West Germany*)
Have you a figure of the costs for the amides in comparison with CMPO?

ANSWER from C. Cuillerdier
Nowadays diamides are not industrially synthesised but the synthesis has been optimised and the price can be reduced in case of industrial development.

QUESTION from Dr. R. Bush (*AEA Technology, UK*)
I am interested in the behaviour of neptunium in your backwash contactor. If Np(VI) is reduced to Np(V) it will backwash; but if it is further reduced to Np(IV) it will not. Have you any information on what proportion of the Np remains extracted in your process, and on whether this depends on the residence time in the contactor?

ANSWER from C. Cuillerdier
During the experiment several phenomena occurred:
- increase of the acidity of the aqueous phase in the back extraction battery due to a perturbation in the flow.
- presence of iron that can reduce Np as Np(IV) when it is reduced as iron II by hydroxylamine nitrate.

So the Np(IV) remaining in the solvent is not a significant feature of the process in normal conditions.

QUESTION from Dr. P.J. Thompson (*AEA Technology, UK*)
Can you comment on the degree of aqueous entrainment or haze formation in your solvent product from the extract mixer-settler?

ANSWER from C. Cuillerdier
Only qualitative observations have been done; the solvent appeared to be clear, no secondary emulsion was formed.

QUESTION from Dr. P. Wilson (*BNFL, UK*)
In the choice of substituents, it seems to be assumed that the best R & R' with R" = H would be the same as with R" analkyl or alkoxyalkyl group. That is what might be expected, but is there any practical evidence to support it?

ANSWER from C. Cuillerdier
No systematic researches were undertaken for a given R". But for instance changing $R=CH_3$ by $R=C_2H_5$ for $R' = C_yH_g$ and a given R" leads to a decrease in distribution coefficient of Am^{3+}.

Paper 10
Extraction of non ferrous metals from hydrometallurgical waste waters. Planning product recycling using separation techniques
F. Mijangos, J.I. Lombrana, F. Varona (*University of Pais Vasco, Spain*) and M. Diaz (*University of Oviedo, Spain*)

QUESTION from Professor B. Waldie (*Heriot-Watt University, UK*)
What was the preferred resin used in the study? I don't think it's mentioned in the paper.

ANSWER from Dr. F. Mijangos
It was mentioned that the selected resin, after a previous experiment with the industrial effluent, was a commercial product containing a amino-di-acetic group supported upon a polystyrene-divinylbenzene matrix.

QUESTION from R. Klaassen (*T N O, The Netherlands*)
Is the resin used commercially available?

ANSWER from Dr. F. Mijangos
Yes, this resin is produced by Bayer AG (Germany) with the commercial name Lewatit TP207.

Paper 11
Flow distribution in fixed bed ion exchange units
D.J. Gunn (*University College of Swansea, UK*)

QUESTION from Dr. L.R. Weatherley (*Heriot-Watt University, UK*)
Do you think that your development can be extended to predict breakthrough curves for the case of axial and radial dispersion?

ANSWER
The effect of radial dispersion is usually to reduce the influence of axial dispersion as shown in Figure 9. This figure shows the combined effect of axial and radial dispersion encapsulated in D_L, the total coefficient of axial dispersion.

QUESTION from Dr. L.R. Weatherley (*Heriot-Watt University, UK*)
What are the likely effects of swelling and sorption in the bed (as will happen during cycling) upon the dispersion model?

ANSWER
One can also add the effect of heat evolution. Once process specific effects such as resin expansion are involved, specific process information is required for the process model, including, for example, a simultaneous differential thermal balance. Intra-particle effects can also be added.

The principal limitation is the quality of available process information, not the capabilities of the model.

Paper 12
Development of CMP-impregnated solid extractant for TRU removal from radioactive liquid wastes
K. Takeshita ***(Institute of Research and Innovation, Japan)***, **S. Matsumoto** ***(Saitama University, Japan)***, **M. Kumagai** ***(Institute of Research and Innovation, Japan)***, **Y. Endo** ***(Ishikawajima-Harima Heavy Industries Co Ltd, Japan)***, **J. Koja** ***(Saitama University, Japan)*** **and Y. Takashima** ***(Institute of Research and Innovation, Japan)***

QUESTION from Dr. D.J. Gunn (*University College of Swansea, UK*)
In conventional ion exchange processes the saturated resin is regenerated for further use. Has Dr. Takeshita considered the process that would include extraction of TRU followed by regeneration of the resin using, for example, nitric acid?

QUESTION from C. Cuillerdier (*Commissariat à L'Energie Atomique, France*)
As there is no regeneration of the extractant and as the CMP is expensive, is this process more interesting economically than a solvent extraction process where the extractant can be recycled?

ANSWER from K. Takeshita
I think that the regeneration of solid extractants will generate the large amount of secondary liquid wastes. If the volume of spent solid extractants can be reduced by a volume-reduction method such as dry distillation, TRU can be recovered effectively and vitrificated compactly. The volume of secondary wastes generated in our proposed process will become smaller than those of conventional extraction processes.

QUESTION from Dr. R.P. Bush (*AEA Technology, UK*)
For particle sizes greater than 0.2 mm, is the rate of absorption of cerium dependent on the degree of cross-linking (ie the percentage of divinyl benzene)?

ANSWER from K. Takeshita
Crosslinkage of SDB resin is an important factor to increase the extraction rate of cerium. I think that by use of low-crosslinked resin, the impregnated amount of CMP increases and the coverage of CMP on pore surface in SDB resin also increases, so that the extraction rate of cerium increases.

Paper 14
Electrokinetic processes for biochemical extractions
L.R. Weatherley ***(Heriot-Watt University, UK)***

QUESTION from Dr. I. Ortiz (*Universidad Del Pais Vasco, Spain*)
Is there any research on the application of electrokinetic techniques to a particular separation technique in your laboratory?

ANSWER
We have several current research programmes examining the application of electrokinetic techniques to liquid/liquid extraction, to two-phase biotransformation and to microencapsulation processes.

Paper 15
Solvent extraction of penicillin - A study of electrostatic spraying behaviour during enhanced extraction in a spray column
L.R. Weatherley, I. Arnott, J.C. Slaughter and I. Campbell (*Heriot-Watt University, UK*)

QUESTION from R. Klaassen (*T N O, The Netherlands*)
You are making a very fine dispersion. Does it split up easily or is there entrainment of organic phase in the aqueous stream coming out of the top of the column?

ANSWER from I. Arnott
There is no evidence of any appreciable carry-over of CH_2Cl_2 into the aqueous raffinate collected at the top of the column. On this subject; there is also no apparent transfer of whole Broth solids into the organic phase.

QUESTION from Dr. P.J. Thompson (*AEA Technology, UK*)
How do you see the scale-up of this technique for an industrial plant?

ANSWER from I. Arnott
The scale-up philosophy is reflected in the current project which concentrates on 'Combined-field extraction' ie a combination of a centrifugal field generated in the organic phase, at the same time applying an electric field to disperse the aqueous phase containing penicillin.

Also, for small volumes of high value biological products it may be possible to use a redesigned column eg countercurrent solvent flow etc.

QUESTION from J. Jenkins (*AEA Technology, UK*)
You showed a graph of the increase of mass transfer rate (product of mass transfer coefficient and interfacial area) with increasing applied voltage. Is this increase of rate due only to a decrease in drop size and increase of interfacial area, or is the mass transfer coefficient enhanced, and what is the relative contribution of the two effects?

ANSWER from Dr. L.R. Weatherley
We think that the enhancement of interfacial area is the major enhancing factor in the electrostatic dispersion region. Undoubtedly, drop acceleration and better phase separation contribute to the overall improvements observed.

QUESTION from Dr. M.A. Hashim (*University of Malaya, Malaysia*)
How do you explain the increase in internal circulation for the charged droplets?

ANSWER from Dr. L.R. Weatherley
The presence of surface charge is thought to reduce the stability of the droplet and enhance oscillation, which would encourage internal droplet turbulence. Electrostatically accelerated single droplets may also experience shear-induced turbulence at the drop surface which may also encourage internal circulation.

Paper 16

Mechanism for the removal of alcohol from beer by emulsion liquid membranes

B.R. Etuk and K. Murray (*Heriot-Watt University, UK*)

QUESTION from E. Brejza (*Imperial College, UK*)
Have the addition of surfactants affected, at all, the physical properties of the alcohol?

The fact that the predicted and experimental results were similar probably implies that they were not influenced by the surfactant. Had they been significantly different in value, would this be due to surfactant dissolving with the alcohol or something else?

ANSWER from Dr. B.R. Etuk
The surfactant used was not soluble (not really!) in the aqueous phase (beer), hence is believed to have virtually no effect on the beer properties after the lm separation (Analysis of beer properties available in Proc. Biochem. 25 (1990) 24).
The use of surfactant, however, was necessary in order to obtain a stable emulsion required for the separation process.

Paper 18

The development of a flowsheet to separate uranium and plutonium present in irradiated oxide fuel

I.S. Denniss and C. Phillips (*BNFL, UK*)

QUESTION from J. Jenkins (*AEA Technology, UK*)
I am interested in your data which indicates that the rate of U/Pu separation is controlled by diffusion rather than by chemical reaction. We came to the same conclusion at Harwell on our work on U extraction as part of the AEA Technology programme on fast reactor fuel reprocessing. What is the source of the data on rate of reaction constants?

ANSWER from Dr. I.S. Denniss
I think that the data was published in the proceedings of the solvent extraction conference at Harwell in 1956.

Paper 19

U/Pu separation in an organic continuously operated pulsed column

J. Schön, H. Schmieder, B. Kanellakopulos, H.-J. Bleyl, D. Ertel and P. Dressler (*Kernforschungszentrum Karlsruhe Gmbh, FRG*)

QUESTION from Dr. I.S. Denniss (*BNFL, UK*)
Does gas build up in the mechanical pulsing system cause variations in the pulse energy in the column?

ANSWER from J. Schön
Yes, I think so. But our pulse control system does not allow a continuous observation of the pulse wave form in the presence of gas, so we cannot give a clear answer as to how high the loss of pulse energy is.

QUESTION from Dr. R. Bush (*AEA Technology, UK)*
What was the temperature of the MINKA runs? (Answer 25°C)

I asked because this may be a lower temperature than is used in the THORP process described by Dr. Denniss and the oxidation/reduction reactors may therefore be significantly slower in MINKA experiments.

ANSWER from J. Schön
We don't use heating of the columns (and no cooling). The system was operating at the normal temperature of the box (22° - 25°C). From other experiments, not shown here, we found great differences in the rate of hydrazine destruction. Now we are looking at the different compositions of this solution. It may be that these or some elements or compounds in the solutions strongly influenced the redox kinetics of hydrazine and therefore Pu and Tc too.

Paper 20
A technetium rejection flowsheet
R. Baker, J.H. Miles and P.T. Roberts (*AEA Fuel Services, UK*)

QUESTION from Dr. P. Thompson (*AEA Fuel Services, UK*)
How did the performance of the Tc rejection flowsheet depend on the dispersion droplet size?

ANSWER from Dr. J.H. Miles
The stirring speed was adjusted to give "good" performance. This implies steady interfaces and solvent and aqueous products uncontaminated by the other phase. Stirring speed was maintained by observing the stroboscopic effect against the fluorescent lighting. We have, however, no information on the drop size and/or stage efficiency at this speed.

COMMENT from H. Schmieder (*Kernforschungszentrum, West Germany*)
We had in 1988 and 1990 with our MILLI plant two successful experiments to improve the decontamination of fission products. On these occasions we also investigated the behaviour of technetium. With an extreme high solvent loading (automatically controlled) we got for Tc DF's of 2.5 to about 20 depending on flowsheet conditions after fission product scrub. The results are published in Radiochimica Acta and Kerntechnik 1989 and will be published in the Proceed. of ISEC'90.

REPLY from Dr. J.H. Miles
I think in your process you have a higher acidity which tends to reject TC; but you have a higher uranium concentration and are working in presence of F.P.Z. which would tend to extract it. I am surprised that you can get the overall effect favourable to rejection.

Paper 21
Some Plutonium (IV) polymers properties in purex process
H. Scoazec, J.Y. Pasquiou and M. Germain (*Commissariat a L'Energie Atomique, France*)

COMMENT from Dr. P. Wilson (*BNFL, UK*)
The author commented that the rates and extents of the Pu polymer formation have been well documented. There is indeed a wealth of literature on the subject, but some evidence that polymerisation depends so strongly on particular circumstances as to require caution in generalisation. For instance, a colleague tried recently to form

plutonium (IV) polymer under conditions expected to produce it in abundance, but failed to obtain any significant amount.

Paper 22
Compounds of Plutonium (IV) with alkylphosphates and phosphate in nuclear reprocessing
H. Scoazec, J. Livet, J.Y. Pasquiou and M. Germain (*Commissariat a L'Energie Atomique, France*)

QUESTION from Dr. I.S. Denniss (*BNFL, UK*)
What was the solvent residence time in each settler?

ANSWER from H. Scoazec
It was about 7 (seven) minutes. According to the scale of this pilot plant, this corresponds to a plutonium production of 36 kg/day for the industrial plant. The production can be increased up to 80 kg/day in the same unit, and back extraction of Pu is better (better F.D. of the solvent) In this case flow rates are increased and residence times lower.

QUESTION from Dr. I.S. Denniss (*BNFL, UK*)
Is there a realistic mechanism to form polymeric plutonium in an industrial process? (Plutonium was poured into hot water in the experiment.)

ANSWER from H. Scoazec
The conditions to form polymeric plutonium in the process are reported in a publication by Costanzo and Biggers (1963). There are many places in the process where it can occur, for example, in the storage tanks for concentrated plutonium, or during concentration of plutonium by evaporation.

QUESTION from Dr. I.S. Denniss (*BNFL, UK*)
The Pu 238 content of the feed used (3.4%) appears to be very high for typical oxide fuel?

ANSWER from H. Scoazec
This is the isotopic composition (3.2% 238Pu) of mixed oxide fuels when Plutonium is recycled in reactors. We intend to reprocess these fuels in the new U.P.2 800 plant of La Hague.

QUESTION from Dr. P.J. Thompson (*AEA Technology, UK*)
Can you comment on the mechanism for complex formation with HDBP for ol- and oxo-forms?

ANSWER from J. Livet
Polymeric Plutonium is a rodlike particle and the complexant (DBP_-) is attached on the outside of the rod by an ionic bond. The evidence from small angle neutron scattering is that the radius of the polymer is increased by the exact length of the extractant molecule.

QUESTION from Dr. P. Wilson (*BNFL, UK*)
Is there any information on the solubility of these compounds in diluents other than hydrogenated propylene tetramer?

ANSWER from J. Livet
My first tests were made in solvents such as "Hyfran 120" essentially n-dodecane. It seems to me that the precipitations were the same. The reason for precipitation if we look at the crystal structures is the formation of Van der Vaal's binding between the polymer chains. To prevent this binding and this precipitation we have to choose a solvent such as Benzene or carbon tetrachloride with an intermediate dielectric coefficient, which destroys this Van der Vaal's binding, or to have strong ligand which fixes to the metal engaged with DBP and prevents the chains' aggregation, such as a maloamide with alkyl chains, which is used to redissolve the metallic DBP precipitates.
(Patented procedure).

Paper 23
Calculation of the dynamic pressure drop in pulsed sieve-plate columns
M. Lorenz (*Technical University of Claustal, FRG*), H. Haverland (*Schott Glaswerke, FRG*) and A. Vogelpohl (*Technical University of Claustal, FRG*)
Paper presented by J. Jenkins (*AEA Technology, UK*)

QUESTION from Dr. P.J. Thompson (*AEA Technology, UK*)
How does the model behave when a solute is present which is extracting into the second phase?

ANSWER from J. Jenkins
Provided that the effect of the presence of the solute on the densities of the two phases is taken into account, the model predicts behaviour well.

QUESTION from Dr. P.J. Thompson (*AEA Technology, UK*)
In practical terms what do you consider the minimum length of column to be for application of this technique?

ANSWER from J. Jenkins
The authors (and the presenter) have applied this technique to lengths of column of 1 m but provided the pressure transmitter is adequately sensitive, lengths of 0.5 m or less could probably be studied.

QUESTION from Dr. P. Wilson (*BNFL, UK*)
The presenter commented that the appearance of inflections on the sinusoidal wave form should have been expected on account of the phase difference between acceleration and velocity. In fact the combination of two sinusoidal wave forms out of phase should itself be sinusoidal but displaced in phase, and the actual waveform suggests the presence of a harmonic. Is this harmonic inherent or due to interference?

ANSWER from J. Jenkins
In fact, the velocity function is a cosine function but the acceleration is a sine function, and the earlier application of the model by Jenkins (Extraction '84) did predict some of the inflections. I suspect the other inflections are due to harmonics rather than interference.

AUTHOR/TITLE/KEYWORD